D0218458

Annual Editions:
World Politics
35/e

Edited by Robert Weiner
University of Massachusetts, Boston

http://create.mcgraw-hill.com

ISBN-10: 1259219275   ISBN-13: 9781259219276

Contents

Preface

In publishing ANNUAL EDITIONS, we recognize the enormous role played by the magazines, newspapers, and journals of the public press in providing current, first-rate educational information in a broad spectrum of interest areas. Many of these articles are appropriate for students, researchers, and professionals seeking accurate, current material to help bridge the gap between principles and theories and the real world. These articles, however, become more useful for study when those of lasting value are carefully collected, organized, indexed, and reproduced in a low-cost format, which provides easy and permanent access when the material is needed. That is the role played by ANNUAL EDITIONS.

The international system continues to exhibit the characteristics of a multipolar system marked by the economic decline of the U.S. in 2013, the rise of China, and the economic difficulties faced by some of the BRICS. Geography, which has received short shrift in recent texts in international relations, is still an important factor in understanding the problems of the international system, as illustrated by the theme of the annual international studies conference in 2014: "Spaces and Places: Geopolitics in an Era of Globalization." The conference is interested in "looking at the juxtaposition of spatiality/distance/place and the geopolitical context." Territorial disputes, rather than no longer being considered as an important source of conflict between state actors, continue to serve as a major source of tension, especially in Asia. Of particular interest are borders, territorial disputes, and the continuing importance of the Westphalian concept of sovereignty in the age of globalization. This is underscored by China's disputes with its neighbors over the sovereignty of islands in the South China Sea. This raises the question as to whether conciliation can be found between the forces of realism and liberal internationalism in the world political system in the second decade of the 21st Century. The collection of articles in this book focus on these and other important developments in the international system.

The articles in this book continue to analyze the structure of the international system from a multipolar perspective, looking at the effect of China's rise, and the United States' decline on the international system as Washington teetered on the verge of default in 2013, Chinese–Russian cooperation, the role of France in maintaining stability in Africa, and the competition between Brazil and Mexico as regional hegemons in Latin America. Articles deal with the process and setbacks of democratization, in Pakistan, and especially in Egypt, almost three years after the Arab Spring; articles focus on the foreign policy of the Obama administration, which has suffered a number of setbacks especially with the shocks administered to the foreign policy of the U.S. by the Snowden revelations; U.S.–Chinese

relations; the implications of the pivot toward Asia; the 50th anniversary of the Cuban missile crisis; and the Israeli–Palestinian conundrum. Conflicts continued in the more volatile regions of the world, as following the U.S. withdrawal from Iraq, the bombings continued there, and as the U.S. prepared to withdraw from Afghanistan in 2014, relations with Pakistan became even more relevant. An important development has been the continuation of terrorism in Africa, as illustrated by the war in Mali, the Somali-based jihadist group Shabab's attack on a Kenyan mall in 2013, and continued chaos in Libya over a year after the ouster of Gadaffi. Arms control and disarmament measures continued to form an important part of the global effort to eliminate weapons of mass destruction, with negotiations scheduled to take place between Iran and the U.S. about Iran's nuclear weapons program, the decision by the Syrian government to place its chemical weapon destruction in the hands of the UN and the Organization for the Prohibition of Chemical Weapons, and the adoption of an Arms Trade Treaty by the UN General Assembly in 2013. Other essays deal with the increasing use of robotic warfare, an assessment of the UN's Millennium Development Goals, and Africa's economic boom. Concluding articles deal with global environmental issues, the effects of climate change on the Amazon forest, and the effects of climate change on food security. I would also like to express my thanks to McGraw-Hill Senior Development Editor Jill Meloy, who has made the completion of this project possible. My thanks also go to the several graduate students who helped me with this project, especially Dan Torres, whose assistance was invaluable in selecting the essays that appear in this book. Thanks also go to Arvind, who suggested a number of possible topics, and to Titilope Osunkojo, Aindrea Benduzek, Victoria Omiti, and John Hashem for their assistance in this project.

Editor

Robert Weiner *Editor*
Robert Weiner
University of Massachusetts—Boston

Dr. Robert Weiner received his PhD in international relations from New York University. He is a center associate at the Davis Center for Russian and Eurasian Studies at Harvard University, and a fellow at the Center for Peace, Democracy, and Development at the University of Massachusetts—Boston. His research interests cover theories and concepts of international relations, Eastern Europe, comparative foreign policy, diplomacy and war, genocide, and the European Union and the emerging democracies. He is the author of *Romanian Foreign Policy at the United Nations* (Praeger, 1984) and *Change in*

Eastern Europe (Praeger, 1994). He also is the author of a number of book chapters, and his articles have been published in such journals as *Orbis, Problems of Postcommunism, The International and Comparative Law Quarterly, Sudost Europa, Demokratizatsiya,* and *The International Studies Encyclopedia.*

Academic Advisory Board

Members of the Academic Advisory Board are instrumental in the final selection of articles for Annual Editions books and ExpressBooks. Their review of the articles for content, level, and appropriateness provides critical direction to the editor(s) and staff. We think that you will find their careful consideration reflected here.

Rafis Abazov
Columbia University

Alton Alade-Chester
San Bernardino Valley College

Tahereh Alavi Hojjat
DeSales University

Januarius Asongu
Santa Monica University

Arthur Cyr
Carthage College

Tahmineh Entessar
Webster University

Beverly Fogelson
Oakland University

Jeremy L. Goldstein
St. George's School

Cindy Jebb
United States Military Academy

Jean-Gabriel Jolivet
Ashford University

Anders M. Kinney
Calhoun Community College

James McGann
University Of Pennsylvania

Joseph B. Mosca
Monmouth University

Kanishkan Sathasivam
Salem State University

Choudhury M. Shamim
California State University, Fullerton

John M. Trayhan
Our Lady of the Lake University

Correlation Guide

The *Annual Editions* series provides students with convenient, inexpensive access to current, carefully selected articles from the public press. **Annual Editions: World Politics, 35/e** is an easy-to-use reader that presents articles on important topics such as the *multipolar international system, democratization, foreign policy, war, international organization and law, human security,* and many more. For more information on *Annual Editions* and other *McGraw-Hill Create™* titles, visit www.mcgrawhillcreate.com.

This convenient guide matches the articles in **Annual Editions: World Politics, 35/e** with **International Politics on the World Stage, Brief, 8/e** by Rourke/Boyer.

International Politics on the World Stage, Brief, 8/e	Annual Editions: World Politics, 35/e
Chapter 1: Thinking and Caring About World Politics	Think Again: The BRICS The Currency of Power: Want to Understand America's Place in the World? Write Economics Back into the Plan
Chapter 2: The Evolution of World Politics	The Global Shift of Power from East to West The Irony of American Strategy
Chapter 3: Levels of Analysis and Foreign Policy	Beyond the Pivot: A New Road Map for U.S.–Chinese Relations Japan's Cautious Hawks: Why Tokyo Is Unlikely to Pursue an Aggressive Foreign Policy The Cuban Missile Crisis at 50: Lessons for U.S. Foreign Policy The Future of United States–Chinese Relations: Conflict Is a Choice, Not a Necessity
Chapter 4: Nationalism: The Traditional Orientation	France in Africa: A New Chapter
Chapter 5: Globalization: The Alternative Orientation	Climate Change and Food Security
Chapter 6: Power, Statecraft, and the National State: The Traditional Structure	Beware Collusion of China, Russia Israel's New Politics and the Fate of Palestine
Chapter 7: Intergovernmental Organizations: Alternative Governance	The Crisis of Europe: How the Union Came Together and Why No Chemical Weapons Use by Anyone: An Interview with OPCW Director-General Ahmed Üzümcü Why UNESCO Is a Critical Tool for Twenty-First Century Diplomacy
Chapter 8: International Law and Human Rights	General Mladic in the Hague: A Report on Evil in Europe—and Justice Delayed Law and Ethics for Robot Soldiers The Showdown: Winners and Losers in Egypt's Ongoing Revolution One Step Forward, Two Steps Back Pakistan on the Brink of a Democratic Transition?
Chapter 9: Pursuing Security	Mutual Assured Production: Why Trade Will Limit Conflict Between China and Japan State of Terror: What Happened When an Al Qaeda Affiliate Ruled in Mali Reduce the Global Nuclear Risk Syria's Long Civil War Talking Tough to Pakistan: How to End Islamabad's Defiance The Cuban Missile Crisis at 50: Lessons for U.S. Foreign Policy Why Iran Should Get the Bomb: Nuclear Balancing Would Mean Stability
Chapter 10: National Economic Competition: The Traditional Road	Shifting Fortunes: Brazil and Mexico in a Transforming Region
Chapter 11: International Economics: The Alternative Road	Africa's Economic Boom Own the Goals: What the Millennium Development Goals Have Accomplished
Chapter 12: Preserving and Enhancing the Biosphere	A Light in the Forest: Brazil's Fight to Save the Amazon and Climate Change Diplomacy Too Much to Fight Over

Topic Guide

All the articles that relate to each topic are listed below the boldface term.

Unit 1

UNIT

Prepared by: Robert Weiner, *University of Massachusetts/Boston*

The Multipolar International System

There continues to be a general consensus among international relations specialists that a multipolar system exists, or if it has not yet solidified, is in the process of emerging with perhaps six or seven centers of power. International systems don't remain the same, but change when the overall distribution of power in the system changes. The most dangerous time is when there is a power transition between rising and declining powers. As far as the position of the United States in the international system is concerned, the declinist school of thought argues that the United States is clearly losing its hegemonic position in the system, especially from an economic point of view; as it suffers from imperial overstretch. This was underscored by the debate in the U.S. government about raising the ceiling on the national debt in 2013, which had soared to about 16.7 trillion dollars. That, along with the shutdown of the U.S. government, certainly weakened the image of the U.S. in foreign capitals as a hegemonic power. The United States' economic disarray also resulted in calls by such countries as China for the replacement of the U.S. dollar as an international reserve currency, especially as the Chinese became concerned over the security of their trilliion-dollar-plus investment in U.S. treasury bonds. Clearly, a key shift of power is occurring in the international system with the rise of China, as the Pacific region becomes more important to the vital interests of the United States than Western Europe. The Obama administration has recognized the change that has occurred in the geopolitical shift in power, as it executes a "pivot" to Asia, downsizes its military presence in Iraq and Afghanistan, and engages in a military buildup in the Pacific. Pivot means a shift in the direction of U.S. policy in the aftermath of the withdrawal of the United States from Iraq and Washington's plans to downsize its military presence in Afghanistan in 2014. The question then arises as to what role the United States can play as a declining power in an emerging multipolar system.

The fluid nature of the multipolar system is underscored by the lack of cohesiveness and slow economic growth of the BRICS, the increasing tension between China and Japan over a set of disputed islands in the South China Sea, the growing collusion between China and Russia in response to the enlargement of NATO and the European Union, the continuing French involvement in Africa, and the competition between Mexico and Brazil in Latin America.

An important dimension of the structure of the emerging multipolar system consists of regional and sub-regional organizations. Particularly important also is the continuing salience of the geographical concept of region, as geography still matters and plays a critical role in understanding the causes of regional conflicts. This was the case of Syria's civil conflict, which continued to rage on in 2013, destabilizing the entire Middle Eastern region. Regionalism therefore forms an important element of the structure of the emerging multipolar system, which is undergoing a process of reconfiguration of power. Regional and sub-regional organizations represent one of the most numerous types of non-state actors in the international system. Until recently, the best example of a successful regional organization was the European Union, although the gloss had now somewhat worn off of the European project due to the Eurozone crisis, which may have called into question the future of the organization itself. Other regions in the world have tried to emulate the European Union, which represents the state-centric "old regionalism", but without too much success, given all of the background and process conditions that need to be met for a successful regional organization which may have been unique in the case of Europe. Nonetheless, the emergence of the New Regionalism, in which—non-state actors may play a more prominent role than the state-centric old regionalism, bears watching.

Prepared by: Robert Weiner, *University of Massachusetts/Boston*

Article

Think Again: The BRICS

Together, their GDP now nearly equals the United States. But are they really the future of the global economy?

Antoine Van Agtmael .

Learning Outcomes

After reading this article, you will be able to:

- Explain some of the problems the Brics face in dominating the economy.
- Explain the key assumptions that have been made about the Brics.

The BRICS Are in a Class by Themselves

YES AND NO.
There is no question that the brics—Brazil, Russia, India, China, and the group's newest member, South Africa—are big. They matter. In terms of population, landmass, and economic size, their pure dimensions are impressive and clearly stand out from those of other countries. Together, they make up 40 percent of the world's population, 25 percent of the world's landmass, and about 20 percent of global GDP. They already control some 43 percent of global foreign exchange reserves, and their share keeps rising.

Jim O'Neill of Goldman Sachs put the spotlight on the rise of the original four of these big new economic powers when he gave them the name brics in 2001, and their collective growth began to soar. But in reality their economic success had been a long time coming. Twenty years before that, when I was at the World Bank's International Finance Corp. (IFC), we were identifying the opportunity to rebrand these countries, which, despite their enormous economic potential, were still lumped together with the world's perennial basket cases as "underdeveloped countries" stuck in the "Third World." At the time, Third World stock markets were simply off the radar screen of most international investors, even though they were starting to grow; I gave them the name "emerging markets." Local investors were already quite active in Malaysia, Thailand, South Korea, Taiwan, Mexico, and elsewhere, as homegrown companies became larger and more export-competitive while

market regulation became more sophisticated. But until the IFC built its Emerging Markets Database and index in 1981, there was no way to measure stock performance for a representative group of these markets, a disabling disadvantage when stacked against other international indices, which were skewed in favor of developed countries such as Germany, Japan, and Australia. This brand-new research on markets and companies provided investors with the confidence to launch diversified emerging-market funds following the success of individual country funds in markets such as Mexico and South Korea.

The brics, however, took much longer to get ready for prime time. Until the beginning of the 1990s, Russia was still behind the Iron Curtain, China was recovering from the Cultural Revolution and the Tiananmen Square unrest, India remained a bureaucratic nightmare, and Brazil experienced bouts of hyperinflation combined with a decade of lost growth. These countries had largely muddled along outside the global market economy; their economic policies had often been nothing short of disastrous; and their stock markets were nonexistent, bureaucratic, or supervolatile. Each needed to experience deep, life-threatening crises that would catapult them onto a different road of development. Once they did, they tapped into their vast economic potential. Their total GDP of close to $14 trillion now nearly equals that of the United States and is even bigger on a purchasing power parity basis.

Here's the problem, however, with asking whether the brics "matter": Big is not the same as cohesive. The brics are part of the G-20, but not a true power bloc or economic unit within or outside it. None is fully accepted as "the" leader even within its own region. China's rise is resented in Japan and distrusted throughout Southeast Asia. India and China watch each other jealously. Brazil is a major supplier of commodities to China and has relied on it for its economic success, but the two powers compete for resources in Africa. Russia and China may have found common cause on Syria, but they compete elsewhere. And though intra-bric commerce is growing rapidly, the countries have not yet signed a single free trade agreement with each other. Then there's South Africa, which formally joined this loose political grouping in 2010. But being a member of the BRICS doesn't make it an equal: South Africa doesn't have the population, the growth, or the long-term economic potential of the other four. Indonesia, Mexico, and Turkey would have been other logical contenders—or South Korea and Taiwan, for that matter, which have comparable GDPs but much smaller populations than the original BRICS.

The BRICS are also nowhere near economically cohesive. Russia and Brazil are way ahead in per capita income, beating China and India by a huge amount—nearly $13,000 compared with China's $5,414 and India's $1,389, according to 2011 IMF data. And their growth trajectories have been very different. What's more, the BRICS face stiff competition from other emerging powerhouses in the developing world. While China and India seemed to have a competitive edge for a while due to their low labor costs, countries like Mexico and Thailand are now back on the competitive map. And while growth in the BRICS seems to be slowing, many African countries are receiving more foreign investment, may be more politically stable, and are at long last moving away from slow or no growth toward much more robust economies.

The Continued Rise of the BRICS Is Inevitable

TRUE, BUT THEIR GROWTH IS SLOWING.
Forecasts by Goldman Sachs and others project China will overtake the United States in GDP before 2030. China, meanwhile, dwarfs the other BRICS, whose combined economic size isn't expected to catch up to China during that period. The BRICS will approach the total size of the seven largest developed economies by 2030, and by the middle of this century they are projected to be nearly double the size of the G-7.

BRICS consumers are also beginning to rival their American counterparts in terms of total purchasing power. More cars, cell phones, televisions, refrigerators, and cognac are now sold in China alone than in the United States. Even with slower growth, the economic engine of the BRICS should be more important than that of the United States or the European Union for most of the 21st century.

Then again, there's no guarantee that the BRICS can maintain their torrid growth rates. Just as their expanding economies took the world by surprise over the past decade, the big shock for the next decade may be that they will grow less quickly than assumed. Japan, South Korea, and Taiwan have already shown that growth rates slow down once a basic level of industrialization has been reached. The unquenchable thirst for "goods" tends to moderate when basic infrastructure is in place and consumers want more health care, education, and free time.

To some extent, this is already occurring. Leading Chinese economists now expect China's annual growth to slow down from rates of 10 to 12 percent to 6 to 8 percent by the end of this decade. Dreams of India reaching sustainable annual growth of 8 percent or more have been lowered to 5 to 6 percent after the country hit an inflation barrier and offshore gas production disappointed. Brazil has also struggled to return to its exuberant

pre-crisis growth, while Russia has been staggered by Europe's economic problems. The projections by Goldman Sachs and others always expected slower growth for the future, but some enthusiasts did not read the footnotes.

The Financial Crisis Was Good for the BRICS

NOT FOR LONG.
The 2008 financial crisis did not emanate from emerging markets. Instead, the BRICS came to the rescue when the United States, Europe, and Japan collapsed due to their overspending, fiscal imprudence, and overreliance on just-in-time production that made them too dependent on a consumer economy that quickly blew up. After the BRICS suffered brief, V-shaped recessions of their own, as swift in their decline as they were in their recovery, the BRICS' demand helped pull the global economy out of its initial slump.

It certainly wasn't clear initially that this was how the crisis would play out. The *Financial Times* warned (and many investors feared) that the banking systems of emerging markets would succumb to the same massive financial problems that plagued the United States and Europe, but Asia and Latin America had learned their lessons from earlier financial crises and put their houses in order. The Chinese had ample reserves for a fiscal stimulus that was not only massive, but, unlike its U.S. counterpart, also disbursed funds quickly. The BRICS' central banks, along with those in other emerging markets, cooperated on global monetary easing. Without it (and without China's quickly disbursed stimulus at home), Western stimulus and easing would have been inadequate and ineffective. With it, demand for commodities stabilized and the world avoided a depression.

These crisis interventions came at a significant cost, however, the full price of which is not yet clear even today. The real estate bubble, which played such a big part in the United States and Southern Europe, didn't burst in the BRICS. Inflation also increased well beyond the comfort zone of central banks in China, India, and Brazil. Although all this did not provoke another crisis, it might have planted the seeds for future problems. Economic history teaches us that the next crisis usually comes from the region where the applause and self-satisfaction were loudest the previous time around. If that holds true, the next economic shock will more likely than not come from the BRICS.

The BRICS Are Unbeatable Competitors

NO.
The BRICS benefited for several decades from cheap labor, higher productivity, massive (but far from universal) investment in infrastructure and education, and a hunger to catch up with wealthier rivals. Their transformation was remarkable: With better-off populations, domestic markets finally became economically attractive, South-South trade exploded, and leading

corporations transformed themselves from second-rate producers of cheap goods into world-class manufacturers of smartphones, semiconductors, software, and planes. China's Lenovo took over IBM's PC business. Brazilian and South African beer companies became leading global brewers. Just as had been the case with the Russians after Sputnik and the Japanese in the 1980s, the BRICS became feared and formidable competitors, even if some of the fears about their rise were exaggerated.

But the story is not over. Cheap, abundant energy from shale gas is attracting new investment in the United States, giving energy-intensive industries a renewed competitive edge. Abundant shale gas could also make Russian Arctic drilling and Brazilian pre-salt production too expensive. Stagnant U.S. wages and soaring pay in China and India are eroding the BRICS' labor-cost advantage, while their seemingly bottomless labor pool has suddenly started emptying out, leaving them with shortages of trained labor.

Mechanization is also allowing the developed world to make a comeback. Increasingly affordable and sophisticated robots can now do what 10 or more human workers did until recently. They work 24 hours a day and do not ask for higher wages or better benefits. Smartphones and tablets may still be made in Asia, but the BRICS lag behind in taking advantage of the productivity gains they bring. As a result, traditional multinationals are fighting back after years of retreat, from General Motors winning the biggest market share in China to General Electric's foray into producing low-cost medical equipment to Nestlé's invention of the wildly successful Nespresso machines, turning high-end coffee from a store-bought luxury into an at-home convenience. The competitive edge may be turning back to the West much faster than we thought.

The BRICS Are the Best Place to Invest

NO LONGER TRUE.
Until 2008, the BRICS performed far better than other emerging equity markets—or developed markets, for that matter. And by a lot: For the five years ending in 2007, investors in the four original BRICS earned an annualized 52 percent return, compared with just 16 percent in the G-7 markets. But in the past five years, through Aug. 31, that figure was −3 percent for the BRICS and −1 percent for the G-7. This was in part a correction to exaggerated expectations, which drove up valuations and currencies to unsustainable levels. It also seems, however, that the BRICS' competitive edge is now being questioned in more fundamental terms. Of course, it makes perfect sense for investors to diversify and not ignore such a huge, successful part of the global economy, but that is different from blind euphoria.

Each of the BRICS is very different, and so are the question marks that accompany their economies. For example, China's wage costs had been so much lower than Mexico's for several decades that Mexico had difficulty competing, despite its closeness to the U.S. market. But that wage gap has closed in recent years—Chinese labor rates have grown from 33 percent of Mexico's in 1996 to 85 percent in 2010—and now investment

is flowing back to Mexico. Even when Indian growth rates went through the roof, bureaucracy, budget deficits, and infrastructure bottlenecks remained serious impediments. Brazil successfully turned around its floundering economy in the 1980s and then benefited from three windfalls: China's thirst for commodities, energy discoveries, and a competitive edge as an agribusiness giant. Now, however, China's slowing economy and the world's shift toward ubiquitous shale gas is changing the picture. Or consider Russia, which, to its peril, has squandered its oil-and-gas weapon by pooh-poohing the potential of shale gas, opening up export opportunities for the United States in Europe.

The BRICS Will Surpass the West

NOT SO FAST.
Yes, the BRICS will remain the main source of growth in tomorrow's world, as they already are today. Together they will dominate the global economy later this century the way Europe and the United States once did.

Just as the pendulum swung far toward the BRICS but then swung back hard in recent years, there are signs of new forms of BRICS competitiveness. Research and development in the BRICS is paving the way for increasingly high-value-added production. Ninety-one percent of U.S. plants are more than a decade old, versus only 43 percent of China's plants, according to a 2007 *Industry Week* survey. While 54 percent of Chinese companies cited innovation as one of their top objectives in the survey, only 27 percent of U.S. respondents did. Chinese telecom equipment-makers are giving more traditional players a run for their money, Indian-made generic drugs are making inroads, Brazilian protein producers dominate world markets, and Russian oligarchs are making smart investments abroad. The BRICS are going through a rough patch right now, yet they're poised for a roaring comeback.

But though the era of American or Western domination may be over, BRICS domination is still some time off. What is already a fact is that the clear delineation between developed and "backward" countries is a thing of the past. Western multinational companies are seeking to expand in the BRICS as growth in their home markets has dried up. Chinese and Indian corporations are building their brands in other emerging markets and the West. More than ever, developed countries' economic fates are tied to those of emerging markets.

Intellectual property remains a strong suit of advanced economies. The United States, Japan, and Germany—just three advanced economies—accounted for 58 percent of patent filings in 2011, according to the World Intellectual Property Organization. But even here the BRICS are catching up: China's applications soared 33 percent in 2011, Russia's filings were up 21 percent, Brazil's 17 percent, and India's 11 percent. Compare that with 8 percent growth for the United States and 6 percent for Germany. Chinese telecommunications equipment giant ZTE Corp. dislodged Japan's Panasonic from the global top spot with 2,826 patent applications. China's Huawei

Technologies was in third place, while a previous American leader, Qualcomm, dropped from third to sixth place in 2011. Why does it matter? Because patents are a key indicator of future economic strength.

Politics Could Be the BRICS' Undoing

TRUE, AND YOU DISREGARD THEM AT YOUR PERIL. The spread of democracy and free markets in much of Asia, Latin America, and Eastern Europe is impressive, but some BRICS have been laggards rather than leaders in this area. Legitimacy in these countries often depends on meeting sky-high expectations for economic success, while political checks and balances remain in their infancy. So forget about all those paeans to "authoritarian capitalism" you read in the op-ed pages. Just because Beijing has a fancy new airport and President Vladimir Putin can bulldoze entire neighborhoods at will doesn't mean China's and Russia's politics give them an edge. Even in democratic India, politics are often overwhelmed by corruption, and politically open Brazil struggles with crippling crime stats and political scandals.

The BRICS may seem stable now, but nobody knows what the future holds. Admiration for oligarchs easily turns into envy and anger. Ubiquitous mobile-phone cameras and instant Internet distribution constrain the use of public force. Under the surface and among the younger generation, pride in economic achievements and a sense of material well-being are now coupled with demands for better health care and national recognition. Increasingly, more is not the answer—citizens of the BRICS want better. Local elites must act adroitly to keep this new mood from developing into a combustible mix. The current generation of leaders in China has not forgotten the lessons of the Cultural Revolution—but the next generations may.

Some tailwinds that have benefited the BRICS these past decades may yet turn into headwinds. For instance, these countries have benefited from relatively low budget allocations to military spending—a fruit of Pax Americana. That could change if conflict broke out on the Indian subcontinent or Iran acquired nuclear weapons. And serious political unrest could easily derail the rise of the BRICS: The Bo Xi-lai case in China, the upheavals following the Arab Spring, and the power blackout in India were recent red flags that showed the dramatic impact of sudden events.

Still, the BRICS are not going anywhere. Sure, they may face tough adjustments getting used to less lofty growth expectations while satisfying more demanding populations. But one way or another it's safe to say: These big emerging economies will put their stamp on the 21st century.

Critical Thinking

1. What factors prevent the BRICS from becoming a cohesive power bloc?
2. Why are the BRICS considered a major economic bloc?
3. Why is the BRIC's economic growth slowing down?
4. Why is the era of U.S. economic domination over?

Create Central

www.mhhe.com/createcentral

Internet References

Brics Forum
    www.BRICSforum.org/.
The Foreign Ministry of China
    www.mfa.gov.cn/eng
Ministry of Foreign Affairs of the Russian Federation
    www.mid.ru/brp.4.nst/main_eng

ANTOINE VAN AGTMAEL, a founder and former chairman of AshmoreEMM, is author of *The Emerging Markets Century*.

Van Agtmael, Antoine. Reprinted in entirety by McGraw-Hill Education with permission from *Foreign Policy*, November 2012, pp. 76–79. www.foreignpolicy.com. © 2012 Washingtonpost.Newsweek Interactive, LLC.

Article

Prepared by: Robert Weiner, *University of Massachusetts/Boston*

Japan's Cautious Hawks: Why Tokyo Is Unlikely to Pursue an Aggressive Foreign Policy

GERALD L. CURTIS

Learning Outcomes

After reading this article, you will be able to:

- Explain the effects of multipolarity on Japanese security.
- Explain why autonomous forces are a factor in Japanese foreign policy.

The Japanese have thought about foreign policy in similar terms since the latter half of the nineteenth century. The men who came to power after the 1868 Meiji Restoration set out to design a grand strategy that would protect their country against the existential threat posed by Western imperialism. They were driven not, as their American contemporaries were, to achieve what they believed to be their manifest destiny nor, like the French, to spread wide the virtues of their civilization. The challenge they faced-and met-was to ensure Japan's survival in an international system created and dominated by more powerful countries.

That quest for survival remains the hallmark of Japanese foreign policy today. Tokyo has sought to advance its interests not by defining the international agenda, propagating a particular ideology, or promoting its own vision of world order, the way the United States and other great powers have. Its approach has instead been to take its external environment as a given and then make pragmatic adjustments to keep in step with what the Japanese sometimes refer to as "the trends of the time."

Ever since World War II, that pragmatism has kept Japan in an alliance with the United States, enabling it to limit its military's role to self-defense. Now, however, as China grows ever stronger, as North Korea continues to build its nuclear weapons capability, and as the United States' economic woes have called into question the sustainability of American primacy in East Asia, the Japanese are revisiting their previous calculations. In particular, a growing chorus of voices on the right are advocating a more autonomous and assertive foreign policy, posing a serious challenge to the centrists, who have until recently shaped Japanese strategy.

In parliamentary elections this past December, the Liberal Democratic Party and its leader, Shinzo Abe, who had previously served as prime minister in 2006-7, returned to power with a comfortable majority. Along with its coalition partner, the New Komeito Party, the LDP secured the two-thirds of seats needed to pass legislation rejected by the House of Councilors, the Japanese Diet's upper house. Abe's victory was the result not of his or his party's popularity but rather of the voters' loss of confidence in the rival Democratic Party of Japan. Whatever the public's motivations, however, the election has given Japan a right-leaning government and a prime minister whose goals include scrapping the constitutional constraints on Japan's military, revising the educational system to instill a stronger sense of patriotism in the country's youth, and securing for Tokyo a larger leadership role in regional and world affairs. To many observers, Japan seems to be on the cusp of a sharp rightward shift.

But such a change is unlikely. The Japanese public remains risk averse, and its leaders cautious. Since taking office, Abe has focused his attention on reviving Japan's stagnant economy. He has pushed his hawkish and revisionist views to the sidelines, in part to avoid having to deal with divisive foreign policy issues until after this summer's elections for the House of Councilors. If his party can secure a majority of seats in that chamber, which it does not currently have, Abe may then try to press his revisionist views. But any provocative actions would have consequences. If, for example, he were to rescind statements by previous governments that apologized for Japan's actions in World War II, as he has repeatedly said he would like to do, he not only would invite a crisis in relations with China and South Korea but would face strong criticism from the United States as well. The domestic political consequences are easy to predict: Abe would be flayed in the mass media, lose support among the Japanese public, and encounter opposition from others in his own party.

In short, chances are that those who expect a dramatic change in Japanese strategy will be proved wrong. Still, much depends on what Washington does. The key is whether the United States continues to maintain a dominant position in East Asia. If it does, and if the Japanese believe that the United States' commitment to protect Japan remains credible, then Tokyo's foreign policy will not likely veer off its current track. If, however, Japan begins to doubt the United States' resolve, it will be tempted to strike out on its own.

The United States has an interest in Japan's strengthening its defensive capabilities in the context of a close U.S.-Japanese alliance. But Americans who want Japan to abandon the constitutional restraints on its military and take on a greater role in regional security should be careful what they wish for. A major Japanese rearmament would spur an arms race in Asia, heighten regional tensions (including between Japan and South Korea, another key U.S. ally), and threaten to draw Washington into conflicts that do not affect vital U.S. interests. The United States needs a policy that encourages Japan to do more in its own defense but does not undermine the credibility of U.S. commitments to the country or the region.

Plus ÇA Change

For many years now, pundits have been declaring that Japan is moments away from once again becoming a great military power. In 1987, no less an eminence than Henry Kissinger saw Tokyo's decision to break the ceiling of one percent of gnp for defense spending, which had been its policy since 1976, as making it "inevitable that Japan will emerge as a major military power in the not-too-distant future." But Japan's defense budget climbed to only 1.004 percent of gnp that year, and it fell below the threshold again the following year. Today, the ceiling is no longer official government policy, but Tokyo still keeps its defense spending at or slightly below one percent of gnp. What is more, its defense budget has shrunk in each of the last 11 years. Although Abe has pledged to reverse this trend, Japan's fiscal problems all but guarantee that any increase in military spending will be modest.

That Japan's military spending has remained where it is points to a larger pattern. Neither the end of the Cold War nor China's emergence as a great power has caused Japan to scuttle the basic tenets of the foreign policy set by Prime Minister Shigeru Yoshida following the end of World War II. That policy stressed that Japan should rely on the United States for its security, which would allow Tokyo to keep its defense spending low and focus on economic growth.

To be sure, Japanese security policy has changed greatly since Yoshida was in power. Japan has stretched the limits of Article 9 of its constitution, which renounces the right to wage war, making it possible for the Self-Defense Forces to develop capabilities and take on missions that were previously prohibited. It has deployed a ballistic missile defense system, its navy patrols sea-lanes in the East China Sea and helps combat piracy in the Gulf of Aden, and Japanese troops have joined UN peacekeeping operations from Cambodia to the Golan Heights. Spending one percent of gnp on its military still gives Japan, considering the size of its economy, the sixth-largest defense budget in the world. And despite the constitutional limits on their missions, Japan's armed forces have become strong and technologically advanced.

Yet the strategy that Yoshida designed so many years ago continues to constrain Japanese policy. Japan still lacks the capabilities needed for offensive military operations, and Article 9 remains the law of the land. Meanwhile, Tokyo's interpretation of that article as banning the use of force in defense of another country keeps Japan from participating more in regional and global security affairs. Abe has indicated his desire to change that interpretation, but he is proceeding cautiously, aware that doing so would trigger intense opposition from neighboring countries and divide Japanese public opinion.

The durability of Yoshida's foreign policy has puzzled not just observers; the architect of the strategy was himself dismayed by its staying power. Yoshida was a realist who believed that the dire circumstances Japan faced after the war left it no choice but to prioritize economic recovery over building up its military power. Yet he expected that policy to change when Japan became economically strong.

The Japanese public, however, saw things differently. As Japan boomed under the U.S. security umbrella, its citizens became content to ignore the left's warnings that the alliance would embroil the country in the United States' military adventures and the right's fears that Japan risked abandonment by outsourcing its defense to the United States. Yoshida's strategy, crafted to advance Japan's interests when the country was weak, became even more popular in good times. And that remains true today: in a 2012 survey conducted by Japan's Cabinet Office, for example, a record high of 81.2 percent of respondents expressed support for the alliance with the United States. Only 23.4 percent said that Japan's security was threatened by its having insufficient military power of its own.

It is worth noting that Japan's opposition to becoming a leading military power cannot be chalked up to pacifism. After all, it would be an odd definition of pacifism that included support for a military alliance that requires the United States to take up arms, including nuclear weapons, if necessary, to defend Japan. Most Japanese do not and never have rejected the use of force to protect their country; what they have resisted is the unbridled use of force by Japan itself. The public fears that without restrictions on the military's capabilities and missions, Japan would face heightened tensions with neighboring countries and could find itself embroiled in foreign wars. There is also the lingering concern that political leaders might lose control over the military, raising the specter of a return to the militaristic policies of the 1930s.

Furthermore, the Japanese public and Japan's political leaders are keenly aware that the country's security still hinges on the United States' dominant military position in East Asia. Some on the far right would like to see Japan develop the full range of armaments, including nuclear weapons, in a push to regain its autonomy and return the country to the ranks of the world's great powers. But the conservative mainstream still believes that a strong alliance with the United States is the best guarantor of Japan's security.

Islands in the Sun

Given Japan's pragmatic approach to foreign policy, it should come as no surprise that the country has reacted cautiously to a changing international environment defined by China's rise. Tokyo has doubled down on its strategy of deepening its alliance with the United States; sought to strengthen its relations with countries on China's periphery; and pursued closer economic, political, and cultural ties with China itself. The one development that could unhinge this strategy would be a loss of confidence in the U.S. commitment to Japan's defense.

It is not difficult to imagine scenarios that would test the U.S.-Japanese alliance; what is difficult to imagine are realistic ones. The exception is the very real danger that the dispute between China and Japan over the Senkaku Islands (known as the Diaoyu Islands in China), in the East China Sea, might get out of hand, leading to nationalist outbursts in both countries. Beijing and Tokyo would find this tension difficult to contain, and political leaders on both sides could seek to exploit it to shore up their own popularity. Depending on how events unfolded, the United States could well become caught in the middle, torn between its obligation to defend Japan and its opposition to actions, both Chinese and Japanese, that could increase the dangers of a military clash.

The Japanese government, which took control of the uninhabited islands in 1895, maintains that its sovereignty over them is incontestable; as a matter of policy, it has refused to acknowledge that there is even a dispute about the matter. The United States, for its part, recognizes the islands to be under Japanese administrative control but regards the issue of sovereignty as a matter to be resolved through bilateral negotiations between China and Japan. Article 5 of the U.S.-Japanese security treaty, however, commits the United States to "act to meet the common danger" in the event of "an armed attack against either Party in the territories under the administration of Japan." Washington, in other words, would be obligated to support Tokyo in a conflict over the islands—even though it does not recognize Japanese sovereignty there.

The distinction between sovereignty and administrative control would matter little so long as a conflict over the islands were the result of aggression on the part of China. But the most recent flare-up was precipitated not by Chinese but by Japanese actions. In April 2012, Tokyo's nationalist governor, Shintaro Ishihara (who resigned six months later to form a new political party), announced plans to purchase three of the Senkaku Islands that were privately owned and on lease to the central government. He promised to build a harbor and place personnel on the islands, moves he knew would provoke China. Well known for his right-wing views and anti-China rhetoric, Ishihara hoped to shake the Japanese out of what he saw as their dangerous lethargy regarding the threat from China and challenge their lackadaisical attitude about developing the necessary military power to contain it.

Ishihara never got the islands, but the ploy did work to the extent that it triggered a crisis with China, at great cost to Japan's national interests. Well aware of the dangers that Ishihara's purchase would have caused, then Japanese Prime Minister Yoshihiko Noda decided to have the central government buy the islands itself. Since the government already had full control over the islands, ownership represented no substantive change in Tokyo's authority over their use. Purchasing them was the way to sustain the status quo, or so Noda hoped to convince China.

But Beijing responded furiously, denouncing Japan's action as the "nationalization of sacred Chinese land." Across China, citizens called for the boycott of Japanese goods and took to the streets in often-violent demonstrations. Chinese-Japanese relations hit their lowest point since they were normalized 40 years ago. Noda, to his credit, looked for ways to defuse the crisis and restore calm between the two countries, but the Chinese would have none of it. Instead, China has ratcheted up its pressure on Japan, sending patrol ships into the waters around the islands almost every day since the crisis erupted.

The United States needs to do two things with regard to this controversy. First, it must stand firm with its Japanese ally. Any indication that Washington might hesitate to support Japan in a conflict would cause enormous consternation in Tokyo. The Japanese right would have a field day, exclaiming that the country's reliance on the United States for its security had left it unable to defend its interests. The Obama administration has wisely reiterated Washington's position that the islands fall within the territory administered by Tokyo and has reassured the Japanese—and warned the Chinese—of its obligation to support Japan under the security treaty.

Second, Washington should use all its persuasive power to impress upon both China and Japan the importance of defusing this issue. Abe could take a helpful first step by giving up the fiction that no dispute over the islands exists. The Senkaku controversy is going to be on the two countries' bilateral agenda whether the Japanese want it there or not. Abe's willingness to discuss it would give China an opening to back down from its confrontational stance and would better align U.S. and Japanese policy.

Terms of Engagement

Barack Obama's election in 2008 initially raised concerns in Tokyo. Ever fearful that the United States' interest in their country is waning, the Japanese worried that the new U.S. president's Asia policy would prioritize cooperation with China above all and give short shrift to Japan. Those apprehensions have been alleviated, however, thanks to the recent tensions in U.S.-Chinese relations, repeated visits to Japan by senior U.S. officials, Japanese appreciation for U.S. support following the March 2011 earthquake and tsunami, and Washington's decision to sign the Association of Southeast Asian Nations' Treaty of Amity and Cooperation and to join the East Asia Summit.

The Obama administration's emphasis on the strategic importance of Asia, symbolized by the use of such terms as "pivot," "return," and "rebalancing," has been dismissed by some as mere rhetoric. But it is important rhetoric, which has signaled Washington's commitment not only to continued U.S. military involvement in the region but also to a much broader engagement in the region's affairs. By any measure, the administration has succeeded in communicating to U.S. allies and U.S. adversaries alike that Washington intends to bolster its presence in Asia, not downgrade it.

What worries Tokyo now is not the possibility of U.S.-Chinese collusion; it is the prospect of strategic confrontation. Japan's well-being, as well as that of many other countries, depends on maintaining both good relations with China, its largest trading partner, and strong security ties with the United States. Given its dependence on Washington for defense and the depth of anti-Japanese sentiment in China, Japan would have little choice but to side with the United States if forced to choose between the two.

But a conflict between China and the United States would not necessarily strengthen U.S.-Japanese relations. In fact, it would increase the influence of advocates of an autonomous Japanese security policy. Arguing that Washington lacked the capabilities and the political will necessary to retain its leading position in East Asia, they would push for Japan to emerge as a heavily armed country able to protect itself in a newly multipolar Asia. To avoid this outcome and to help maintain a stable balance of power, Washington needs to temper its inevitable competition with China by engaging with Beijing to develop institutions and processes that promote cooperation, both bilaterally and among other countries in the region.

You Can't Always Get What You Want

In assessing the current Japanese political scene and the possible strategic course that Tokyo might chart, it is important to remember that a right-of-center government and a polarized debate over foreign policy are nothing new in Japan's postwar history. Abe is one of the most ideological of Japan's postwar prime ministers, but so was his grandfather Nobusuke Kishi, who was a cabinet minister during World War II and prime minister from 1957 to 1960. Kishi wanted to revise the U.S.-imposed constitution and to undo other postwar reforms; these are his grandson's goals more than half a century later.

But Kishi was also a pragmatist who distinguished between the desirable and the possible. As prime minister, he focused his energies on the latter, negotiating with the Eisenhower administration a revised security treaty that remains the framework for the U.S.-Japanese alliance today. For Abe as well, ideology will not likely trump pragmatism. The key question to ask about Japan's future is not what kind of world Abe would like to see but what he and other Japanese leaders believe the country must do to survive in the world as they find it.

If Tokyo's foreign policy moves off in a new direction, what will drive it there is not an irrepressible Japanese desire to be a great power. Although some Japanese politicians voice that aspiration, they will gain the support of the public only if it

becomes convinced that changes in the international situation require Japan to take a dramatically different approach from the one that has brought it peace and prosperity for decades.

The Japanese public remains risk averse; nearly 70 years after World War II, it has not forgotten the lessons of that era any more than other Asian nations have. And despite changes in the region, the realities of Japanese politics and of American power still favor a continuation of Japan's current strategy: maintaining the alliance with the United States; gradually expanding Japan's contribution to regional security; developing security dialogues with Australia, India, South Korea, and the Association of Southeast Asian Nations; and deepening its engagement with China. China's growing economic clout and military power do present new challenges for Tokyo and Washington, but these challenges can be met without dividing Asia into two hostile camps. If Japanese policy changes in anything more than an incremental manner, it will be due to the failure of Washington to evolve a policy that sustains U.S. leadership while accommodating Chinese power.

Will the Abe government chart a new course for Japanese foreign policy? Only if the public comes to believe that the threat from China is so grave and the credibility of the United States' commitment to contain it is so weakened that Japan's survival is at stake. But if rational thinking prevails in Beijing, Tokyo, and Washington, the approach that has made Japan the linchpin of the United States' security strategy in Asia, stabilized the region, and brought Japan peace and prosperity is likely to persist.

Critical Thinking

1. Under what circumstances would Japan rely on itself for its own defense?
2. What is the relationship between Japanese security and the U.S. pivot to Asia?
3. What is the Chinese–Japanese territorial dispute about?
4. What is meant by a new multipolar Asia?

Create Central

www.mhhe.com/createcentral

Internet References

ASEAN (Association of Southeast Asian Nations)
www.aseansec.org
Japan's Ministry of Foreign Affairs
www.mofa.go.jp

GERALD L. CURTIS is Burgess Professor of Political Science at Columbia University.

Prepared by: Robert Weiner, *University of Massachusetts/Boston*

Article

Beware Collusion of China, Russia

LESLIE H. GELB AND DIMITRI K. SIMES

Learning Outcomes

After reading this article, you will be able to:

- Explain what dual containment is.

- Explain why the U.S. should pay attention to Russian and Chinese fundamental interests.

Visiting Moscow during his first international trip as China's new president in March, Xi Jinping told his counterpart, Vladimir Putin, that Beijing and Moscow should "resolutely support each other in efforts to protect national sovereignty, security and development interests." He also promised to "closely coordinate in international regional affairs." Putin reciprocated by saying that "the strategic partnership between us is of great importance on both a bilateral and global scale." While the two leaders' summit rhetoric may have outpaced reality in some areas, Americans should carefully assess the Chinese-Russian relationship, its implications for the United States and our options in responding.

The Putin-Xi summit received little attention in official Washington circles or the media, and this oversight could be costly. Today Moscow and Beijing have room for maneuver and a foundation for mutual cooperation that could damage American interests.

Specifically, the two nations could opt for one of two possible new courses. One would be to pursue an informal alliance to counter U.S. power, which they see as threatening their vital interests. This path might prove difficult, given competing interests that have burdened relations between Russia and China in the past. Still, stranger things have happened in history between two nations that confront similar challenges. But there is a second possibility. They could play a game of triangular diplomacy similar to the Nixon/Kissinger strategy of the 1970s. In this scenario, Moscow and Beijing could dangle the prospect of a potential alliance or ad hoc cooperative arrangement with the other to gain leverage over Washington and put the United States at a bargaining and power disadvantage.

So far, Russian-Chinese ties appear in large part to be an unintended consequence of American policies aimed at other objectives. Thinking about unintended consequences in foreign policy has never come easily to U.S. policy makers, particularly since the end of the Cold War, when the pursuit of democratic and humanitarian triumphalism has virtually become a form of political correctness among both Republicans and Democrats. Though the wars in Iraq and Afghanistan eventually produced a modest degree of soul-searching, the excitement of the Arab Spring—and the external pressure of the interventionist impulses of Britain and France in particular in Libya and Syria—seems to have cut short this much-needed introspection about what works and what doesn't in U.S. foreign policy.

It is ironic that some European countries that are unable to pursue minimally sound economic policies, or to effectively integrate exploding immigrant populations, have developed the irresistible temptation to promote Europe as a model for the rest of the world—if, of course, the United States supplies the muscle. Taking into account their own history, it is especially curious that these Europeans should not recognize the increasingly apparent reemergence globally of traditional power politics at the expense of their social-engineering vision of peace through democracy.

In fact, the future in many ways now resembles the past, with competing power centers and clashing values. As historian Christopher Clark writes in his magisterial work on the origins of World War I, "Since the end of the Cold War, a system of global bipolar stability has made way for a more complex and unpredictable array of forces, including declining empires and rising powers—a state of affairs that invites comparison with the Europe of 1914." While his stark comparison may seem excessive, there is reason for concern that the current multipolar confusion could once again evolve into two loose alliances or ad hoc alignments increasingly at odds with one another.

America's conventional wisdom virtually dismisses the possibility of a global realignment set in motion by China and Russia, which feel threatened by American and European policies and by having to function in the world's Western-made system. And, whatever the likelihood of a lasting alliance between the two based on their particular strategic interests and values, even a temporary tactical arrangement could have a huge and lasting impact on global politics. Remember the short-lived Molotov-Ribbentrop Pact, which in less than two years had dramatic consequences for the world on the eve of World War II. Hardly anyone in London or Paris could conceive of such a diplomatic development.

True enough, much stands in the way of a genuine Chinese-Russian alliance: a history of mutual mistrust; the combination of China's sense of superiority and Russia's imperial nostalgia; China's declining need for Russian technology, including military hardware; Russia's wariness of substantial Chinese investment in Siberia's energy development; and the fact that in the long run, China and Russia alike need more from the United States and the European Union than from each other.

Nevertheless, Chinese and Russian leaders will measure these very important differences against fundamental interests that Beijing and Moscow have in common. First and foremost, both face challenges to the very legitimacy of their rule as well as serious challenges from restless ethnic and religious minorities. Accordingly, they are highly sensitive to outside influence in their political systems. And make no mistake, what U.S. and European politicians consider noble efforts to promote freedom and democracy look like hostile efforts at regime change to Chinese and Russian leaders. Foreign guidance on governance to countries with different histories, traditions and circumstances is rarely welcomed, particularly by proud major powers.

Second, despite the fact that Russia's leaders played a critical part in destroying the Soviet Union, the West generally has treated Russia as heir to the USSR's policies and objectives. Thus did NATO expand to incorporate not only the former Warsaw Pact but also the three Baltic states. And it has declared its intent to admit Ukraine and Georgia. More broadly, in almost every dispute between Russia and other former Soviet states, even with the authoritarian and repressive Belarus, the United States and the European Union have sided with Moscow's opponents. This creates an impression that the West's top priorities, long after the Cold War, include not merely containing Russia but also transforming it.

Similarly, the United States has supported China's neighbors in nearly all disputes with that country, including territorial disputes. This is not only the case with respect to traditional U.S. allies, such as Japan and the Philippines, but also with Vietnam, which is no more democratic than China and represents a painful episode in American history. The Obama administration's pivot, while weak on substance, has contributed to China's narrative of encirclement. From an American standpoint, these moves make sense, and many Asian nations welcome the pivot. But Beijing predictably sees it as a threat. Thus, it isn't surprising that during President Barack Obama's recent two-day summit with Xi Jinping at Rancho Mirage, California, the Chinese leader kept the atmospherics positive but evaded any concessions on major issues currently separating the two countries.

China and Russia want to break out of what appears to many in both countries as a new "dual containment" policy, and they also wish to reshape a global political and economic system they see as created by and for the United States and the West to their own benefit. Russian and Chinese leaders instantly see their nations disadvantaged when they hear that they should be "responsible stakeholders" in supporting decisions made in Washington and Brussels, when they see the World Bank and the International Monetary Fund operating largely as instruments of Western policy, and when they experience the United States and the European Union regularly orchestrating the international financial system to advance their own interests. More important, all this stimulates a desire to reshape the global rules to accommodate their power and their aspirations. A number of emerging regional powers seem to share these sentiments.

Make no mistake, what U.S. and European politicians consider noble efforts to promote freedom and democracy look like hostile efforts at regime change to Chinese and Russian leaders.

No wonder leading Russian commentator Andranik Migranyan asks rhetorically whether, notwithstanding many common Russian and U.S. interests, there might be "a greater convergence in Russian and Chinese interests on the matter of containing Washington's arrogant and unilateral foreign policy that attempts to dominate the world."

Similar concerns are seen in Beijing and Moscow when the United States pushes them on hot-button issues such as Syria, Iran and North Korea. Certainly, pushing is the right course for Washington. The United States needs their help on these matters, and China and Russia do have their own worries about these countries. But those worries are not necessarily equal to America's, and they have other important priorities to consider. Accordingly, they don't feel comfortable being yoked to American interests, especially when they don't see much effort by Washington to engage in genuine give-and-take or to significantly accommodate their interests in these troubled lands.

Many in Washington seem to believe that notwithstanding the frustrations and ambitions of Chinese and Russian decision makers, they inevitably will wish to avoid rocking the boat in their relations with the United States and the European Union. The European Union is China's number-one trading partner, while the United States is number two—and Russia comes in at number nine. Likewise, the European Union is Russia's top trade partner, with China a distant second. The United States is number four on Russia's list, after Ukraine. China and Russia also have a huge stake in the stability of the euro and particularly the dollar, since a large share of their central-bank reserves is held in these currencies. And China's holdings of U.S. debt give Beijing a big interest in America's fiscal solvency.

Despite these close economic ties, however, history demonstrates that economic interdependence only goes so far in preventing international conflict. Indeed, U.S.-Japanese economic interdependence actually contributed to tensions before World War II. Likewise, before World War I, Britain and Germany were each other's top trading partners. Russia and Germany were economically intertwined before they went to war against each other in 1914—and also before Germany's invasion of the USSR in June 1941. The decisions to go to war in these cases clearly demonstrate that economic interests may be quickly subordinated to national-security concerns and domestic political priorities when disagreements reach the boiling point.

This is why it is a mistake to assume that Washington or Brussels can essentially continue to set the global agenda and decide upon international actions. China and Russia alike agree with the United States and the European Union that it would be better to see Iran and North Korea without nuclear weapons and to avoid Taliban rule in Afghanistan. From Moscow's perspective as well as Beijing's, however, these mutual interests are secondary when set against their efforts to retain influence in Central Asia or East Asia and particularly their desire for stability at home.

Looking to the future, we cannot know the precise consequences of a Sino-Russian alliance if one should emerge. Among other factors, the results would depend on the durability of the arrangement, the strength of the conflicting interests pushing Beijing and Moscow apart, and the magnitude of the pressure from the United States and its allies pushing them closer together. Nevertheless, the Cold War was not so long ago that Americans cannot envision a polarized world, with increasing diplomatic stalemate or worse.

Regarding Iran, imagine if China and Russia offered Tehran security guarantees or promised to rebuild its nuclear infrastructure after a U.S. or Israeli attack. In Syria, we already see the results of having Russia on the other side and China sitting on the fence. Or imagine Chinese support for guerrillas in the Philippines or Kremlin encouragement of Russian-speaking minorities in Latvia and Estonia. If U.S. relations with Russia and China sour, these nightmares can't be excluded.

Russia and particularly China already are steadily increasing and modernizing their military capabilities. For now, Washington is responding with caution to avoid the appearance of overreacting. But picture what might happen if those militaries continue to grow and maneuver worldwide, especially in cooperation with each other. It isn't that war would become likely between the West and these other superpowers. But tensions and conflicts could grow; hot spots could further fester, à la Syria. Great-power animosities would seriously complicate international efforts at crisis management. This all would make international life that much more uncomfortable, if not also dangerous. It certainly would create a specter of miscalculation, escalatory pressure and sense of crisis. And there would be nasty consequences for U.S. prospects for prosperity.

A world of a Sino-Russian alliance or even triangular diplomatic games is certainly not inevitable. But it is a risk the West must be much more aware of. Moreover, making it less likely does not require surrender or appeasement. The United States, Europe, Japan, South Korea, and numerous other allies and friends around the world have enough power and leverage to discourage leaders in Beijing or Moscow who might set aside their own conflicts and seek to disadvantage the United States and the West. But a tough-minded yet prudent American foreign policy based on the world as it actually is would realistically evaluate the interests of other powers and take them into account in order to reduce the risk of provoking a counterbalancing global coalition. Thus, U.S. foreign policy should pay more attention to the benefits of working with Russia and China and taking into account their fundamental interests. Obviously, U.S. leaders must stand their ground on matters of national concern. But more cooperation with Russia and China should be on their minds. Such cooperation is not a reward for good behavior. It's the best and perhaps the only way to defuse crises and reduce international stalemate; it is also a fundamental U.S. national interest.

Critical Thinking

1. Does the future resemble the past with contesting power centers and clashing interests?
2. What is meant by triangular diplomatic games?
3. What is meant by the U.S. policy of dual containment?
4. What is the relationship between the U.S. pivot and Russian–Chinese collusion?

Create Central

www.mhhe.com/createcentral

Internet References

U.S. State Department
www.state.gov
NATO
www.nato.int/

LESLIE H. GELB is president emeritus of the Council on Foreign Relations, a former senior official in the State and Defense Departments, and a former *New York Times* columnist. He is also a member of *The National Interest's* Advisory Council. DIMITRI K. SIMES is president of the Center for the National Interest and publisher of *The National Interest.*

Article

Prepared by: Robert Weiner, *University of Massachusetts/Boston*

The Global Power Shift from West to East

CHRISTOPHER LAYNE

Learning Outcomes

After reading this article, you will be able to:

- Explain what is meant by the Declinists school of thought.
- Explain what is meant by Pax Americana.
- Discuss the new order that is replacing the old order.

When great powers begin to experience erosion in their global standing, their leaders inevitably strike a pose of denial. At the dawn of the twentieth century, as British leaders dimly discerned such an erosion in their country's global dominance, the great diplomat Lord Salisbury issued a gloomy rumination that captured at once both the inevitability of decline and the denial of it. "Whatever happens will be for the worse," he declared. "Therefore it is our interest that as little should happen as possible." Of course, one element of decline was the country's diminishing ability to influence how much or how little actually happened.

We are seeing a similar phenomenon today in America, where the topic of decline stirs discomfort in national leaders. In September 2010, Secretary of State Hillary Clinton proclaimed a "new American Moment" that would "lay the foundations for lasting American leadership for decades to come." A year and a half later, President Obama declared in his State of the Union speech: "Anyone who tells you that America is in decline . . . doesn't know what they're talking about." A position paper from Republican presidential candidate Mitt Romney stated flatly that he "rejects the philosophy of decline in all of its variants." And former United States ambassador to China and one-time GOP presidential candidate Jon Huntsman pronounced decline to be simply "un-American."

Such protestations, however, cannot forestall real-world developments that collectively are challenging the post-1945 international order, often called *Pax Americana,* in which the United States employed its overwhelming power to shape and direct global events. That era of American dominance is drawing to a close as the country's relative power declines, along with its ability to manage global economics and security.

This does not mean the United States will go the way of Great Britain during the first half of the twentieth century. As Harvard's Stephen Walt wrote in this magazine last year, it is more accurate to say the "American Era" is nearing its end. For now, and for some time to come, the United States will remain primus inter pares—the strongest of the major world powers—though it is uncertain whether it can maintain that position over the next twenty years. Regardless, America's power and influence over the international political system will diminish markedly from what it was at the apogee of *Pax Americana.* That was the Old Order, forged through the momentous events of World War I, the Great Depression and World War II. Now that Old Order of nearly seven decades' duration is fading from the scene. It is natural that United States leaders would want to deny it—or feel they must finesse it when talking to the American people. But the real questions for America and its leaders are: What will replace the Old Order? How can Washington protect its interests in the new global era? And how much international disruption will attend the transition from the old to the new?

The signs of the emerging new world order are many. First, there is China's astonishingly rapid rise to great-power status, both militarily and economically. In the economic realm, the International Monetary Fund forecasts that China's share of world GDP (15 percent) will draw nearly even with the United States share (18 percent) by 2014. (The United States share at the end of World War II was nearly 50 percent.) This is particularly startling given that China's share of world GDP was only 2 percent in 1980 and 6 percent as recently as 1995. Moreover, China is on course to overtake the United States as the world's largest economy (measured by market exchange rate) sometime this decade. And, as argued by economists like Arvind Subramanian, measured by purchasing-power parity, China's GDP may already be greater than that of the United States.

Until the late 1960s, the United States was the world's dominant manufacturing power. Today, it has become essentially a rentier economy, while China is the world's leading manufacturing nation. A study recently reported in the *Financial Times* indicates that 58 percent of total income in America now comes from dividends and interest payments.

Since the Cold War's end, America's military superiority has functioned as an entry barrier designed to prevent emerging powers from challenging the United States where its interests are paramount. But the country's ability to maintain this barrier faces resistance at both ends. First, the deepening financial crisis will compel retrenchment, and the United States will be increasingly less able to invest in its military. Second, as ascending powers such as China become wealthier, their military expenditures will expand. The *Economist* recently projected that China's defense spending will equal that of the United States by 2025.

Thus, over the next decade or so a feedback loop will be at work, whereby internal constraints on United States global activity will help fuel a shift in the distribution of power, and this in turn will magnify the effects of America's fiscal and strategic overstretch. With interests throughout Asia, the Middle East, Africa, Europe and the Caucasus—not to mention the role of guarding the world's sea-lanes and protecting United States citizens from Islamist terrorists—a strategically overextended United States inevitably will need to retrench.

Further, there is a critical linkage between a great power's military and economic standing, on the one hand, and its prestige, soft power and agenda-setting capacity, on the other. As the hard-power foundations of *Pax Americana* erode, so too will the United States capacity to shape the international order through influence, example and largesse. This is particularly true of America in the wake of the 2008 financial crisis and the subsequent Great Recession. At the zenith of its military and economic power after World War II, the United States possessed the material capacity to furnish the international system with abundant financial assistance designed to maintain economic and political stability. Now, this capacity is much diminished.

All of this will unleash growing challenges to the Old Order from ambitious regional powers such as China, Brazil, India, Russia, Turkey and Indonesia. Given America's relative loss of standing, emerging powers will feel increasingly emboldened to test and probe the current order with an eye toward reshaping the international system in ways that reflect their own interests, norms and values. This is particularly true of China, which has emerged from its "century of humiliation" at the hands of the West to finally achieve great-power status. It is a leap to think that Beijing will now embrace a role as "responsible stakeholder" in an international order built by the United States and designed to privilege American interests, norms and values.

These profound developments raise big questions about where the world is headed and America's role in the transition and beyond. Managing the transition will be the paramount strategic challenge for the United States over the next two decades. In thinking about where we might be headed, it is helpful to take a look backward—not just over the past seventy years but far back into the past. That is because the transition in progress represents more than just the end of the post-1945 era of American global dominance. It also represents the end of the era of Western dominance over world events that began roughly five hundred years ago. During this half millennium of world history, the West's global position remained secure, and most big, global developments were represented by intracivilizational power shifts. Now, however, as the international system's economic and geopolitical center of gravity migrates from the Euro-Atlantic world to Asia, we are seeing the beginnings of an intercivilizational power shift. The significance of this development cannot be overemphasized.

The impending end of the Old Order—both *Pax Americana* and the period of Western ascendancy—heralds a fraught transition to a new and uncertain constellation of power in international politics. Within the ascendant West, the era of American dominance emerged out of the ashes of the previous international order, *Pax Britannica*. It signified Europe's displacement by the United States as the locus of global power. But it took the twentieth century's two world wars and the global depression to forge the transition between these international orders.

Following the end of the Napoleonic wars in 1815, at the dawn of the Industrial Revolution, Britain quickly outstripped all of its rivals in building up its industrial might and used its financial muscle to construct an open, international economic system. The cornerstones of this *Pax Britannica* were London's role as the global financial center and the Royal Navy's unchallenged supremacy around the world. Over time, however, the British-sponsored international system of free trade began to undermine London's global standing by facilitating the diffusion of capital, technology, innovation and managerial expertise to emerging new centers of power. This helped fuel the rise of economic and geopolitical rivals.

Between 1870 and 1900, the United States, Germany and Japan emerged onto the international scene more or less simultaneously, and both the European and global power balances began to change in ways that ultimately would doom *Pax Britannica*. By the beginning of the twentieth century, it had become increasingly difficult for Britain to cope with the growing number of threats to its strategic interests and to compete with the dynamic economies of the United States and Germany.

The Boer War of 1899–1902 dramatized the high cost of policing the empire and served as both harbinger and accelerant of British decline. Perceptions grew of an ever-widening gap between Britain's strategic commitments and the resources available to maintain them. Also, the rest of the world became less and less willing to submit to British influence and power. The empire's strategic isolation was captured in the plaintive words of Spenser Wilkinson, military correspondent for the *Times:* "We have no friends, and no nation loves us."

Imperially overstretched and confronting a deteriorating strategic environment, London was forced to adjust its grand strategy and jettison its nineteenth-century policy of "splendid isolation" from entanglements with other countries. Another consideration was the rising threat of Germany, growing in economic dynamism, military might and population. By 1900, Germany had passed Britain in economic power and was beginning to threaten London's naval supremacy in its home waters by building a large, modern and powerful battle fleet. To concentrate its forces against the German danger, Britain allied with Japan and employed Tokyo to contain German and Russian expansionism in East Asia. It also removed America as a potential rival by ceding to Washington supremacy over the Americas and the Caribbean. Finally, it settled its differences with France and Russia, then formed fateful de facto alliances with each against Germany.

World War I marked the end of *Pax Britannica*—and the beginning of the end of Europe's geopolitical dominance. The key event was American entry into the war. It was Woodrow Wilson who called the power of the New World "into existence to redress the balance of the Old" (in the words of the early nineteenth-century British statesman George Canning). American economic and military power was crucial in securing Germany's defeat. Wilson took the United States to war in 1917 with the intent of using American power to impose his vision of international order on *both* the Germans and the Allies. The peace treaties that ended World War I—the "Versailles system"—proved to be flawed, however. Wilson could not persuade his own countrymen to join his cherished League of Nations, and European realpolitik prevailed over his vision of the postwar order.

Although the historical wisdom is that America retreated into isolationism following Wilson's second term and Warren Harding's return to "normalcy," that is not true. The United States convened the Washington Naval Conference and helped foster the Washington naval treaties, which averted a United States naval arms race with Britain and Japan and dampened prospects for increased great-power competition over influence in China. America also played a key role in trying to restore economic, and hence political, stability in war-ravaged Europe. It promoted Germany's economic reconstruction and political reintegration into Europe through the Dawes and Young plans that addressed the troublesome issue of German reparations. The aim was to help get Europe back on its feet so it could once again become a vibrant market for American goods.

Then came the Great Depression. In both Europe and Asia, the economic cataclysm had profound geopolitical consequences. As E. H. Carr brilliantly detailed in his classic work *The Twenty Years' Crisis, 1919–1939,* the Versailles system cracked because of the growing gap between the order it represented and the actual distribution of power in Europe. Even during the 1920s, Germany's latent power raised the prospect that eventually Berlin would renew its bid for continental hegemony. When Adolf Hitler assumed the chancellorship in 1933, he unleashed Germany's military power, suppressed during the 1920s, and ultimately France and Britain lacked the material capacity to enforce the postwar settlement. The Depression also exacerbated deep social, class and ideological cleavages that roiled domestic politics throughout Europe.

In East Asia, the Depression served to discredit the liberal foreign and economic policies that Japan had pursued during the 1920s. The expansionist elements of the Japanese army gained sway in Tokyo and pushed their country into military adventurism in Manchuria. In response to the economic dislocation, all great powers, including the United States, abandoned international economic openness and free trade in favor of economic nationalism, protectionism and mercantilism.

The crisis of the 1930s culminated in what historian John Lukacs called "the last European war." But it didn't remain a European war. Germany's defeat could be secured only with American military and economic power and the heroic exertions of the Soviet Union. Meanwhile, the war quickly spread to the Pacific, where Western colonial redoubts had come under intense military pressure from Japan.

World War II reshaped international politics in three fundamental ways. First, it resulted in what historian Hajo Holborn termed "the political collapse of Europe," which brought down the final curtain on the Eurocentric epoch of international politics. Now an economically prostrate Western Europe was unable to defend itself or revive itself economically without American assistance. Second, the wartime defeats of the British, French and Dutch in Asia—particularly the humiliating 1942 British capitulation in Singapore—shattered the myth of European invincibility and thus set in motion a rising nationalist tide that within two decades would result in the liquidation of Europe's colonies in Asia. This laid the foundation for Asia's economic rise that began gathering momentum in the 1970s. Finally, the war created the geopolitical and economic conditions that enabled the United States to construct the postwar international order and establish itself as the world's dominant power, first in the bipolar era of competition with the Soviet Union and later as the globe's sole superpower following the 1991 Soviet collapse.

Periods of global transition can be chaotic, unpredictable, long and bloody. Whether the current transitional phase will unfold with greater smoothness and calm is an open question.

Thus do we see the emergence of the new world order of 1945, which now represents the Old Order that is under its own global strains. But we also see the long, agonizing death of *Pax Britannica,* which had maintained relative global stability for a century before succumbing to the fires of the two world wars and the Great Depression. This tells us that periods of global transition can be chaotic, unpredictable, long and bloody. Whether the current transitional phase will unfold with greater smoothness and calm is an open question—and one of the great imponderables facing the world today.

As the United States emerged as the world's leading power, it sought to establish its postwar dominance in the three regions deemed most important to its interests: Western Europe, East Asia and the Middle East/Persian Gulf. It also fostered an open international-trading regime and assumed the role of the global financial system's manager, much as Britain had done in the nineteenth century. The 1944 Bretton Woods agreement established the dollar as the international reserve currency. The World Bank, International Monetary Fund, and the General Agreement on Tariffs and Trade fostered international commerce. The United Nations was created, and a network of American-led alliances established, most notably NATO.

It is tempting to look back on the Cold War years as a time of heroic American initiatives. After all, geopolitically, Washington accomplished a remarkable double play: while avoiding great-power war, containment—as George F. Kennan foresaw in 1946—helped bring about the eventual implosion of the Soviet Union from its own internal contradictions. In Europe, American power resolved the German problem, paved the way for Franco-German reconciliation and was the springboard for Western Europe's economic integration. In Asia, the United States helped rebuild a stable and democratic Japan from the ashes of its World War II defeat. For the trilateral world of *Pax Americana*—centered on the United States, Western Europe and Japan—the twenty-five years following World War II marked an era of unprecedented peace and prosperity. These were remarkable accomplishments and are justly celebrated as such. Nevertheless, it is far from clear that the reality of the Cold War era measures up to the nostalgic glow in which it has been bathed. Different policies might have brought about the Cold War's end but at a much less expensive price for the United States.

The Cold War was costly in treasure and in blood (the most obvious examples being the wars in Korea and Vietnam). America bears significant responsibility for heightening postwar tensions with the Soviet Union and transforming what ought to have been a traditional great-power rivalry based on mutual recognition of spheres of influence into the intense ideological rivalry it became. During the Cold War, United States leaders engaged in threat inflation and overhyped Soviet power. Some leading policy makers and commentators at the time—notably Kennan and prominent journalist Walter Lippmann—warned against the increasingly global and militarized nature of America's containment strategy, fearing that the United States would become overextended if it attempted to parry Soviet or communist probes everywhere. President Dwight Eisenhower also was concerned about the Cold War's costs, the burden it imposed on the United States economy and the threat it posed to the very system of government that the United States was supposed to be defending. Belief in the universality of American values and ideals was at the heart of United States containment strategy during most of the Cold War, and the determination to vindicate its model of political, economic and social development is what caused the United States to stumble into the disastrous Vietnam War.

Whatever questions could have been raised about the wisdom of America's Cold War policies faded rapidly after the Soviet Union's collapse, which triggered a wave of euphoric triumphalism in the United States. Analysts celebrated America's "unipolar moment" and perceived an "end of history" characterized by a decisive triumph of Western-style democracy as an end point in human civic development. Almost by definition, such thinking ruled out the prospect that this triumph could prove fleeting.

But even during the Cold War's last two decades, the seeds of American decline had already been sown. In a prescient—but premature—analysis, President Richard Nixon and Secretary of State Henry Kissinger believed that the bipolar Cold War system would give way to a pentagonal multipolar system composed of the United States, Soviet Union, Europe, China and Japan. Nixon also confronted America's declining international financial power in 1971 when he took the dollar off the Bretton Woods gold standard in response to currency pressures. Later, in 1987, Yale's Paul Kennedy published his brilliant *Rise and Fall of the Great Powers,* which raised questions about the structural, fiscal and economic weaknesses in America that, over time, could nibble away at the foundations of United States power. With America's subsequent Cold War triumph—and the bursting of Japan's economic bubble—Kennedy's thesis was widely dismissed.

Now, in the wake of the 2008 financial meltdown and ensuing recession, it is clear that Kennedy and other "declinists" were right all along. The same causes of decline they pointed to are at the center of today's debate about America's economic prospects: too much consumption and not enough savings; persistent trade and current-account deficits; deindustrialization; sluggish economic growth; and chronic federal-budget deficits fueling an ominously rising national debt.

Indeed, looking forward a decade, the two biggest domestic threats to United States power are the country's bleak fiscal outlook and deepening doubts about the dollar's future role as the international economy's reserve currency. Economists regard a 100 percent debt-to-GDP ratio as a flashing warning light that a country is at risk of defaulting on its financial obligations. The nonpartisan Congressional Budget Office (CBO) has warned that the United States debt-to-gdp ratio could exceed that level by 2020—and swell to 190 percent by 2035. Worse, the CBO recently warned of the possibility of a "sudden credit event" triggered by foreign investors' loss of confidence in United States fiscal probity. In such an event, foreign investors could reduce their purchases of Treasury bonds, which would force the United States to borrow at higher interest rates. This, in turn, would drive up the national debt even more. America's geopolitical preeminence hinges on the dollar's role as reserve currency. If the dollar loses that status, United States primacy would be literally unaffordable. There are reasons to be concerned about the dollar's fate over the next two decades. United States political gridlock casts doubt on the nation's ability to address its fiscal woes; China is beginning to internationalize the renminbi, thus laying the foundation for it to challenge the dollar in the future; and history suggests that the dominant international currency is that of the nation with the largest economy. (In his piece on the global financial structure in this issue, Christopher Whalen offers a contending perspective, acknowledging the dangers posed to the dollar as reserve currency but suggesting such a change in the dollar's status is remote in the current global environment.)

Leaving aside the fate of the dollar, however, it is clear the United States must address its financial challenge and restore the nation's fiscal health in order to reassure foreign lenders that their investments remain sound. This will require some combination of budget cuts, entitlement reductions, tax increases and interest-rate hikes. That, in turn, will surely curtail the amount of spending available for defense and national security—further eroding America's ability to play its traditional, post-World War II global role.

Beyond the United States financial challenge, the world is percolating with emerging nations bent on exploiting the power shift away from the West and toward states that long have been confined to subordinate status in the global power game. (Parag Khanna explores this phenomenon at length further in this issue.) By far the biggest test for the United States will be its relationship with China, which views itself as effecting a restoration of its former glory, before the First Opium War of 1839–1842 and its subsequent "century of humiliation." After all, China and India were the world's two largest economies in 1700, and as late as 1820 China's economy was larger than the combined economies of all of Europe. The question of why the West emerged as the world's most powerful civilization beginning in the sixteenth century, and thus was able to impose its will on China and India, has been widely debated. Essentially, the answer is firepower. As the late Samuel P. Huntington put it, "The West won the world not by the superiority of its ideas or values or religion . . . but rather by the superiority in applying organized violence. Westerners often forget this fact; non-Westerners never do."

Certainly, the Chinese have not forgotten. Now Beijing aims to dominate its own East and Southeast Asian backyard, just as a rising America sought to dominate the Western Hemisphere a century and a half ago. The United States and China now are competing for supremacy in East and Southeast Asia. Washington has been the incumbent hegemon there since World War II, and many in the American foreign-policy establishment view China's quest for regional hegemony as a threat that must be resisted. This contest for regional dominance is fueling escalating tensions and possibly could lead to war. In geopolitics, two great powers cannot simultaneously be hegemonic in the same region. Unless one of them abandons its aspirations, there is a high probability of hostilities. Flashpoints that could spark a Sino-American conflict include the unstable Korean Peninsula; the disputed status of Taiwan; competition for control of oil and other natural resources; and the burgeoning naval rivalry between the two powers.

These rising tensions were underscored by a recent Brookings study by Peking University's Wang Jisi and Kenneth Lieberthal, national-security director for Asia during the Clinton administration, based on their conversations with high-level officials in the American and Chinese governments. Wang found that underneath the visage of "mutual cooperation" that both countries project, the Chinese believe they are likely to replace the United States as the world's leading power but Washington is working to prevent such a rise. Similarly, Lieberthal related that many American officials believe their Chinese counterparts see the United States-Chinese relationship in terms of a zero-sum game in the struggle for global hegemony.

An instructive historical antecedent is the Anglo-German rivalry of the early twentieth century. The key lesson of that rivalry is that such great-power competition can end in one of three ways: accommodation of the rising challenger by the dominant power; retreat of the challenger; or war. The famous 1907 memo exchange between two key British Foreign Office officials—Sir Eyre Crowe and Lord Thomas Sanderson—outlined these stark choices. Crowe argued that London must uphold the *Pax Britannica* status quo at all costs. Either Germany would accept its place in a British-dominated world order, he averred, or Britain would have to contain Germany's rising power, even at the risk of war. Sanderson replied that London's refusal to accommodate the reality of Germany's rising power was both unwise and dangerous. He suggested Germany's leaders must view Britain "in the light of some huge giant sprawling over the globe, with gouty fingers and toes stretching in every direction, which cannot be approached without eliciting a scream." In Beijing's eyes today, the United States must appear as the unapproachable, globally sprawling giant.

In modern history, there have been two liberal international orders: *Pax Britannica* and *Pax Americana*. In building their respective international structures, Britain and the United States wielded their power to advance their own economic and geopolitical interests. But they also bestowed important benefits—public goods—on the international system as a whole. Militarily, the hegemon took responsibility for stabilizing key regions and safeguarding the lines of communication and trade routes upon which an open international economy depend. Economically, the public goods included rules for the international economic order, a welcome domestic market for other states' exports, liquidity for the global economy and a reserve currency.

As United States power wanes over the next decade or so, the United States will find itself increasingly challenged in discharging these hegemonic tasks. This could have profound implications for international politics. The erosion of *Pax Britannica* in the late nineteenth and early twentieth centuries was an important cause of World War I. During the interwar years, no great power exercised geopolitical or economic leadership, and this proved to be a major cause of the Great Depression and its consequences, including the fragmentation of the international economy into regional trade blocs and the beggar-thy-neighbor economic nationalism that spilled over into the geopolitical rivalries of the 1930s. This, in turn, contributed greatly to World War II. The unwinding of *Pax Americana* could have similar consequences. Since no great power, including China, is likely to supplant the United States as a true global hegemon, the world could see a serious fragmentation of power. This could spawn pockets of instability around the world and even general global instability.

The United States has a legacy commitment to global stability, and that poses a particular challenge to the waning hegemon as it seeks to fulfill its commitment with dwindling resources.

The fundamental challenge for the United States as it faces the future is closing the "Lippmann gap," named for journalist Walter Lippmann. This means bringing America's commitments into balance with the resources available to support them while creating a surplus of power in reserve. To do this, the country will need to establish new strategic priorities and accept the inevitability that some commitments will need to be reduced because it no longer can afford them.

These national imperatives will force the United States to craft some kind of foreign-policy approach that falls under the rubric of "offshore balancing"—directing American power and influence toward maintaining a balance of power in key strategic regions of the world. This concept—first articulated by this writer in a 1997 article in the journal *International Security*—has gained increasing attention over the past decade or so as other prominent geopolitical scholars, including John Mearsheimer, Stephen Walt, Robert Pape, Barry Posen and Andrew Bacevich, have embraced this approach.

Although there are shades of difference among proponents of offshore balancing in terms of how they define the strategy, all of their formulations share core concepts in common. First, it assumes the United States will have to reduce its presence in some regions and develop commitment priorities. Europe and the Middle East are viewed as less important than they once were, with East Asia rising in strategic concern. Second, as the United States scales back its military presence abroad, other states need to step up to the challenge of maintaining stability in key regions. Offshore balancing, thus, is a strategy of devolving security responsibilities to others. Its goal is burden shifting, not burden sharing. Only when the United States makes clear that it will do less—in Europe, for example—will others do more to foster stability in their own regions.

Third, the concept relies on naval and air power while eschewing land power as much as possible. This is designed to maximize America's comparative strategic advantages—standoff, precision-strike weapons; command-and-control capabilities; and superiority in intelligence, reconnaissance and surveillance. After all, fighting land wars in Eurasia is not what the United States does best. Fourth, the concept avoids Wilsonian crusades in foreign policy, "nation-building" initiatives and imperial impulses. Not only does Washington have a long record of failure in such adventures, but they are also expensive. In an age of domestic austerity, the United States cannot afford the luxury of participating in overseas engagements that contribute little to its security and can actually pose added security problems. Finally, offshore balancing would reduce the heavy American geopolitical footprint caused by United States boots on the ground in the Middle East—the backlash effect of which is to fuel Islamic extremism. An over-the-horizon United States military posture in the region thus would reduce the terrorist threat while still safeguarding the flow of Persian Gulf oil.

During the next two decades, the United States will face some difficult choices between bad outcomes and worse ones. But such decisions could determine whether America will manage a graceful decline that conserves as much power and global stability as possible. A more ominous possibility is a precipitous power collapse that reduces United States global influence dramatically. In any event, Americans will have to adjust to the new order, accepting the loss of some elements of national life they had taken for granted. In an age of austerity, national resources will be limited, and competition for them will be intense. If the country wants to do more at home, it will have to do less abroad. It may have to choose between attempting to preserve American hegemony or repairing the United States economy and maintaining the country's social safety net.

The constellation of world power is changing, and United States grand strategy will have to change with it. American elites must come to grips with the fact that the West does not enjoy a predestined supremacy in international politics that is locked into the future for an indeterminate period of time. The Euro-Atlantic world had a long run of global dominance, but it is coming to an end. The future is more likely to be shaped by the East.

> **American elites must come to grips with the fact that the West does not enjoy a predestined supremacy in international politics that is locked into the future for an indeterminate period of time.**

At the same time, *Pax Americana* also is winding down. The United States can manage this relative decline effectively over the next couple of decades only if it first acknowledges the fundamental reality of decline. The problem is that many Americans, particularly among the elites, have embraced the notion of American exceptionalism with such fervor that they can't discern the world transformation occurring before their eyes.

But history moves forward with an inexorable force, and it does not stop to grant special exemptions to nations based on past good works or the restrained exercise of power during times of hegemony. So is it with the United States. The world has changed since those heady days following World War II, when the United States picked up the mantle of world leadership and fashioned a world system durable enough to last nearly seventy years. It has also changed significantly since those remarkable times from 1989 to 1991, when the Soviet Union imploded and its ashes filled the American consciousness with powerful notions of national exceptionalism and the infinite unipolar moment of everlasting United States hegemony.

But most discerning Americans know that history never ends, that change is always inevitable, that nations and civilizations rise and fall, that no era can last forever. Now it can be seen that the post-World War II era, romanticized as it has been in the minds of so many Americans, is the Old Order—and it is an Old Order in crisis, which means it is nearing its end. History, as always, is moving forward.

Critical Thinking

1. Do you agree that the great Power transition underway is dangerous? Why or why not?

2. Did the United States retreat into isolationism after the First World War? Why or why not?

3. How did World War II reshape international politics?

4. Do you agree with the Declinist school of thought as presented in Layne's article? Why or why not?

5. Is China a stakeholder in the international system? Why or why not?

Create Central

www.mhhe.com/createcentral

Internet References

African Union
www.africa-union.org
Non-Aligned Movement (NAM)
www.nam.gov.ir

CHRISTOPHER LAYNE is professor and Robert M. Gates Chair in National Security at Texas A & M University's George H. W. Bush School of Government and Public Service. His current book project, to be published by Yale University Press, is After the Fall: International Politics, United States Grand Strategy, and the End of the Pax Americana.

Layne, Christopher. From *The National Interest*, May/June 2012, pp. 21–31. Copyright © 2012 by National Interest. Reprinted by permission.

Article _____   Prepared by: Robert Weiner, *University of Massachusetts/Boston*

France in Africa: A New Chapter?

STEPHEN W. SMITH

Learning Outcomes

After reading this article, you will be able to:

- Explain why the new reality in the post-Cold War era is continued French intervention in Africa.
- Discuss the legacy of France in Africa.

Are the French fighting Islamist terrorists in Mali to thwart a threat against Europe and the wider Western world? Or has *la Françafrique,* that is, the incestuous postcolonial body politic uniting the ex-metropolis and its former African colonies, arisen from its grave?

On January 11, 2013, the very day that French President François Hollande announced his country would go to war against jihadists in Mali's north, the government and rebels in the Central African Republic signed a power-sharing agreement dictated to them by neighboring states, namely Chad. Despite 600 French soldiers deployed in the Central African Republic for the protection of French expatriates, Paris's erstwhile tutelary shadow was conspicuously absent from that crisis. A regional peacekeeping force, along with some 400 South African soldiers sent to the rescue of beleaguered President François Bozizé, had been dispatched to either steady or tip the balance of power in Bangui, the capital, where France used to rule the roost. In March, Bozizé was toppled without Paris lifting a finger.

So, is this sufficient evidence to conclude that "Africa's gendarme" has finally come out of the cold war? That France has gradually become a rational-choice mid-level power player in francophone Africa, its former preserve? Or, on the contrary, does Mali prove that, where and when it matters, Paris still slides back into its postcolonial rut to assert its *présence* south of the Sahara?

The Wager

One way of addressing these questions is to rewind the complex Franco-African history to the defining moment that was the wager between Kwamé Nkrumah and Félix Houphouët-Boigny, respectively Ghana's champion of independence and the leader of Ivory Coast, at the time still a French colony. On April 7, 1957, Nkrumah visited his Ivorian neighbor, who had

not deigned to attend the ceremony, only a month ago, when the Black Star had replaced the Union Jack and Nkrumah had proclaimed his country "forever free." During three days in Abidjan, the Ghanaian hailed the "political kingdom"— his millenarian vision of independence, heedless of the harsh realities of underdevelopment—as the sine qua non of Africa's emancipation. Wherever Nkrumah appeared, he was frantically acclaimed.

Houphouët-Boigny, then an elected member of the French government, waited until the last day of the visit to respond in public. His falsetto jarred with Nkrumah's sonorous voice as much as the message he carried. "Your experience is rather impressive," he declared. "But on account of the human relationship between the French and the Africans, and because in the twentieth century people have become interdependent, we consider that it would perhaps be more interesting to try a new and different experience than yours, and unique in itself, one of a Franco-African community based on equality and fraternity." This was to be called "the wager," as Houphouët-Boigny suggested they meet again a decade later to compare results.

Ten years later, toppled by a coup d'état, Nkrumah was living in exile in Guinea, the state run by his francophone alter ego, Ahmed Sékou Touré, who had said "no" to Charles de Gaulle's proposal of a Franco-African community, preferring "freedom in poverty to riches in slavery." For his part, Houphouët-Boigny had become, at Ivory Coast's independence in 1960, the president of a country well on its way to superseding Ghana as the world's most important cocoa producer—and overall to turning into an economic "miracle"—while the Black Star sank amid instability and mismanagement.

However, in the 1970s and 1980s, the 50,000 French expatriates running the Ivorian state and economy—five times more than under colonial rule—gave the "miracle" a hollow ring. The dependence on France, arguably a neo– rather than ex-colonial power, was there for all to see. In its own way, Ivory Coast's independence was as nominal as the beggar's choice next door.

Who won the wager? Strictly speaking, within the bet's 10-year limit, Houphouët-Boigny carried the day. Yet, inasmuch as the only way to learn how to play the harp is to play the harp, Ghana at least made its own mistakes and, since the 1990s, seems to have learned from them. This is small comfort for a generation of Ghanaians after independence who grew up in misery and chaos, without much schooling and health care or a functioning state. But it is also little comfort for the Ivorian youth of the past 20 years—years marred by a putsch, a civil war, and an outbreak of xenophobia—to know that their parents had enjoyed a better life before.

In fact, the unexpected winner of the wager might be the analyst, who keeps moving the time horizon and asking the same question, and who realizes that the Franco-African relationship has never been, simplistically, the association of a French rider and his African horses. Though in varying and mostly unequal proportions, there has always been agency on either side.

Whatever the final word on Houphouët-Boigny's choice, it was his choice. Initially affiliated with the French Communist Party, and abhorred by the colonial establishment in Ivory Coast, he eventually offered the colonizer a deal. He became coauthor of a political project that he summarized, in a speech in 1973, as *la Françafrique*.

La Françafrique

Though never in name, the Franco-African state existed both de jure and de facto. Legally, French decolonization—conducted under the informal motto *partir pour mieux rester* ("leaving so as to better stay behind")—resulted in a web of bilateral defense and aid agreements subsumed under the label *Coopération*. Since 1960, when the bulk of French colonies in sub-Saharan Africa—15 countries—acceded to independence over the course of a summer, the ministry in Paris in charge of Africa has been *la Coop*, short for *ministère de la Coopération*. It used to be subdivided into military and civilian departments.

For decades, a French general managed a latticework of defense agreements, which were all similar in writing, with France guaranteeing the territorial integrity of each former colony and pledging to train the local army in exchange for the free right to station in the African country, or transition through it, any number of soldiers. In secret clauses, France was furthermore granted a right of first purchase of the raw materials that its former possessions exported, from uranium in Niger to cocoa or coffee in Ivory Coast to oil in Cameroon, Gabon, or Congo-Brazzaville. In turn, Paris committed itself to intervene militarily not only against an external threat to an African partner, but also "in case of internal turmoil." This amounted to life insurance for friendly regimes.

On the civilian side, so-called *coopérants*—technical assistants or advisers—were dispatched by the thousands to prop up shaky state apparatuses south of the Sahara. Education, health care, and the higher echelons of administration were flooded with *coopérants*, who enjoyed a comfortable lifestyle in the tropics while making a pile of the common currency in former French Africa, the CFA franc. (Unchanged after independence, the acronym evolved from meaning *Colonies françaises d'Afrique* to *Communauté financière d'Afrique*.) In wealthier countries such as Ivory Coast, Cameroon, or Gabon, their salary and housing were paid for by the host country. In poor states, namely in the Sahel, France ploughed back into Africa part of the benefit it reaped from the system as a whole.

Generally speaking, the Franco-African state functioned like a federation, with Paris as its overbearing center in charge of defense, diplomacy, and the macro-management of the Franco-African economy. But even weak members of the "family"—a paternalistic term widely used in the heyday of *la Françafrique*—gained leverage from the collective bargaining power of francophone Africa (in the 1970s, the former Belgian colonies were integrated into the Franco-African state). A food crisis in Niger would become a test of France's wider commitment to its strategic depth in Africa, much as Libyan leader Colonel Muammar el-Qaddafi's repeated attempts to take over Chad obliged Paris to intervene time and again in a country that, before the discovery of oil, produced as much national wealth as a provincial town of 200,000 inhabitants in France.

The Franco-African state was part of the Gaullist aspiration for *grandeur* despite France's defeat and occupation in the Second World War. Some 250,000 African soldiers had fought Hitler's Germany for *la France libre*. Victory achieved, de Gaulle sent them back home as colonial subjects. But the lesson was not lost on him. When he returned to power in 1958, four years after France's loss of Indochina and in the midst of the Algerian war for independence, he entrusted the decolonization of sub-Saharan Africa to the leader of a World War II resistance network, Jacques Foccart.

"Our positions in Algeria have been squandered by plentiful mistakes, bloodshed, and suffering," de Gaulle told him. "Only Black Africa is left, and here the decolonization must succeed as a friendship, with us accompanying the peoples of these countries. This is what I ask you to be in charge of." Foccart, born in 1913 in Guadeloupe, became the fulcrum of the Franco-African state, not only under de Gaulle but also under his successors, from Georges Pompidou to Jacques Chirac. Until his death in 1997, at the age of 84, Foccart embodied the continuity of the postcolonial pact between France and its former colonies south of the Sahara.

Elite cooperation was the cement of the Franco-African state. As a result of its colonial policy of "assimilation," Paris had nurtured a sub-Saharan elite of "black Frenchmen" (women were rarely part of the happy few), who had bought into the normative universalism of French culture and politics. Quite a number of them had been elected to the French parliament at a time when "colored people" in parts of the United States were not allowed to use the same public facilities as whites. Houphouët-Boigny had been a minister of health in France, and the protagonist of a much disputed health care reform—without any public reference ever made to his skin color or his origin in a colony.

So was this a genuine partnership? Or was it, as Senegal's poet-president Léopold Sedar Senghor once claimed in a spark of ire, "Kollaboration"—intelligence-sharing with the enemy and a sellout of the African masses under neocolonial rule? There is room for debate. However, if the lot of ordinary Africans is accepted as a yardstick, it is not a foregone conclusion that Guineans under Sékou Touré were sold down the river less than Ivorians under Houphouët-Boigny.

After the Funeral

It is hazardous to date the demise of the Franco-African state, but three events in 1994 epitomized its disintegration. First, an unprecedented devaluation of the CFA franc brought down the monetary wall around the Franco-African enclave economy. Second, genocide in Rwanda left blood on the escutcheon of Africa's gendarme, who was accused of having sided with the Hutu planners and perpetrators of the 1994 genocide against the Tutsis. Finally, that year a state funeral was held for Houphouët-Boigny, the godfather of *la Françafrique* and, in tandem with Foccart, the co-manager of the Franco-African state.

The Franco-African state was laid to rest in Houphouët-Boigny's native village, and in the presence of its remaining dramatis personae: two generations of French and African heads of state, prime ministers, mercenaries, merchants, and minions. The octogenarian Foccart attended in a wheelchair. By his side, in the front pew, sat Jean-Christophe Mitterrand, who, in the early 1990s, was in charge of Africa for his terminally ill father, François Mitterrand, the Socialist president. *Papa m'a dit* ("daddy-told-me"), as Mitterrand's son was nicknamed, did not even attempt to reform a system whose fate was linked to the cold war.

This crucial fact is easily overlooked by observers, often French, who tend to concentrate on colonial history and its aftereffects: The bipolar postwar world order was the geopolitical sine qua non of the Franco-African state. During the cold war, the French army acted as an auxiliary of the "free world," intervening no fewer than 39 times in sub-Saharan Africa to change the course of history with a few hundred men. (Most often this meant saving or removing a president.) During this period, the state-tethered business that France conducted south of the Sahara prospered on sweetheart deals, which Western competitors such as the United States and Germany countenanced as remuneration of Africa's gendarme. Meantime, Paris remained dependent on the old imperialist hallucinogen known as *La plus grande France*, "Greater France."

In this regard, the fall of the Berlin Wall meant cold turkey for France. It also meant that francophone Africa became a competitive market within a global economy. The cold war's end led to the collapse of France's comfortable trade surplus with sub-Saharan Africa, which had been on the order of 2 billion euros a year—that is, between two and three times the net benefit from France's far greater volume of trade with the United States.

Under pressure to adapt, French companies parlayed their wealth of African experience. In 1980, the proportion of France's overseas capital investment directed to Africa had stood at 35 percent, compared with 29 percent for the United Kingdom and 19.5 percent for West Germany. By 1995, Britain's share had dropped to 3.8 percent and Germany's to 2.4, while France remained assertive, and exposed, at 30.4 percent. However, most French investments had been redirected to non-francophone countries, namely Nigeria, Angola, Kenya, and South Africa. (Since the 1990s, French investment in Africa has drawn level with other industrialized countries. For the past five years, it has remained consistently under 3 percent of total foreign investment.)

Surreptitiously, *la Françafrique* changed its size and boundaries on the continent. No longer was it the postcolonial territory staked out by Houphouët-Boigny. Increasingly it was a new frontier for high-risk high-reward venture capital, in particular in the oil industry. There was still substantial overlap between the old and the new in countries like Gabon, Congo, or Cameroon. But at the very moment that *la Françafrique* (a term Foccart had never used) entered the media-driven public debate in France—that is, at the end of the 1990s, after Foccart's death—the word blurred the lines between two quite distinct realities: the late Franco-African state, on the one side, and, on the other, the new and wider but also shallower Franco-African business relationship. Since both were rife with corruption and shady dealings, they subsequently fed off each other to compound a single, rather monstrous entity.

Look Back in Anger

La Françafrique, Houphouët-Boigny's eulogistic description of the Franco-African Atlantis, which had hitherto traded solely among insiders of African affairs, now plunged into the French public sphere like a meteorite. It became incendiary, a vituperation of the past. And there was indeed much that deserved the flames. After decades of complacent silence along the lines of "what is good for France's standing in the world is also good for Africa," French public opinion at the end of the 1990s began to assess what the Franco-African state had really been about.

In hindsight, the phenomenon added up to a sum of scandals. Ignominious African dictators such as Zaire's Mobutu Sese Seko and Jean-Bédel Bokassa, the "emperor" of the Central African Republic, had been hosted and toasted in Paris. Mercenary adventures to overthrow "unfriendly regimes," such as Bob Denard's attempted invasion of Marxist Benin in 1977, and his repetitive coups d'état in the Comoro Islands throughout the 1980s, had been sanctioned as undercover operations by the French state.

Elf Aquitaine, the state-owned French oil company, had turned corruption into something more than standard in-house practice—it became a giant sewer system linking the political leadership in Paris to the petroleum-producing capitals of Africa. And finally, reviving memories of colonial "pacification," the French army had intervened in favor of the Hutu regime in Rwanda, which was held responsible for genocide.

Not much historical context survived France's look back in anger. For example, it overlooked the fact that, throughout the cold war, only one African leader—the Mauritian prime minister Seewoosagur Ramgoolam in 1982—relinquished power in the wake of an electoral defeat. Rather, French media gobbled up the claim by African opposition leaders (who typically failed to field a single candidate to challenge the incumbent head of state) that France highhandedly "installed" its best friends—all "dictators"—in power in Africa. Not only were Senghor or Houphouët-Boigny equated to Mobutu or Bokassa, but bloody tyrants like Uganda's Idi Amin, Equatorial Guinea's Francisco Macías Nguema, or Nigeria's Sani Abacha—none of them of French making—were passed over in silence.

By the same token, Africa's much-reviled gendarme was not given credit for the fact that he had not only protected France's satraps. Between 1960 and 1990, roughly 40,000 people are estimated to have died as a result of internecine violence in former French Africa, half of them in Chad. By comparison, about 2 million died in former British Africa, another 2 million in former Belgian Africa, 1.2 million in the former Portuguese colonies, and another million in the residual category that includes Ethiopia, Somalia, Liberia, and Equatorial Guinea. A different indicator, which corrects for demographic imbalances, corroborates the (nonetheless ambivalent) value of the *pax franca:* For the same period between 1960 and 1990, the number of "victims of repression or massacres" stands at 35 per 10,000 inhabitants in ex-French Africa; 790 in postcolonial Anglophone Africa; 3,000 in the Belgian Congo, Rwanda, and Burundi; and a staggering 4,000 in the former Portuguese colonies, which did not achieve independence until the mid-1970s.

In the emotional heat of the French public's soul-searching exercise, it all became a blur. Even more so as new grist, since the late 1990s, has constantly been added to the mill. Another French oil company, Total, was accused of bribery in Nigeria (scant surprise, we had seen that before). A French arms manufacturer, Thales, allegedly corrupted South African leader Jacob Zuma and his entourage (how would you expect otherwise, given France's track record in Africa?). A multi-billion-dollar arms shipment, paid for in Angolan oil, had transited France (the deal was promptly dubbed "Angolagate" because the middleman, Pierre Falcone, had handed out cash to politicians and opinion makers in Paris to peddle influence).

But, wicked as these practices most likely were, what had they to do with the defunct Franco-African state? Certainly Nigerians and Zuma did not need France's colonial past to teach them business practices, good or bad. And Angola, a former Portuguese possession, was not part of France's postcolonial playing field, level or atilt. Why would a corrupt French businessman be a simple criminal if he operated, say, in Brazil, but a living testimony to France's opprobrious postcolonial presence in Africa, even if he acted outside of France's historical zone of influence on the continent?

The New Reality

The Franco-African state is dead and buried. Since the end of the cold war, there have been no more commercial safe havens in Africa—China's spectacular rise attests to this, and other countries such as India and Turkey are following in its wake. No more geopolitical rent is paid to Africa's gendarme who, not incidentally, has closed three of his six permanent bases on the continent since 1989. Today, anywhere between 500,000 and a million Chinese are living in Africa, while the number of French expatriates has been halved since the fall of the Berlin Wall, from about 200,000 to 100,000.

And this overall figure masks a migration to other, often non-francophone countries, as well as the fact that a high proportion of the French in Africa have dual citizenship. A stark example is Ivory Coast: From 50,000 in the mid-1980s, the number of

French has fallen to 8,000, of whom only an estimated 1,200 are not Franco-Lebanese or Franco-Ivorians. At the same time, *la Coopération* has been steadily dismantled: The number of technical assistants in sub-Saharan Africa has dropped below 1,000 from around 6,500 in the early 1990s; there were 925 French military advisers on the continent in 1990, only 264 by 2008. That year the French budget for military assistance reached a low-tide mark of 60 million euros, down by 50 percent since the mid-1990s.

It was not only changing geopolitics that demolished the Franco-African state. Demography and democracy also helped. To stick to Ivory Coast as an example, its population has increased six-fold since 1960, rising from 3.5 million to 21 million. The country's main city, Abidjan, has grown even faster, from a modest town of 180,000 inhabitants to a lagoon megalopolis of an estimated 4 million—22 times as many. Meanwhile, the number of Parisians has risen from 7 to 10 million, that is, by 30 percent, while the French population overall has increased by 40 percent, from 45 to 64 million.

In 2004, when the then-Ivorian president Laurent Gbagbo unleashed his "young patriots" against the French expatriates in Abidjan, there was little the French army could do to protect them in the urban maze. It was a matter of sheer numbers. For the same reason, Paris has had to share the stewardship of its monetary zone in Africa with the European Union and international financial institutions such as the World Bank and the International Monetary Fund. The days when Paris could singlehandedly bail out francophone Africa with its budgetary aid are long gone.

Long gone, too, is the inbred elite connivance between Paris and the francophone capitals in Africa. The Franco-African microcosm is swept by winds of change. Today, most young French do not take gap-years to cross the Sahara and venture into black Africa; they travel in Europe, Asia, or the Americas. On their side, talented or privileged young Africans study in the United States, Great Britain, Germany, or Canada instead of France. All else being equal, their choice between the Sorbonne and Stanford is more often than not decided in favor of California.

Democracy, even under its widespread pseudo-form of competitive authoritarianism, has broken the monolithic mold of the ruling class in Africa. Some African politicians still view Paris as a strategic constituency, but others are busy currying favor with Washington if not Beijing. In any event, incumbency no longer equates with eternity the way it has in the past. Power changes hands. Again, this holds true not only for Africa. Houphouët-Boigny, who ruled Ivory Coast from 1960 to 1993, found a similarly permanent interlocutor in Foccart, who was irremovable as the chief "Africa hand" since France's Fifth Republic made no room for an alternation in power before 1981—and then only to reveal that the Socialists had rallied to the Gaullist idea of a Greater France. (Would it be outrageous to conclude that, in this regard, France resembled its colonies turned into "independent" states? It too was a "democracy" in which the key governing personnel would not be voted out of power.)

Successive French presidents—from Mitterrand to Hollande via Chirac and the post-Gaullist Nicolas Sarkozy—all pledged to issue *la Françafrique's* death certificate. But with the exception of Hollande, for whom the jury is still out, they failed to deliver. Apparently, the Franco-African state was too exquisite a corpse to let go. Paris no longer called the shots in francophone Africa, but why acknowledge this sobering reality against the powerful make-believe of *grandeur?*

On the African side, after decades of postcolonial guilt-peddling, why admit to one's own limited capacities when the French were still there to blame? Since the mid-1990s, anti-French feelings had run high in the former colonies and, as Gbagbo demonstrated in Ivory Coast, this ground-swell of anger provided a potent political resource for African leaders touting a "second independence." All the more so as French public opinion had sunk deep into postcolonial shame, as France mutated into a disconsolate ex-empire averting its eyes from the present to lick the wounds of the past. Once again, just as during the cold war but now for solipsistic reasons, the realities on the ground—the "Africa of the Africans"—did not count for much.

The Gendarme Returns

This period, however, may be coming to an end. For one, self-punishing remorse seems to have run its course in Paris. In a more realistic assessment, France views francophone Africa again as a crucial echo chamber for its international pretensions, not to mention the continent's raw materials which, *merci la Chine,* are more coveted than ever. Paris is also warming up to the realization that, with 1 billion inhabitants, Africa is an awakening giant. Even if only a fraction of its population escapes the poverty trap, the continent will be an important market where France could benefit from comparative advantages, among them *la Francophonie* and culturally shaped consumption patterns.

Moreover, with nearly 10 percent of its population originating from Africa, the former metropolis has a vested human interest in the emerging continental powerhouse. Otherwise, "black France" might turn into an even bigger problem than it revealed itself to be in the fall of 2005 when "race riots" erupted in the poor outskirts of Paris.

Sarkozy, under whom the "parallel diplomacy" of semi-official intermediaries with Africa was allowed a comeback,

intervened in 2011 in Ivory Coast to pick the presidential winner, Alassane Ouattara, a personal friend of his. Tipping the balance of power in Abidjan made political sense, however, inasmuch as stability in Ivory Coast remains a condition of stability in francophone West Africa. Moreover, the international community was eager for Paris to do the job once the United Nations had officially "certified" Gbagbo, the incumbent since 2000, as the loser of a free and fair election.

Though no longer Africa's gendarme, France is still a policeman with a big stick on the continent. The action in Mali illustrates that, post-Gaullist or neo-Socialist, French presidents are again prepared to intervene with the overwhelming support of public opinion. Still, the French could be right about their new, "post-neocolonial" assertiveness in general, and nevertheless wrong in Mali, where the military confrontation with jihadists risks tardily buying into former US President George W. Bush's "global war on terrorism." Be that as it may, it would be wrong for reasons owing nothing to *la Françafrique.*

Critical Thinking

1. What is meant by La Françafrique?
2. What is the crtique of French policy in Africa during the Cold War?
3. What is the new reality of the French relationship with Africa?

Create Central

www.mhhe.com/createcentral

Internet References

The French Foreign Ministry
www.diplomatie.gouv.fr/en

Mali's Embassy in the United States
www.Maliembassy.US

Office of the Coordinator for Counterterrorism
www.state.gov/s/ct

STEPHEN W. SMITH, a former Africa editor of Le Monde, is a visiting professor of African and African-American studies at Duke University.

Smith, Stephen W. From *Current History*, May 2013, pp. 163–168. Copyright © 2013 by Current History, Inc. Reprinted by permission.

Prepared by: Robert Weiner, *University of Massachusetts/Boston*

Article

Shifting Fortunes: Brazil and Mexico in a Transformed Region

MICHAEL SHIFTER AND CAMERON COMBS

Learning Outcomes

After reading this article, you will be able to:

- Explain why Brazil and Mexico are the two most important countries in Latin America.
- Discuss the changes in Brazilian and Mexican narratives.

Just three years ago, Brazil, the country that long marketed itself as the home of samba and soccer, was all the rage in global affairs. Adorning its cover with Rio de Janeiro's Christ statue shooting up like a rocket, the *Economist* proclaimed in November 2009 that Brazil was "taking off." The first among the BRICS (Brazil, Russia, India, China, and South Africa), the preeminent power bloc of developing nations, Brazil was, it seemed, realizing its enormous potential. It was enjoying strong economic growth. It was aggressively reducing levels of poverty—and even inequality—while fostering a vibrant democracy. If the twenty-first century was really to be the Century of the Americas, Brazil was poised to lead the way. Its star power had few peers.

While Brazil basked in effusive praise, Mexico, the other Latin American titan, was portrayed in markedly negative, sometimes even macabre, terms. The media described unremitting, drug-fueled violence that, in the most extreme depictions, threatened the very integrity of the Mexican government. Indeed, in a US government report, Mexico was viewed as possibly becoming a "failed state." Spreading criminality was compounded by a severe recession in 2009. The economy contracted by over 6 percent, largely the result of a financial crisis originating in the country to which Mexico was inextricably tied: the United States.

In 2013, the narratives of the two countries that account for over half of Latin America's population and almost two-thirds of its economic output have nearly been inverted. In the past two years, Mexico's economic growth has exceeded Brazil's; its violence, though still severe, declined markedly in 2012; and the onset of a new administration in Mexico City committed to pursuing far-reaching reforms has injected a measure of optimism regarding the country's outlook.

In sharp contrast to Mexico's moderate growth of roughly 4 percent, Brazil's economy expanded by just 1 percent in 2012. Among other constraints, vulnerability resulting from Brazil's heavy dependence on commodity sales—chiefly to an economically slowing China—became increasingly evident. In addition, international media accounts have focused on escalating violence in São Paulo and police raids into the *favelas* of Rio de Janeiro. And, as the World Bank has reported, it is far more difficult to do business in Brazil: Investors have become increasingly frustrated with a cumbersome public sector.

The ostensibly shifting fortunes of both countries caution against hyperbole and argue for nuance. The dominant characterizations of three years ago were overdrawn, and so are many today. Brazil and Mexico represent distinctive approaches to governance and development, and their positions in the Western Hemisphere and the world are appreciably different. Until now, their separate paths have only rarely intersected.

Brazilians and Mexicans are keenly conscious of each other's performance, making comparisons inevitable. Still, it would be a stretch to describe the relationship as a rivalry in the strictest sense. Brazil and Mexico seldom clash over major economic questions (making a recent dispute over auto sales especially stand out), and their policies are hardly antagonistic. Rather, the two countries often jockey for position, power, and influence in a Latin America that in recent decades has been utterly transformed.

By most measures, both nations have advanced in remarkable ways. But they also face similar, long-term challenges—not unlike those confronting the United States—including high inequality and deficient infrastructure and education systems.

The Brasília Consensus

Latin America has long been a veritable laboratory of political experimentation, yet over the past decade a formula for success embodied by the Brazilian experience has taken hold and exerted broad appeal. The ingredients consist of a commitment to economic growth through fiscal discipline, a significant concern for poverty reduction, and a deepening of democracy. Progress in each sphere can be attributed in part to a succession of effective political leaders, each with a remarkable personal story: Fernando Henrique Cardoso (1995–2002), Luiz Inácio "Lula" da Silva (2003–2010) and, currently, Dilma Rousseff.

All have, under different circumstances, sought to balance and integrate these three ingredients. Cardoso's chief accomplishments were taming Brazil's chronic inflation and initiating a more vigorous social agenda. Under Lula, and especially during his second term, Brazil "took off" in impressive fashion. The combination of strong growth and the expansion of a conditional cash transfer program, *Bolsa Família,* helped lift some 20 million Brazilians out of poverty during Lula's eight years in office. Rousseff has adhered to the broad outlines of her predecessors' approaches, though she has had to confront difficult challenges—both external and domestic—that have led to slower economic growth.

Few, if any, countries can match such competence and continuity in political leadership. Solid macro-economic management, combined with a profound concern for the social agenda and a penchant for political negotiation, has produced extraordinary results in policy making. Brazil's model was particularly tested by the 2009 financial crisis, which, defying many expert predictions, it weathered remarkably well. Its economy, today the world's seventh largest, grew by 7.5 percent in 2010.

Brazil has also been at the forefront of the development of biofuels (ethanol in particular). And with its new oil finds and large reserves of shale gas, the country appears poised to become a major global energy power. The Brazilian government plays an active, interventionist role in the economy that, coupled with a strong private sector, has resulted in a hybrid, public-private approach toward economic development. This approach is reflected in Brazil's biggest and best-known companies, such as the mining juggernaut Vale, the aerospace conglomerate Embraer, and, of course, the oil giant Petrobras.

Brazil's record on its domestic agenda over the past decade has enabled it to play a more influential, high-profile role in global and regional affairs—fulfilling its long-held feelings of *grandeza* (greatness). The country will surely be in the spotlight in coming years as it hosts the 2014 World Cup and 2016 Summer Olympics. At the global level, Brazil has been notably assertive in forums such as the Group of 20 (Mexico and Argentina are the other Latin American members) and the World Trade Organization.

Its BRICS membership and ambition to become a permanent member of the United Nations Security Council (Brazil has been a non-permanent member 10 times since 1946) have opened doors for its presidents and top officials across the globe. Under Lula, buoyed by progress on the domestic front, Brazil pursued a more aggressive foreign policy, seeking to perform a mediating role in the Middle East while developing a joint proposal with Turkey (much to Washington's displeasure) for dealing with Iran's nuclear program. Showcasing Brazil's commendable HIV/AIDS efforts, Lula also pursued more robust south-south diplomacy, including deeper cooperation with Africa.

Brazil has sought to enhance its global position with greater engagement in regional affairs. Regarding itself much more as a "South American" than a "Latin American" nation, Brazil has principally concentrated on its immediate geographic sphere of influence, while showing limited interest in Central and North America, which are viewed as the United States' strategic prerogative. Indeed, Brazil's distance and independence from Washington have been key to its aspiration to be taken more seriously as a global actor.

In 1991, Brazil jointly spearheaded the effort to create Mercosul (whose original members also included Argentina, Uruguay, and Paraguay), which aimed to eventually become a common market. This organization has since become less credible on trade questions, diverging from its original purpose and acquiring a more political cast. For many, this perception crystallized when Venezuela became a Mercosul member through the back door—the result of a hasty suspension of Paraguay (whose senate had been blocking Venezuela's entry) following President Fernando Lugo's (equally hasty) impeachment in June 2012.

Also on the regional front, building on what Cardoso began years before, Brazil (with Venezuela's backing) in 2008 launched the Union of South American Nations (UNASUL). This organization assumes an explicitly political function; it has, for example, established an affiliated South American Defense Council. In 2011, in a move that partly reflected the evolving Brazil-Mexico relationship, the Community of Latin American and Caribbean States (CELAC) got off the ground, encompassing all of the hemisphere's governments—with the exception of Canada and the United States.

Although Brazil has had little choice but to devote increased attention to hemispheric concerns in recent years, it has accorded even greater precedence to its global priorities.

Mexico's Trajectory

Geography may not be destiny, but Mexico's location just south of the United States has—for better or worse—significantly shaped its economic, security, and demographic situation. Roughly 80 percent of Mexico's exports, chiefly manufactures, go to the United States—compared with less than 20 percent of Brazil's. Two decades ago, the United States, Mexico, and Canada adopted the North American Free Trade Agreement (NAFTA), which has substantially boosted trade and investment among the three countries. Today Mexico is the second-largest trading partner of the United States, slightly ahead of China. Some experts predict that in six to eight years, Mexico could overtake Canada as the United States' principal trading partner, as connections across the Rio Grande continue to deepen on all fronts.

Under the administration of President Felipe Calderón, which ended in December 2012, the main story of the past six years has been Mexico's drug-fueled violence. It has taken some 60,000 Mexican lives. It has also hurt the country's economy (investment and tourism in particular) and, of course, its international image. Major US publications have frequently invoked sensationalist phrases to describe the country: "deepening drug-war mayhem," "reign of terror," "criminal anarchy." When Calderón took office in 2006, he called on the military to play a central role in carrying out the drug war, a move questioned by some Mexican and international observers.

As with trade, the links with the United States in the drug war are profound—but in this case, far less benign. As the world's largest consumer of cocaine and the primary provider of the smuggled arms used in drug-related killings, the United States bears an enormous responsibility for the criminal violence wracking Mexico. The Mérida Initiative, a program of cooperation between Mexico and the United States to combat drug-fueled violence, has been in place since 2008. While security collaboration has never been higher, the level of US support falls short of the magnitude of the challenge, and implementation of the initiative has been plagued by delays and inefficiencies. At the same time, the United States has been unable or unwilling to tackle contentious and difficult domestic reforms related to drug policy and gun control that could help allay the situation.

In recent years, Mexico's security crisis and even stronger ties with its northern neighbor have restricted its ability to pursue a more multifaceted foreign policy, as it once did during the Central American civil conflicts of the 1980s through the Contadora Group, a regional initiative. Instead, it has focused on deepening cooperation on a wide-ranging agenda with the United States, and on trade promotion worldwide. Mexico today has free trade agreements with some 40 countries, compared with the United States' 18.

Before 2010, Mexico's trade boom had been accompanied by lackluster economic growth, which encouraged significant out-migration to the north. This was largely the result of failure to pursue long-needed energy, fiscal, and education reforms, together with economic problems in the United States. For many years, the effect of China's rising role in the region was also considerably less helpful than it was in the case of Brazil. China, when its labor costs were lower, dealt a blow to Mexico's manufacturing sector, causing scores of factories to close and many jobs to be lost. Hit hard by the financial crisis and understandably wearied by the frustrating drug war, Mexico dropped to a low point in 2009 as pessimism intensified. At the same time, hype about Brazil was nearing its zenith.

Turning the Tables?

In the last three years of the Calderón administration, the outlook in Mexico began to brighten. Most notably, the economy began climbing out of its deep hole, with growth approaching 4 percent in 2011 and 2012. The US recovery, however anemic, gave Mexico a needed boost. With labor costs increasing in China, foreign investment in Mexico began to increase, and exports became more competitive.

A new narrative has emphasized an expanding middle class (by some estimates, reaching nearly 50 percent of the population); a reduction in poverty levels (the conditional cash transfer program, though often associated with Brazil, actually began in Mexico in the 1990s); a public mood increasingly ready for real reforms; and, according to an April 2012 study produced by the Pew Research Center, net zero migration to the United States.

Further, although the overall security situation may not be on a sustained path of improvement, levels of violence declined in 2012. Some success stories have even begun to emerge: Ciudad Juárez, long a site of unprecedented levels of violence, recently witnessed a sharp drop in homicides.

The election of Enrique Peña Nieto of the Institutional Revolutionary Party (PRI) in July 2012 helped crystallize the perceptions of positive change. Although the PRI had dominated Mexico's political system for more than seven decades—and its highly authoritarian old guard has hardly disappeared—many argue that Mexican society has become far more democratic since the 1990s. There is little prospect of recreating the old PRI model in the context of an increasingly autonomous and assertive Congress, judiciary, civil society, and media.

Since the beginning of the new administration and announcement of the cabinet on December 1, 2012, there have been clear signs that Peña Nieto is moving quickly to undertake a broad reform agenda. Observers are relatively sanguine about the prospects for reform that would open Mexico's energy sector, with its significant endowments of oil and gas, to some form of private sector participation.

Even during the lame duck session of Congress, labor reform designed to create a more flexible and formal workforce was adopted with ample support. And, within the framework of a new political pact among the country's three main parties, Peña Nieto has left little doubt that he plans to introduce badly needed education reforms, which would likely mean a fierce battle with the country's largest and most powerful teachers' union.

Just as Mexico has shown signs of an upswing, Brazil seems to be facing mounting difficulties. A slump in key European markets and the economic slowdown in China, together with indications that its huge domestic market has become overly leveraged, requires significant adjustments in Brazil's economic policy. Some argue, moreover, that Brazil has begun to reach limits on rapid growth unless it pushes through pending reforms to boost productivity across all sectors.

For example, a recent *Foreign Affairs* article entitled "Bearish on Brazil," by Morgan Stanley's Ruchir Sharma, criticized the country for being overly dependent on commodity exports and not investing enough in infrastructure or manufacturing. The World Economic Forum's 2012–2013 Global Competitiveness Report ranked Brazil 107th worldwide in quality of overall infrastructure—far behind Mexico at 65th. Unlike Brazil, Mexico has recently been lauded for diversifying its exports over the past two decades: In the 1980s, 10 percent of Mexico's total exports were manufactures; now that figure stands at over 75 percent.

The automobile sector starkly illustrates the relative weight each country gives to exports versus serving its internal market. In 2011, the year for which the most updated data are available, Mexico produced 2.6 million vehicles and exported 2.1 million of them. Brazil, meanwhile, produced 3.4 million vehicles, but exported just 0.54 million of them.

Also, though investment is high in both countries, the so-called "Brazil cost" (the additional expense of goods due to insufficient infrastructure, high taxes and interest rates, and an excessively onerous bureaucracy) renders doing business more difficult. According to the World Bank, it takes an average of 119 days to start a new business in Brazil—the fifth-longest wait in the world. Mexico, in contrast, is one of the best countries in the region for conducting business. Although none of these problems is new to Brazil, they were largely eclipsed by more positive features high-lighted during its widely touted "takeoff" phase.

Brazil's economic setbacks have begun to constrain the country's foreign policy projection. After a period of frenetic worldwide diplomacy at the end of the Lula period, Rousseff has assumed a more subdued global and regional profile. At the same time, some of Brazil's longstanding problems, such as corruption, criminal violence, and strains on the justice system, are bound to draw even more media scrutiny in the run-up to the World Cup.

After the United States, Brazil is the world's second largest consumer of cocaine, and although there have been improvements and success stories in reducing urban violence, its homicide rate is higher than Mexico's (23 versus 18 per 100,000 people, according to the United Nations). Brazil's stepped-up antidrug policies in relation to coca-producing neighbors like Bolivia and Peru reveal how worried national authorities are about this problem.

Nevertheless, though Mexico's star has recently brightened—while Brazil's has faded a bit—it is unwise to embrace any sweeping, zero-sum interpretations about the region's two most important countries. Both nations are protean, susceptible to natural economic cycles, shifting global tendencies, and national vicissitudes. Moreover, though continued progress is likely in terms of both countries' national development and global role, their enduring challenges are significant. They are also strikingly similar.

Two Peas in a POD

The litany of problems facing Brazil and Mexico is by now familiar. While crime and security issues have different manifestations in each country, they remain serious challenges that demand more effective and credible police forces and justice systems. Deteriorating infrastructure remains an obstacle to robust development, though the Rousseff administration has recently undertaken a notable initiative with public and private sector support to address this challenge, especially in advance of the World Cup and Olympics.

Low productivity stems in part from the poor quality of education; although access to schooling in both countries has significantly increased, the amount that students are actually learning remains dismal by global standards. The Program for International Student Assessment, a project of the Organization for Economic Cooperation and Development, in 2009 ranked Brazil 52nd out of 64 countries for reading and 56th for math; Mexico fared only slightly better, ranking 47th for reading and 49th for math.

In terms of fiscal issues, both governments have displayed admirable discipline, yet taxation in Brazil is too high and regressive. Mexico, in contrast, taxes too little and makes up for it by drawing on oil profits, undercutting the vitality of its energy sector.

The problems of inequality and corruption deserve special mention. Although recent studies by the economists Nora Lustig and Luis López Calva show that there have been modest reductions in inequality in both Brazil and Mexico, both countries remain among the worst performers in the Americas.

In 2009, for example, the top 20 percent of Brazilian earners accounted for 59 percent of the nation's income, while the bottom 20 percent's share was merely 3 percent. Mexico is quite similar: That year, the top 20 percent received 54 percent of earnings, compared with the bottom quintile's meager 4.7 percent. The still enormous gaps between rich and poor have important political and economic implications. In coming years, progress on fiscal and education reforms will be critical to achieving a more equitable distribution of income.

There are few rigorous and reliable measures of corruption, though governments in both countries (and throughout much of Latin America) have identified it as a priority. According to the 2012 Corruption Perceptions Index produced by the monitoring group Transparency International, Brazil fares better than Mexico. It placed 9th among the 26 Latin American and Caribbean countries, while Mexico came out 16th. Not surprisingly, the issue has drawn considerable media attention in both countries.

Since Rousseff assumed office, she has wasted no time demonstrating her intolerance for corruption, and has fired seven of her ministers. The so-called *mensalão* scandal, which involved millions of dollars in bribes to members of Congress during the Lula administration, has resulted in the most significant trial in the country's modern history, even sending the former president's chief of staff to prison. A new "clean slate" law, forbidding individuals with criminal records from running for office in Brazil, barred some 1,200 candidates nationwide in 2012.

In Mexico, meanwhile, *The New York Times* has done much to expose the misbehavior of the giant retailer Wal-Mart, which reportedly offered bribes on a regular basis to Mexican officials for favors such as zoning approvals. The recently installed Peña Nieto administration has already set up a new anticorruption commission.

In sum, while it is tempting to extrapolate from short-term swings in each country, it is more productive to examine longer-term trends. Brazil and Mexico exhibit distinctive policy approaches, largely attributable to their geographical positions and divergent political legacies. On balance, the tendencies in each are heartening and point to more significant influence in global affairs, sustained growth, social progress, and political openness. Dramatic retrogressions in each country seem improbable. Nonetheless, the pending challenges remain serious and, unless effectively tackled, will impede continued progress for both nations and the region as a whole.

Looking Ahead

Forecasts predict Mexico and Brazil to be among the world's top ten economies by 2020; indeed, some projections rank them within the top five by mid-century. Both countries' influence in regional and global affairs is bound to grow in coming decades. In a hemisphere in considerable flux, Brazil and Mexico's decisions about their economic policies and governance models will be critical in shaping the future direction of the Americas.

For Mexico, the strategic alignment with the United States is likely to endure. No matter what strains and tensions emerge over sensitive bilateral questions that invariably arise in a close relationship, the increasingly profound connections between the two countries militate against any significant rupture. Migratory flows over the 2,000-mile border are bound to accelerate, not diminish, though one hopes they will occur under a more sensible and realistic legal framework. Current and subsequent Mexican administrations may well seek to diversify their relationships in the hemisphere and throughout the world, but the bilateral agenda will require continued attention and effort.

To be sure, such close ties carry some risks for Mexico. The country's dependence on the United States for such a large percentage of its exports, remittances, and tourist dollars is of some concern. The costs to Mexico of the 2008-09 financial crisis underscore the hazards of close reliance on a single partner. At the same time, such deep linkages confer clear advantages. Most obviously, the US anchor gives Mexico proximity to the world's largest economy and provides a huge market for its goods and labor. Mexico's solid industrial sector is the result in part of competing with countries such as China for market share in the United States.

If there is serious immigration reform in the United States—and reform of the energy sector in Mexico—the prospects for an even stronger bilateral relationship will increase. These were the most sensitive concerns when NAFTA was negotiated and approved two decades ago. If there is progress on both fronts, the opportunities for deepening integration will multiply.

Also critical will be whether a major new trade initiative, the Trans-Pacific Partnership (TPP), actually takes hold. In addition to Australia, New Zealand, a handful of Asian countries, and Mexico, the TPP prospectively includes the United States, Canada, Peru, and Chile. If such a scheme were to carry weight and prosper, it would have profound implications for cementing economic ties between the more open Latin American and Asian economies and ensuring that the Pacific remains the world's most dynamic market.

It is unclear how Brazil and the more closed Mercosul countries would fit into such a scenario. While the TPP and other initiatives to more formally link Latin American countries with Asia hold enormous commercial promise, they risk further accentuating already discernible strains in hemispheric economic relations. A major question moving forward is the possibility of a more pronounced bifurcation in the hemisphere—with discrete US– and Brazil-led blocs—and the growing challenge of achieving broader hemispheric cooperation on a common agenda.

Brazil does, however, enjoy considerable advantages as it carries out its global and regional strategies. The country is economically diverse (its trade, for instance, is fairly evenly distributed among China, the United States, Europe, and Latin America); its resource endowment is formidable; and it has substantial political initiative and ample room for maneuver.

With almost half of South America's population, Brazil has a massive internal market fostered by extending new lines of credit, which has resulted in an expanding middle class. While many bemoan Brazil's protectionist bent, its robust consumption may allow policy makers to expand the economy—albeit perhaps not as efficiently—with limited exposure to foreign competitors. And Brazil's independence from the United States, as it has historically, will likely enable it to more effectively pursue multiple options on the global stage.

While Brazil and Mexico have forged ahead independently of one another, it would be a mistake to rule out the possibility of greater economic and political cooperation between the two in coming years. Certainly, Brazil's decision in 2012 to impose high tariffs on Mexican automobiles did little to inspire confidence among business leaders. But with the prospect of energy sector reform in Mexico—and given that Peña Nieto has cited Petrobras as a possible model to follow—serious bilateral cooperation on that issue could be in store.

In addition, with Brazil's disappointing growth in 2012, and an increasingly problematic commercial relationship with its neighbor Argentina, the Rousseff government may find Mexico's expanding market attractive.

Strategy and Oscillations

Over the past dozen years, Latin America has undergone dramatic changes. The region has become stronger economically, more independent politically, and more assertive in global affairs. Deepening fiscal problems and debilitating political polarization within the United States have only reinforced this trend by limiting Washington's sway. The result has been a region increasingly emboldened, yet also fragmented and moving in a number of directions at once. In this context, the relative success of Mexico and Brazil likely will influence how other countries choose to navigate an unpredictable global marketplace and the demands of an emergent middle class.

In a hemisphere in such flux, it is scarcely surprising that the apparently changing fortunes of Brazil and Mexico have given rise to commentaries about which will occupy the number one spot. In February 2012, Mexico's former foreign minister Jorge Castañeda claimed in an op-ed in *El País*, entitled "The Mexico-Brazil Rivalry," that once Mexico's drug war wound down—and Brazil's poor infrastructure and communications were exposed during the World Cup—the Brazilian government's marketing of a twenty-first century miracle would be effectively debunked.

In August 2012, *Forbes* asserted, "Sorry Brazil, Mexico Is Better." The title of a *New York Times* piece on the G-20 meeting held in Los Cabos, Mexico, in June 2012, was telling: "World Leaders Meet in a Mexico Now Giving Brazil a Run for Its Money." The *Economist* and *Financial Times* blogs also have noted that Mexico seems to be pulling ahead of Brazil, with the latter instructing readers, "Forget the BRIC countries or even just Brazil . . . it's all about Mexico now."

The so-called "Latin rivalry" provides grist for the media mill and makes for a compelling narrative. But tracking the supposed competition to see which one of the region's twin powers is up and which is down serves as little more than a distraction and is shortsighted. For the United States, the fundamental question is how to build a more productive, long-term relationship with both countries. These relationships demand commitments from Washington as serious and sustained as with any top-tier nation in the world.

Regardless of the choices they make, Brazil's and Mexico's underlying trajectories should be matched with a correspondingly farsighted mindset among senior US policy makers—a mindset that remains focused on strategic considerations, no matter the eye-catching oscillations of the moment.

Critical Thinking

1. Discuss the competition that exists between Mexico and Brazil.
2. Why should Brazil be considered a South American rather than a Latin American country?
3. What explains Brazil's economic slowdown over the last several years?

Create Central

www.mhhe.com/createcentral

Internet References

Community of Latin American and Carribean States (CELAC)
www.celac.gov.ve/

Mercosur
www.mercosur.int

MICHAEL SHIFTER, a Current History contributing editor, is president of the Inter-American Dialogue and an adjunct professor of Latin American studies at Georgetown University's School of Foreign Service. Cameron Combs is a program assistant at the Inter-American Dialogue.

Unit 2

UNIT

Prepared by: Robert Weiner, *University of Massachusetts/Boston*

Democratization

Global democratization has generally proceeded in waves. The Arab Spring, in which the long-standing dictatorial regimes of Mubarak and Gadaffi were overthrown in the Middle East by uprisings based on popular revolutions, could be viewed as one of the latest examples of global democratization that were transformed into revolutions. But the process of democratization may depend on such factors as the method of exit from the previous regime and system, the political culture of the society that is undergoing the transition to democracy, and the legacy of the past, or path dependency. It is not that easy to graft the model of liberal democracy—a model that developed over the course of several hundred years in the United Kingdom and the United States—onto a political culture that differs from the Western cultural concept of democracy. A liberal democracy includes such elements as a free press, a rule of law state, and free and periodic elections. The impression given that democracy in general is continuing to spread is belied by the fact that there have been some significant regressions in some cases. Analyses by Freedom House and the Bertelsman Foundation show that there actually has been a freeze empirically in the spread of democracy and a qualitative decline in democracy as well. In a report entitled "Democratic Breakthroughs in the Balance," published in 2013, Freedom House reported 90 free countries, while 27 countries declined and 16 countries gained in being viewed as free. According to Freedom House, "This is the seventh consecutive year that shows more declines than gains worldwide." The difficulties in making the transition to democracy were underscored by the removal of President Mohamed Morsi from power by the Egyptian military in a coup in 2013. This resulted in mass demonstrations by Morsi's followers in the Moslem Brotherhood, which was ruthlessly crushed by the Egyptian military with nearly a thousand demonstrators reportedly killed. Morsi was removed because he was engaged in rooting out corruption in the Egyptian military, concentrating too much power in the hands of the Moslem Brotherhood, and failed to deal with the serious economic problems that Egypt faced. Morsi was placed under arrest, and faced a possible criminal trial. As the military continued to arrest the leaders of the Moslem Brotherhood, the head of the Egyptian Constitutional Court was appointed as the interim president of the country, the constitution was suspended, and a commission to work on the development of a new constitution was created. The military leadership promised to hold elections in the future. The reaction of the U.S. was extremely cautious at first, and the Obama administration did not even want to characterize what had happened as a military coup. Not only did Egypt face difficulties in making the transition to democracy, but problems in consolidating democracy were also experienced in Tunisia and Libya, and efforts to overthrow an undemocratic regime in Syria were bogged down in a protracted civil conflict in which the revolutionaries were not assured of victory.

Article

Prepared by: Robert Weiner, *University of Massachusetts/Boston*

The Showdown

Winners and losers in Egypt's ongoing revolution.

Peter Hessler

Learning Outcomes

After reading this article, you will be able to:

- Explain the failure of the Arab Spring in Egypt.
- Discuss the role of the Moslem Brotherhood in Egypt.

On July 1st, after the Egyptian military issued its ultimatum, and crowds of celebrating people began to stream into Tahrir Square, I walked down the street to the headquarters of Tamarrod, a political-activist group. A neat line of Apache helicopters flew overhead, trailing Egyptian flags. Everywhere around me, citizens were chanting, "The people and the Army are one hand!" The previous month had been full of rumors, and the last time that the Minister of Defense, General Abdel Fattah el-Sisi, had issued a public statement, a week earlier, it had been cryptic enough that it was welcomed by both the opponents and the supporters of President Mohamed Morsi. But this time the military's intentions couldn't have been clearer. The armed forces had given the President forty-eight hours to respond to the protesters' demands, and if he didn't, they said, "it will be incumbent upon us . . . to announce a road map for the future."

All of this had started with Tamarrod, which means Rebellion. In late April, five activists, who ranged in age from twenty-two to thirty, had come up with the idea of a petition campaign that rejected Morsi's Presidency. He had been elected less than a year earlier, the first Egyptian leader to be chosen after the revolution of 2011 toppled the regime of Hosni Mubarak. Morsi was a longtime member of the Muslim Brotherhood, the Islamist organization that had been banned and repressed by the old regime; he was briefly jailed during the revolution. He ran for President against a former Air Force commander, and there had been concerns about whether the election would be stolen, since the military was governing Egypt during the transition period. But Morsi won a vote that was widely hailed as free and fair, and crowds spontaneously gathered in Tahrir to celebrate what seemed to be an important step toward democracy.

His first year in office turned out to be disastrous. He governed with an autocratic style, refusing to offer concessions to opponents and making little effort to gain new allies. In November, he issued a Presidential decree that temporarily granted him powers beyond the reach of any court; the decree allowed a Brotherhood-dominated assembly to draft a new constitution, which most experts described as badly flawed. During the spring, Morsi's government threatened to shut down some television stations that broadcast opposition views, and a few prominent critics were investigated on charges of insulting the President. But the worst problems were economic. The administration was incapable of meeting obligations for a much needed loan from the International Monetary Fund, and Egyptians suffered through frequent power blackouts. By the end of June, drivers in Cairo had to wait in line for four or more hours to get gasoline.

In this climate, the Tamarrod campaign rapidly gained support. On June 24th, when I first visited the downtown apartment that the group used as an office, activists claimed to have collected more than fifteen million signatures, each on a separate sheet of paper. The forms had been piled three feet high in old plastic shopping bags and cardboard boxes. About twenty volunteers were sorting papers, and people constantly arrived with more; at one point, a blind man appeared at the door with a big stack of them. Organizers seemed uncertain what to do with all this material—they talked vaguely about having it certified by the Supreme Constitutional Court, or maybe by the United Nations. They acknowledged that there was no legal basis for impeachment.

The petitions had an amateurish quality. When I asked why they didn't include a space for phone numbers and e-mail addresses, so that the organizers could build a network for future activism, a nineteen-year-old volunteer said, "You expect me to call fifteen million people?" (Another organizer admitted that this had been a mistake, but that now it was too late.) Organizers told me that there was no external funding for the campaign; volunteers paid for photocopies of the petition with their own money. The apartment they worked out of was on the sixth floor of a grungy building on Tahrir Street, and had been loaned to them by an organization called the National Association for

Change. This group had mounted a petition campaign, in 2010, to pressure Mubarak for political reform, contributing to the climate that led to his downfall, and its volunteers were active in the Tamarrod campaign. But many political figures were slow to take Tamarrod seriously. On June 18th, Anne Patterson, the U.S. Ambassador to Egypt, gave a speech in which she dismissed the group's efforts. "Some say that street action will produce better results than elections," she said. "To be honest, my government and I are deeply skeptical."

Nevertheless, organizers seemed confident. "We know our Army," Doa Khalifa, a member of Tamarrod's central committee, told me on June 24th. "We know they will side with us." On the last day of June, Tamarrod called for demonstrations, which turned out to be even larger than the ones that had toppled Mubarak. Huge crowds filled Tahrir and the streets around Cairo's Presidential Palace, and there were protests in every major Egyptian city, with plans to hold sit-ins until Morsi left. A military source said that as many as fourteen million people had participated, out of a population of eighty-three million. Tamarrod gave a final count for the petition campaign: 22,134,465. The day after the protests, the military issued its ultimatum.

That afternoon, the Tamarrod headquarters were almost empty. Three young volunteers sat in a darkened conference room, watching the Muslim Brotherhood's television channel. On one wall of the office, somebody had hung a picture of Morsi behind bars, with the inscription "wanted—Escaped from Prison."

I asked one of the men, a twenty-three-year-old accountant named Mohamed Nada, if he was pleased with the military's statement. "Happiness will come after the road map," he said, through a translator. "But this is a step. We were expecting it— although it happened faster than our expectations. We said that it was going to take more than a week, maybe ten days. El-Sisi surprised us."

They were convinced that Morsi, who had a reputation for stubbornness, would never negotiate, and that the military would remove the President. "Tamarrod is only sixty days old," Mohamed Belkeemi, another volunteer in his twenties, said. "The Muslim Brotherhood has been around for more than eighty years. And yet it was destroyed by this sixty-day-old organization."

Nada grinned. "Anwar Sadat said that the Muslim Brotherhood should be put on trial for political stupidity," he said.

"Failed rulers are much more dangerous than corrupt rulers," Belkeemi said. "The corrupt ones survive because they have to be functional enough to continue to profit. But the failed ones don't survive. That's why Mubarak lasted for thirty years, but Morsi only a year."

Some of the stacks of petitions had toppled over and fanned across the dirty floor. I asked Nada what the organization planned to do with them. "After el-Sisi's statement, these papers won't merely go into the database," he said. "They will go into a museum. This is a historic day." Then he said, "I have a personal question for you. What do you think of the Revolution of June 30th?"

Earlier that day, I had stopped by the Muslim Brotherhood's national headquarters, in Moqattam, a neighborhood situated on a limestone escarpment that rises high above Cairo. The Brotherhood was founded in Egypt, in 1928, and became an Islamist organization that hoped to depose the country's colonial overlords and spark a religious renaissance across the Middle East. In its early years, the Brotherhood had engaged in political violence, but it eventually rejected such tactics. In the seventies and eighties, though, a pattern emerged in which radical Brothers left the group to join more extreme organizations. The Brotherhood was officially banned in 1954, but some of its activities were tolerated under Mubarak, and it had maintained an unassuming office in central Cairo.

Soon after the revolution, though, the Brotherhood had opened an impressive new headquarters in Moqattam. At the time, the building seemed to represent a fresh start, but now it looked ready for demolition. On the night of June 30th, it had been attacked and gutted by men with guns and Molotov cocktails. Armed Brothers had been holed up inside, and a gunfight broke out, in which eight were killed and dozens were injured. A foreign diplomat who had studied videos of the attack told me that she didn't believe it was spontaneous. "The guys who attacked were seasoned," she said. "Even though they were under fire, they were very cool, very calm. I'm sure that was funded. That could have been police or intelligence personnel in plain clothes. They were probably funded by the old regime. The old regime is still here." She believed that the Brotherhood's building had been a symbolic target, because the headquarters of the National Democratic Party, Mubarak's old party, had been gutted during the revolution of 2011. The N.D.P. had since been disbanded, but former adherents continued to play a role in the political scene.

When I arrived at the Brotherhood's headquarters, inspectors from the prosecutor's office were examining bullet holes in the walls of buildings across the street. At the back gate, a police officer directed his men to search for bullet casings amid the rubble. He found one and held it up. "Machine gun," he said. "Foreign-made." I asked him why the police hadn't been there the previous evening. "Because the Interior Ministry announced that we would not defend their headquarters," he said. He showed no particular interest when a private security guard from a neighboring building wandered over and mentioned that he had donated two cans of gasoline to the arson.

The building's furnishings had been looted, and now people were carrying out piles of documents. Somebody showed me a Brotherhood booklet that described the glories of the postrevolution period: "Who could have expected Dr. Morsi's success after just twenty-eight days of campaigning?" An Egyptian journalist asked me to look at a stack of English documents that he had found inside. They consisted of dozens of pages of correspondence involving the United States Special Inspector General for Iraq Reconstruction; later, I learned that they had come from an unclassified audit of Iraq's Police Development Program. I had no idea why the Brothers had been reading U.S. government audits while their country's economy was collapsing, but now it would probably be fodder for new Egyptian conspiracy theories. For months, people had claimed that the American government supported the Brothers. At protests, it became common to see photographs of Ambassador Patterson, a big red "X" drawn across her face. On the Tamarrod petition, one of seven reasons listed for rejecting Morsi was that "you are subservient to the Americans." Such notions were simplistic, but it was true that the Obama Administration had seemed reluctant to criticize Morsi during his aggressive moves of the past year.

Periodically, people drove by to look at the ruined building, waving Egyptian flags and honking their car horns. A striking number of these celebrants were women, who often seem to have particularly strong reactions to the male-dominated Brotherhood. "Do you have any rooms to rent?" one older woman called out as she stopped in front of the building. The policemen laughed and cheered. "By the honor of our martyrs," she shouted, "we'll bring them down!"

The Brotherhood has always had many enemies, but it also has a reputation for successful grass-roots organizing and charitable work, especially in provincial cities and towns. And the Brothers are known for their financial integrity, which is one reason that they performed so well in post-revolution elections—voters were disgusted by the corruption of the Mubarak regime. The Freedom and Justice Party, the Brotherhood's political wing, won every vote in the new Egypt, although its margin of victory diminished over time. Even as criticism mounted, there was never a sense that the Brothers were engaged in profiteering.

But the Brotherhood's culture remained profoundly undemocratic. It has a rigid hierarchy, with a number of figures occupying higher positions than Morsi's had been. Many Egyptians were disturbed when these unelected leaders issued public statements during national crises. Brothers who held public office continued to function with the secrecy and the insularity of a banned group, and they seemed incapable of trusting outsiders. When I talked to people who had left the Brotherhood, they always emphasized how friends in the organization immediately cut off all contact. This response might have made sense when the Brotherhood was semi-underground, but, in a competitive political environment, former associates could have been a valuable source of information, perspective, and constructive criticism.

Within the organization, there's a tradition of strict obedience to superiors. This makes the Brotherhood highly disciplined but also unresponsive to change. Leaders struck me as out of touch, perhaps because subordinates told them what they wanted to hear. As their popularity plummeted in Cairo, Brotherhood leaders often said that the situation in the capital was an anomaly, and that their national network still reported strong support in places like Upper Egypt, which is home to almost forty per cent of the population. During the spring, I spent time in the region, where I decided to visit a series of offices of the Muslim Brotherhood and the Freedom and Justice Party. I planned to start at the governorate level, descend to the district, and finally reach a pair of villages called el-Araba and Beni Mansour.

"Those who say that the Freedom and Justice Party has lost popularity in the street are only talking from air-conditioned offices in Cairo," Youssef el-Sharif, the top Party official in the governorate of Sohag, told me. "We have a presence all the way down to the neighborhood level. There's a Party office in el-Araba." Next, an official at the Brotherhood headquarters in the district of el-Balyana also assured me that I'd find an office in el-Araba. But, when I pressed, he acknowledged that el-Balyana was the lowest organizational level, and that there wasn't a single member in el-Araba or in Beni Mansour, which have a total population of seventeen thousand. In el-Balyana, which is home to more than three hundred and fifty thousand people, I was told that there were only a hundred and fifty Brothers.

The el-Balyana official also claimed that there had been no drop in popularity, and he said that the Brotherhood often sponsored activities and medical drives in the villages. In el-Araba and surrounding villages, though, many people complained bitterly about Morsi and the Brotherhood. They said that they had voted for him out of a desire to have a religious man in office, but they had no specific affection for the organization; the Brotherhood seemed to have won because of weak competition, not strong support. And it had no presence at the grass-roots level. Nobody could tell me about a Brotherhood activity being sponsored in the villages, other than past voter drives. "People just come before the elections, and then you never see them again," a farmer said. "I won't vote for the Brotherhood again."

Late on the evening of July 2nd, the day before the military's ultimatum was set to expire, I called Gehad el-Haddad, a spokesman for the Brotherhood. I asked if the President was negotiating with General el-Sisi.

"There is no dialogue with el-Sisi," el-Haddad told me. "He is an arm of the state. He follows the leadership of his President, Mohamed Morsi." Like other Brothers, el-Haddad seemed to be operating on a theoretical level—he insisted that Morsi's election made him more powerful than the Army, despite the ultimatum. El-Haddad continued, "The outcome has to come from democratic means, and not from one part of the population using the military to bully others."

I asked what the Brothers would do if the Army tried to remove Morsi from office.

"We will stand between the tanks and the President," he said.

The foreign diplomat I talked to had told me that some prominent Brotherhood leaders had approached sympathetic Egyptian academics who sometimes functioned as informal advisers. They had recommended that Morsi offer concessions quickly and publicly, as a way of gaining the moral high ground. But their advice had been ignored. "They still believe that since they were elected they should not step down," the diplomat told me. "It's denial."

Shortly before midnight, Morsi appeared on Egyptian television. He wore a black suit and a black tie, and he spoke calmly. It was the most Presidential I had ever seen him, and the speech was angry and defiant. "If the price of protecting legitimacy is my blood, I'm willing to pay it," he said. The statement seemed to energize both sides, and later that night, near Cairo University, a gunfight broke out between Morsi's supporters and his opponents. The violence killed eighteen, including one Tamarrod activist.

The next day, somebody hung two large banners from the top floor of the shell of the N.D.P. headquarters. The building is near Tahrir Square, and for more than two years it has sat there untouched, a reminder of the violence of the revolution. One banner read "Tamarrod," and the other read "End the Rule of the Brotherhood: Morsi = Mubarak." Three kids had climbed up to the tenth floor and were trying to remove the last air-conditioning unit from the building. Originally, there had been more than a hundred; from a distance, the empty steel frames looked like staples punched into the concrete. The kids seemed to be having a hard time getting the last one out. There was less than an hour until the ultimatum expired, and it was a good bet that cops were too preoccupied to worry about N.D.P. air-conditioners.

In Egypt, members of the military, the police, and other security forces do not have the right to vote. This was first established in the 1923 constitution, and the theory was that officers and soldiers should remain politically neutral. But one can also interpret such a law, especially in a military dictatorship, as an implied acknowledgment that the Army gains its power through something other than direct democracy. Since the revolution, there has been talk about changing the law, which was recently deemed unconstitutional by the courts. Egypt has an estimated two million security personnel, and, in 2012, Morsi won the Presidential election by 882,751 votes. He probably would have lost if soldiers and officers had the franchise.

In fact, Muslim Brotherhood members are banned from both the police and the Army, which have always had a secular culture. The basic terms of Morsi's election—no soldiers or officers voting, no Brothers in the Army or the police—made for a sensitive transition. And yet there seemed to have been little caution on the part of the President. One of his first acts in office, in August, 2012, was to purge the military of its top leaders, some of whom were in their seventies and had been contemporaries of Mubarak.

Morsi appointed as Minister of Defense el-Sisi, who, at fifty-seven, was a full generation younger than the man he replaced. El-Sisi was known to be devout and sympathetic to political Islamism, but soon there were signs of tension. In December, during the battle over the new constitution, Brothers and other Islamists attacked peaceful protesters at the Presidential Palace, and they interrogated and tortured citizens. Later, Brothers told me that they defended the palace because they couldn't trust the police. But their actions infuriated the security forces, and throughout December I saw clear signs that officers disliked protecting Morsi. During this period, el-Sisi invited all parties to talks but was reportedly rebuffed by Morsi, who refused to negotiate.

As the national crisis worsened this spring, one of the biggest complaints from Egyptians concerned reduced security. The police were notably absent, especially in routine tasks like traffic management; they seemed uninspired to work under the new regime. In mid-June, while the Tamarrod campaign was gaining strength, Morsi appointed a member of the Islamist group al-Gama'a al-Islamiyya as the governor of Luxor. Although the group had renounced violence by 2003, in 1997 it organized a terrorist attack in Luxor that killed fifty-eight tourists. The memory of that massacre remains strong for public-security officers in the region. After Morsi's appointment, the protests were so large that the new governor couldn't enter the state building, and he resigned without ever sitting at his desk.

The Brotherhood had always claimed that institutional prejudices were stacked against it, and this was certainly accurate. But the Brothers antagonized old enemies, and made new ones—even among the Salafis, the more conservative Islamists. In incidents like the fight over the constitution, the Brotherhood cited its electoral victories as justification for undemocratic behavior—a defense that alienated liberals who otherwise might have opposed military intervention. As the threat of a coup arose, Brotherhood spokesmen talked about the terrible precedent that such an action would set for democracy in Egypt. But their own behavior had also set a terrible precedent, and after a certain point such arguments became largely theoretical. When you evaluated the situation on the ground—the way people felt, the way people acted—the past mattered much more than the future. Since the revolution, Egypt may have aspired to democracy, but it retained many features and institutions that were highly undemocratic. In order to survive in such an environment, a political player had to be nimble, pragmatic, and connected to reality.

And he had to understand how quickly all the threats could converge. Tamarrod represented the popular mood and the new democratic currents; the ultimatum represented entrenched institutions like the military; the attack in Moqattam possibly represented the vengeance of the defeated Mubarak regime. When I spoke with Mohamed Kadry Said, a retired major general who leads the military-studies unit at the Al-Ahram Center for Political and Strategic Studies, in Cairo, he told me that he was impressed by the energy of the Tamarrod activists. "And also there is some smell of connection between them and the military," he said. He believed that the military had followed the progress of the activists closely, coördinating its own activities accordingly. Tamarrod organizers told me and other journalists that they had no contact with the armed forces. But one of their activists told Reuters that she had quit during the last week of June because she sensed a sudden infiltration of security agents.

I met with Said on the day before the ultimatum expired. He seemed optimistic; he believed that this could be a new start for Egypt. I asked what would happen to Morsi.

"They will put him in some place," he said. "They will find a reason."

I asked if he thought there would be violence.

"Yeah," he said softly.

At nine o'clock on July 3rd, five hours after the deadline expired, el-Sisi appeared on national television. He had always been a cipher. Like many Egyptian military leaders, he had studied at the U.S. Army War College, in Carlisle, Pennsylvania. Robert Springborg, an expert on the Egyptian military, has noted that at the war college el-Sisi wrote a paper arguing against a fundamental separation of powers, because Islam should be the reference point for all government branches. This idea is very much in line with Brotherhood ideology; it also directly contradicts the constitutional reality in Egypt. El-Sisi had never cast a vote in his life. On television, he looked every bit as calm as the President had the night before.

At the Presidential Palace, where tens of thousands had gathered, there was a hush while people listened to radios and phones. When el-Sisi announced the suspension of the constitution, a roar went up. After that, it was impossible to hear the rest. El-Sisi explained that an interim President would come from the judiciary, and then Egypt would hold a national election. He said that the military had been forced to take this step because of the national crisis and Morsi's refusal to negotiate.

Afterward, I left the celebrating crowds and headed to Rabaa al-Adawiya Square, in the Cairo suburb of Nasr City. Thousands of Morsi supporters had been camped out there for days, to show their strength, especially on television. In post-revolutionary Egypt, this has become a crude instrument of democracy: news channels train live cameras on major protests, so viewers can gauge the numbers. On the day of an important political event, the screen might show as many as six sites at once. Often there's no commentary—you watch the crowds in silence, like a mall cop with his closed-circuit screens.

Since midafternoon, the Army had been steadily surrounding the area around Rabaa al-Adawiya Square with troops and barriers. Earlier in the evening, Hassan, my translator, and I hadn't been allowed in, but now we took a back road where soldiers let us through. An Apache helicopter rumbled overhead; the streets were empty. Nasr City is situated in the eastern desert, and it has the ambitious, overbuilt feeling of many distant Cairo suburbs. Large stretches consist of military installations—it would be hard to pick a worse place for a face-off against the Army. But, up until the end, Morsi's rank-and-file supporters hadn't seemed to realize what was happening. The night before, men had told me that the Army would definitely back the President, and a speaker was rallying them with a story about a woman having had a vision of the angel Gabriel flying over Nasr City.

Hassan and I reached a Brotherhood checkpoint, where men sat in formation, wearing construction helmets and clutching simple weapons: clubs, nightsticks, nunchakus. They looked stunned, and they examined our I.D.s without saying much. Down the street, we talked with a thirty-year-old civil engineer and Brotherhood member named Ahmed el-Hawat. He carried a long club, but he had a friendly, gentle manner, and his helmet made him look like a building engineer on site—glasses, button-down shirt, slightly overweight. He told us that all of Egypt's Islamist television channels had been shut down

at the moment of el-Sisi's statement. While we were speaking, another supporter wandered over, weeping, and began to pound on a parked car.

"We will see more blood in all the streets of Egypt," el-Hawat said. There were reports that Mohammed Badie, the top Brotherhood leader, had been arrested near the Libyan border. I asked el-Hawat if he thought the people here might be arrested, too.

"It's very possible," he said. "This does not really concern me. Our message and our goal is to have this country rise up." He continued, "If we are assaulted—we have a huge number of people, and we are armed. This is the simplest tool that I'm carrying, but there are other weapons."

I asked him what they were.

"There are machine guns, shotguns, nine-millimetre pistols," he said. "Every one of us is already a time bomb."

Then he pointed out that the roof of the building across the street was full of Army snipers. Their silhouettes were clear against the night sky: prone at the edge of the roof, guns at the ready. "They showed up a couple of hours before the military issued its statement," el-Hawat said.

I decided that it was time to wrap up this conversation. But then an angry roar came from the crowd down the street, and somebody opened fire with an automatic weapon. Hassan and I turned and walked away fast, heads down. More gunfire: probably somebody was just shooting into the air, but I didn't look back to check. After we had walked for a few minutes, there was another explosion, this time from up ahead—fireworks. The Presidential Palace was three miles from the square; that was all the distance that separated the victors and the losers. Later, I rode through streets that were thronged with Cairenes celebrating Morsi's ouster. When I got home, I thought about the men who were cornered in Nasr City, and I turned on the news, but those live feeds had been removed. All the Islamist channels were blank.

That week, in most parts of Cairo it was hard to tell that anybody was upset about what had happened. People hung national flags above their homes and shops, and the mood was happier than it had been for months. There weren't any lines at gas stations, and the electricity wasn't cut off—such problems seemed to magically disappear. Anybody in uniform had a smile on his face. If you wanted to find anger, you had to search for it. But for two years the violence in Cairo had always been highly localized—problems usually occurred around Tahrir, while life went on as usual in the rest of the city.

The Morsi supporters remained in Nasr City. After a day, the military relaxed the checkpoints and the snipers vanished, and it was clear that the authorities had decided to allow the protest to continue. As time passed, the group changed in subtle ways. Before Morsi's ouster, almost everybody I met was a Brotherhood member, and the energy often flagged; organizers were constantly running through the crowd, shouting at protesters to gather and chant so that it would look better on television. But now I encountered more people without a Brotherhood affiliation, and they seemed more energized. They had chosen to be there—it wasn't just a matter of obeying an order from a superior. Many seemed particularly upset because the activities of the movement were no longer televised. They tended to be older than the Tahrir crowd, mostly in their forties, and often they were lower-middle-class men from the provinces. They had painted new graffiti all around Nasr City: "No to the Coup"; "Down with the Agents of America"; "Hey el-Sisi You Coward, You're the Agent of the Americans."

On July 5th, protesters crowded in front of the headquarters of the Republican Guard, in Nasr City, where they believed that Morsi was being held. Soldiers fired on the crowd, killing four. A video showed one of the men being shot after making no clear provocation. I spoke with a man named Mohamed Ibrahim Ahmed, who said that he had been standing next to one of the protesters who died. Ahmed was a retired Air Force veteran from the city of Mansoura. "I'm not a Muslim Brotherhood member—you can't join if you're in the military," he said. "But I voted for Morsi. It's my right to fight to defend my vote." He told me that the military shouldn't interfere in politics, and he believed that the Americans were behind everything. I asked about the speech in which Ambassador Patterson had criticized Tamarrod.

"Would she say openly that she's the one who supported all those protests?" Ahmed said. "So of course she said the opposite."

A group of soldiers stood behind a nearby wall, and passersby yelled, "You're our family! Don't betray us!" I walked down the road to the Rabaa al-Adawiya mosque, where Badie, the top Brotherhood leader, was scheduled to deliver a speech before the evening prayers. Both the Egyptian and the foreign press had reported Badie's arrest, but now his voice boomed over the loudspeakers: "I have not been arrested, and I'm not running away! Those are lies! I'm free!" An Apache helicopter approached and hovered low over the mosque; Badie's voice grew louder. "Film us, man, film us!" he shouted. "I say to the great Army of Egypt: the people and the Army are one hand! Leave, el-Sisi, leave!"

The crowd roared, chanting pro-Army slogans. Men around me began to weep. In Egypt, the faith in the military runs so deep that people called on it for assistance even after their leader had been removed in a coup. Badie said that Morsi supporters should occupy every public square in Egypt, and, after the speech, men marched on Tahrir and other parts of Cairo. When they encountered anti-Morsi crowds, street fighting broke out, and it continued into the early morning. Ten people were killed.

At the time of Morsi's election, an Egyptian video producer who disliked the Brotherhood had told me that it was the best thing that could happen, because people would quickly learn the true face of the organization. Now that the Brotherhood had been overthrown, it reverted to occupying what seemed like a much more natural role. For decades, it had been in the opposition, and most leaders had been hounded and imprisoned; men like Badie were brilliant at rallying supporters during a tough time. In Cairo, everything had been neatly reversed. The volatile protests had shifted from Tahrir to Nasr City, where a different group called on the same Army for help. During the

spring, the Brotherhood had tried to intimidate journalists and television stations; now the Islamist media were being silenced. Even the Americans had apparently switched sides. It had been all of six days since the Tamarrod demonstration.

On July 8th, a fight broke out in front of the Republican Guard headquarters, and fifty-one civilians were killed, along with one soldier and two police officers. Two days later, the new government issued arrest warrants for Badie and other Brotherhood leaders, accusing them of inciting violence. It was the worst single episode of bloodshed in Cairo since Mubarak had been removed from office. Each side blamed the other: protesters claimed that they had been attacked while praying, but the military said that it had been fired on first.

That evening, I stopped by the old Tamarrod office. The group no longer had a formal headquarters; the premises were once again occupied by the National Association for Change. I talked to Ahmed Salama, a young man who had also worked on the Tamarrod campaign. He blamed the morning's violence on the Brotherhood—he believed that its leaders were inciting aggression. "Their families are not in these clashes," he said. "It's the poor people." After the massacre, statements by the Brotherhood called for an "uprising." That afternoon, Jay Carney, the White House press secretary, condemned what he described as "explicit calls to violence made by the Muslim Brotherhood." Obama Administration officials had made few comments about the week's events, and they still refused to say whether they considered the Army's intervention a coup—a designation that could end the $1.5 billion in U.S. support that Egypt receives every year. But new sources of funding had quickly appeared. The day after the massacre of the protesters, Saudi Arabia and the United Arab Emirates pledged eight billion dollars in aid, a reflection of their distaste for the Brotherhood.

In Cairo, a judge named Adli Mansour had been named interim President, and a transition process had been mapped out, with parliamentary and Presidential elections to be held in the coming months. But an argument had already erupted over the first suggestion for prime minister, Mohamed ElBaradei, a liberal who was eventually rejected after complaints by the Salafis. The Islamists had agreed with Morsi's ouster, and their presence was critical to any political consensus; soon they expressed concerns about the proposed constitutional committee, which was to consist of a small group of legal experts. The Brotherhood still refused to participate in any negotiations, insisting that Morsi remained the legitimate President. Nobody would say for certain where he was.

From a distance, the labels seemed to be all that mattered: Islamists and secularists, military and civilian, democracy and dictatorship. There was a great deal of talk about how to define a coup. But such discussions implied a coherence that simply wasn't relevant in Egypt, where so much of day-to-day life has always been makeshift and improvised. The most important actor in all these events, el-Sisi, was a deeply religious man who had defended the principles of political Islam from a leafy campus in Carlisle, Pennsylvania. In Cairo, though, he acted as a pragmatist. Theories tend to evaporate in a crisis, leaving behind elemental human qualities: stubbornness, frustration, fear, anger, want. The question in Egypt is whether the next elected government will remember these forces, or chase another abstract idea.

At the former Tamarrod office, Salama, the organizer, told me that he was optimistic. "We were like kids in democracy and revolution," he said, remembering the movement of 2011. "We are becoming mature now." Three days earlier, he had been wounded in the leg by a shotgun pellet while defending Tahrir Square against Morsi supporters. There was a scar on his head from a wound he'd received while fighting Mubarak's forces. His sinuses had been damaged by tear gas at the end of 2011, when he had protested the military's first transitional government. But Salama still believed in military guidance for the time being—he pointed out that it had taken Brazil almost a quarter of a century to transition to full civilian rule.

In Egypt, more than half the population is under the age of twenty-five, and in the capital a generation of young people are spending their formative years in the ongoing revolution. It's hard to tell what they're learning—some of the educated ones, like Salama, are engaged in organizing, but many are there for the thrill of the fight. Opposition has always been the easiest stance; until now, no one has been able to build something that works for a positive change. Tamarrod, for all its energy, has an uncertain future. "Tamarrod was an idea instead of a political organization," Salama explained. He didn't believe that it should become a political party or movement; in his description, it almost seemed a kind of performance art. "It's a quick idea that comes and you do something," he said. "It's a tool." He rolled a cigarette and smoked it, ignoring a sign on the wall that said "Smoking Outside the H.Q. Only." The place was a firetrap: stacks of petitions now rose six feet high around us. Salama told me that most of the papers were hidden in four locations around the city, but no journalist or independent group had verified their existence. It didn't seem important to the organizers: from their perspective, that idea had served its purpose, and now it was time for something else.

Critical Thinking

1. Why was Morsi removed by the Egyptian military?
2. What is Tamarod?
3. Under what conditions can democracy be consolidated in Egypt?

Create Central

www.mhhe.com/createcentral

Internet References

Al Jazeera
   http://english.alijazeera.net

Arab Net
   www.arab.net

Middle East Online
   www.middle-east-online.com/english

Prepared by: Robert Weiner, *University of Massachusetts/Boston*

Article

Pakistan on the Brink of a Democratic Transition?

C. Christine Fair

Learning Outcomes

After reading this article, you will be able to:

- Discuss the role of the military in Pakistani politics.
- Explain what the future stability of Pakistan depends on.

For perhaps the first time in recent memory, Pakistan's battered democracy could prove naysayers wrong. A national election is expected in May 2013. If it produces a government, it will mark the second time that such a democratic transition of power has taken place. Also, the government that stepped down in March became the first democratically elected administration to serve out a more-or-less complete term. Unfortunately, events of recent months have marred both the tenure of the outgoing administration and the constitutionality of the transition. And even under the best of circumstances, the myriad challenges awaiting the new government will be daunting.

Since 2008, Pakistan has been governed (in the loosest sense of the word) by the Pakistan Peoples Party (PPP). The PPP government came into power via reasonably free and fair elections held after General Pervez Musharraf stepped down as president. The PPP cobbled together a fraught coalition, which included its archrival, the Pakistan Muslim League-N (PML-N), led by former Prime Minister Nawaz Sharif. However, the unprecedented partnership between the two long-time foes collapsed in the summer of 2008 due to serious disagreement surrounding the reinstatement of a controversial Supreme Court chief justice, Iftikhar Muhammad Chaudhry.

President Musharraf had removed Chaudhry from office in 2007 after the chief justice publicly opposed government actions such as the extra-judicial execution and/or apprehension ("disappearing") of Pakistani citizens, as well as a series of privatizations in which Musharraf's cronies purchased public assets at below-market rates. The PPP, for its part, was not anxious to reinstate Chaudhry, whom the party feared would reverse key Musharraf-era legislation that had allowed party members to contest elections in the first place.

Most important, Chaudhry opposed the National Reconciliation Ordinance (NRO), an executive order issued by Musharraf in 2007. The NRO, which was brokered by the Americans, sought to reconcile President Musharraf and then–PPP leader Benazir Bhutto by granting legal amnesty to an array of PPP politicians. The amnesty was necessary for these politicians to contest elections scheduled for late 2007 and to hold political office despite previous allegations of criminal wrongdoing.

The NRO did not extend such amnesty to the PPP's political foes. Washington put its heft behind the NRO because it was anxious to keep Musharraf on as president, valuing him as a counterterrorism partner, but recognized that his government lacked legitimacy. US officials hoped that if Bhutto became prime minister, her legitimacy would bolster Musharraf, who would remain as president.

Bhutto was assassinated in December 2007. Widespread belief that Musharraf was somehow behind her killing dashed his electoral ambitions and those of his party, the Pakistan Muslim League (Quaid-e-Azam) (PML-Q). Musharraf had assembled this party early in his presidency with defectors from mainstream parties. The PML-Q had formed a government following army-rigged elections in 2002 and had rubber-stamped Musharraf's policies.

By the summer of 2008 it was clear that the PPP would not reinstate Chaudhry of its own accord. Both to punish the PPP and to force the government to a breaking point, the PML-N withdrew from the governing coalition and threw its support behind a mass movement of lawyers (the so-called Lawyers Movement) who had mobilized to demand Chaudhry's reinstatement.

As protesters under the banner of Sharif's "Long March" joined forces with the Lawyers Movement, the cities of Lahore and Islamabad were seized with tumult. In an effort to defuse the standoff, Army Chief Ashfaq Parvez Kayani brokered a settlement between the PML-N and the PPP that culminated in the justice's reinstatement in early 2009. Chief Justice Chaudhry since then has had an enduring soft spot for the PML-N and its leadership, given their role in restoring his job.

Chaudhry, true to his word, vacated the NRO later in 2009. With their amnesty suspended, PPP politicians again became vulnerable to prosecution. The Supreme Court selectively employed the various corruption charges against President Asif Ali Zardari and others, thereby ensuring that the PPP government remained weak and under constant threat. Curiously, the court did not order the prosecution of members of other political parties who also had criminal allegations pending against them and who never had amnesty in the first place.

The army doubtless supported the court's harassment of the PPP. The military has long regarded the PPP government as obsequious to the Americans and embarrassingly inept. Worse, from the generals' perspective, the PPP leadership has repeatedly sought to interfere in the army's affairs. For example, in April 2008, President Zardari announced that Pakistan would pursue a "no first use" nuclear policy. This directly contravened the army's nuclear deterrent posture, which privileges the option of nuclear first use to dissuade India from any military action on Pakistani soil. In July 2008, Zardari again riled the military when he announced that the army's premier spy agency, Inter-Services Intelligence (ISI), would come under civilian control. Needless to say, both of these plans evaporated under the army's outrage.

The PPP government also encouraged US legislation in 2009 that would require Washington to withhold security assistance if the army interfered in the governance of the country. In response, the ISI orchestrated a disinformation campaign regarding the bill to whip up public denouncements of American meddling in Pakistani affairs.

However, lacking a viable alternative to the PPP, the military did not want the government to fall. As odious as the PPP is to the generals, they are also wary of the PML-N, in part because of Sharif's alleged role in the attempted assassination of Musharraf in 1999. A new political option was needed.

Savior in the Wings?

The search for just such an alternative marked the next period of Pakistani politics. The army is suspected of backing the political rise of Imran Khan and his Pakistan Tehreek-e-Insaf party (PTI). Khan is a much-beloved former cricket star and known lothario who left the game to take up charity work and a political career in Pakistan. Until 2011, he had made few political inroads, but in that year he burst onto the national stage in what he called a tsunami, drawing huge crowds of men and women of all ages.

Khan railed against government corruption. He mobilized public antipathy toward American drone strikes, demanding an independent foreign policy that did not rely on the United States. He made a number of preposterous promises to "end" graft and expand the tax base. But he avoided taking a considered and principled stand on the Pakistan Taliban, a criminal and terrorist network that has killed tens of thousands of Pakistanis. Khan favors negotiation with the Pakistan Taliban, rather than military confrontation.

Some of the Supreme Court's most vigorous activism took place during the height of Khan's popularity. For example, in the fall and winter of 2011, the Supreme Court hounded Zardari over "Memogate"—a scandal surrounding the alleged role played by Husain Haqqani, Pakistan's ambassador to the United States, in securing American assistance to help Pakistan's civilian politicians control the army. Had Zardari in fact attempted such a maneuver, it would have been completely consistent with Pakistan's long-flouted constitution, which places authority over the military in the hands of the civilian government. But the Supreme Court, egged on by the PML-N, cast Haqqani as a traitor and forced him out of office.

In June 2012, the high court ousted Prime Minister Yousaf Raza Gilani after he refused to pursue corruption charges against President Zardari. Some worried then that the government would fall, but once again, the PPP and its leadership proved more robust and wily than its critics had believed. Raja Pervez Ashraf soon assumed Gilani's position. And as Khan's tsunami dwindled to a trickle, the court also ceased its efforts.

The PPP-led government limped on after the ousting of Gilani and exiling of Haqqani. It even garnered accolades at home for its handling of a November 2011 tragedy in which US and NATO forces killed 23 Pakistani soldiers at an army outpost on Pakistan territory. The government shut down the supply routes that the Americans use to transport war matériel to Afghanistan. Months later, Pakistan reopened the routes when it became apparent that Washington was forging a logistical strategy that cut Pakistan out of the loop entirely.

In September 2012, US Secretary of State Hillary Rodham Clinton designated the Jalaluddin Haqqani network as a foreign terrorist organization, a decision that would have been inconceivable had the supply routes remained open without official interruption. (The Islamist insurgent group, which has fought US-led forces in Afghanistan, has long enjoyed the support of elements within the Pakistani security establishment.) In fact, Pakistan watchers speculated that the US government might even declare Pakistan a state sponsor of terror. However, when Pakistan reopened the supply routes, it once again established itself as a critical part of US policies in Afghanistan and South Asia.

The Soft Coup

In early 2013, just as it looked as if the government was going to achieve the unprecedented feat of serving out its five-year term without a coup or other extra-constitutional proroguing of the parliament, Muhammad Tahir-ul Qadri, a Canadian religious scholar, arrived on the Pakistani political scene. Qadri's rise was meteoric. Even though he had previously advised Musharraf and President General Mohammad Zia ul-Haq (the Islamist coup maker who governed Pakistan from 1978 until his death in 1988), Qadri was virtually unknown among Pakistanis. Yet almost immediately upon his arrival he began to draw some of the largest crowds Pakistan had seen in years.

Some Pakistanis, however, were wary that this latest savior had been thrust into Pakistani living rooms, video parlors, and tea stalls with the help of the army. Qadri traveled in a "mobile command center," a highly fortified container on wheels, which moved about while Qadri addressed crowds from within. Some cynical Pakistanis joked that the scholar enjoyed a "martyrdom-proof box" while ordinary Pakistanis and even public officials risked terrorist violence without the security of a fortified conveyance.

Qadri's shenanigans, despite their absurdity, nearly brought down the government. He held massive sit-ins in Islamabad, virtually shutting down the capital. Qadri, like Khan, complained about corruption and political dysfunction, and he hinted that an army-backed technocracy might be the way forward. His suggestions should have been dismissed as preposterous, but many saw the army's hand in the matter, and indeed believed that the military had manufactured the drama to legitimize a coup.

Others, such as this author, doubted a hard coup was ever in the making. After all, should the army seize power yet again, Pakistan would face sanctions and international ostracism. Given Pakistan's precarious economic predicament, ongoing security operations throughout the country, and deep popular distaste for direct army rule, it is unlikely that the men in khaki sought to seize direct power.

With Qadri's hijinks in full swing, the Supreme Court swiftly stepped into action. In January 2013 it ordered the arrest of Prime Minister Ashraf and 15 others under various charges of corruption. Qadri and his mob of peaceful miscreants eventually departed the stage—but only after extorting a raft of extra-constitutional concessions. Most egregiously, Qadri insisted that, in advance of this year's election, the government step down and appoint a caretaker administration in consultation with himself and the army by March 16, 2013. This date, which was agreed to, was critical if rarely noted: It was two days before the government's term was set to expire. By coercing the administration to end its term two days early, Qadri and his army allies subtly undermined the claim that this government completed its lawful term. Coincidentally, when Ashraf acquiesced to Qadri's demands, the court again retracted its PPP-specific talons.

While many breathed a sigh of relief that a hard coup had been averted, others understood that the army had in fact engineered a soft coup. Alas, few Pakistanis—and fewer international observers—noticed, much less cared about, the orchestrated miscarriage of constitutionality perpetrated by the Qadri-army combine. After all, Qadri, a Canadian (and thus legally barred from contesting elections in Pakistan), using street theater and with the presumed assistance of the military, had forced an elected government to yield. By some measures, the upcoming election has already been tainted—likely the army's goal in the first place. This is not a military ready to retreat to its garrison without a fight.

An Army with a Country

While much is at stake for Pakistan's nascent democracy, the army also has a great deal at risk: namely, its ability to run roughshod over democracy. The army has long promoted itself as the only institution able to protect Pakistan from domestic and foreign foes. In its attempts to prove its own efficacy, it has exploited inter-party rivalries to sow discord and maximize political incompetence: The worse the politicians appear, the more noble and competent the army seems.

The army has used its privileged place in Pakistani society to demand the lion's share of the budget and to pursue risky policies toward Afghanistan and India. The army has also attracted international opprobrium for its history of sponsoring nuclear proliferation through, among other means, the "procurement networks" of Abdul Qadeer Khan, the father of Pakistan's atomic bomb.

Yet the military's domestic political influence may be diminishing. The PPP government this March became the second Pakistani administration to serve out a (roughly) complete term. The first to do so was the one elected in 2002 under Musharraf's military government. Thus the PPP government became the first to serve a full term (minus the two extra days for a caretaker administration) under an entirely civilian dispensation.

It is true that during the past five years the PPP has become known for industrial-strength corruption, a refusal to expand Pakistan's tax base by imposing industrial and agricultural taxes on parliamentarians and their patronage networks, the failure to mitigate chronic power and gas shortages, and an inability to manage pervasive security problems or end violence against religious and ethnic minorities. Despite these shortcomings, the PPP government's achievements have been as notable and unprecedented as they have been unremarked upon in the international media.

After all, the outgoing parliament has passed more legislation than any previous Pakistani legislature, and it has taken important strides toward becoming more active in foreign and defense policy, an area long dominated by the army. While the parliament walks a fine line with the military establishment, the Pakistani electorate has become accustomed to seeing the army's authority publicly questioned, and it now expects politicians to be active in crafting security policy. Equally groundbreaking is the fact that President Zardari was the first sitting president to devolve extensive presidential powers to the prime minister and central government power to the provinces.

With a second constitutional transition in the offing, the army knows that its wings are slowly being clipped. But if the next government comes to power without a strong mandate, the army may be able to manage the political process and even take over the government if it chooses. This may explain the military's recent indirect manipulations of the civilian government. The army knows that Pakistanis will not tolerate a coup: As irked as citizens are by the current state of governance, surveys by the Pew Global Attitudes Project, WorldPublicOpinion. org, and this author consistently show that Pakistanis prefer even a flawed democracy to an armored state.

The best that the army can hope for is that no party attracts a majority or even a robust plurality of votes in the upcoming election. Without a clear majority, the resulting government will have to be formed from a fractious coalition that may include political rivals. From the army's point of view, the ideal new government is one that is shaky, ineffective, and unable to exert control over the armed forces or influence defense policy or key foreign policies (especially toward India, the United States, and Afghanistan).

Divided Electorate

The army may get what it wishes for. A February 2013 poll fielded by the Sustainable Development Policy Institute (SDPI) for the *Herald* (a Pakistani monthly magazine) suggests that the upcoming election will be close, and that a hung parliament is the most likely outcome. The poll found that among those respondents who claimed to be registered to vote, 29 percent said they would vote for the PPP, 24.7 percent for the PML-N, and 20.3 percent for Imran Khan's PTI. Several regional and ethnic parties barely broke into double digits.

While the poll found little variation in these figures across age groups and genders, it did find that support for the parties varies by region. For example, the PTI fares better in urban areas, whereas the PPP and PML-N have more support in rural areas. Sindhis and Seraiki speakers are most likely to vote for the PPP, while Punjabis and Hindko speakers are more inclined toward the PML-N. The PTI draws support from Hindko speakers and many Pashtuns.

Not only are Pakistanis deeply divided along party lines, they do not agree which of the issues confronting the state are the most important. Respondents were given a list of issues and asked to select the most pressing problems facing the country. While poverty, corruption, power crises, illiteracy, and extremism were the most common choices, no issue garnered more than 17 percent of the responses. Responses differed according to the respondent's socio-economic status, place of residence (rural or urban), and level of education.

Pakistanis are even more deeply conflicted when it comes to their nation's foreign policies, including relations with the United States, India, Afghanistan, and China. The United States and Pakistan have a long and tortured history together. While both sides have frequently been disappointed in the alliance, the last decade has been particularly challenging. The SDPI asked respondents whether Pakistan should have a "strong alliance" with the United States. Despite public outrage over drones and other unpopular American policies, respondents were ambivalent, with nearly one-third answering "yes," another third "no," and the remainder "maybe."

Pakistanis are similarly divided about their country's relations with India. One of the PPP government's greatest accomplishments was offering India "most favored nation" trade status. (India had offered Pakistan the same status in 1996 and a reciprocal arrangement is now being implemented.) Respondents surveyed by the SDPI were not terribly enthusiastic about this breakthrough. In fact, a plurality of interviewees believed Pakistan should not have made the offer (43 percent), with 28 percent agreeing with the move and another 29 percent undecided.

The PPP government also tried to make overtures to Afghanistan. Policy makers have emphasized that they would like a cooperative relationship with Pakistan's western neighbor, even though the army backs a more interventionist approach. According to the SDPI poll, Pakistanis are equally divided about how best to pursue relations with Kabul. When asked whether Pakistan should "actively promote a government favorable to its own interests in Afghanistan," roughly equal percentages of respondents answered "yes" (33 percent), "no" (35 percent), and "maybe" (32 percent).

Despite all of the anti-American fulmination in Pakistan, Pakistanis do not appear to be ready to kick the Americans out. Survey respondents were asked to select the countries most beneficial to Pakistan from a list that included China, India, Iran, Russia, Saudi Arabia, and the United States, as well as countries associated with the South Asian Association for Regional Cooperation and "Muslim countries" in general. China proved most popular, with 15 percent of the respondents identifying it as the "most beneficial." But the other countries and groups of countries, including the United States, polled similarly, at roughly 11 to 13 percent: statistically, more or less a dead heat.

Pakistani views on the parties' foreign policy approaches are similarly fragmented. When asked which party would "handle Pakistan's foreign affairs in the best possible manner," respondents were divided: 30 percent and 29 percent respectively identified the PPP and the PML-N. As on domestic issues, Khan's PTI claimed the support of 21 percent of those surveyed. The other parties all drew support in the single digits.

Interestingly, when asked to say which party would handle Pakistan's foreign affairs in the worst possible manner, respondents were most likely to nominate the PPP for this ignominious distinction (38 percent). The PML-N came in second place, with 18 percent, and the MQM (Mutehida Quami Mahaz, a party generally representing Pakistan's Urdu-speaking migrants from India) ranked third with 12 percent.

Options in a Hung Parliament

The results of the *Herald* and SDPI survey imply two important realities for the upcoming election. First, no single party can dominate the polls. Second, Pakistanis are deeply divided about what kind of policies their country should pursue at home and abroad and what the new government's priorities should be. Thus, whichever party forms the government will have to rely on a coalition, and it will have to navigate a fissiparous electorate whose priorities vary by province and ethnicity more than according to gender or age.

Seeing the writing on the wall, seasoned politicians are abandoning their parties for new opportunities. Such side-switching is not unusual for Pakistani politics, and has given rise to a widespread epithet: *lotas*. The lota is the water pot that Pakistanis use for toilet hygiene. In the course of its use, the lota must tip back and forth, similar to the way in which some politicians move back and forth between parties without shame or consequences.

The survey data make clear that, to remain in power, the PPP will have to cling to its current allies and try to offset defections by enticing rival candidates with significant vote banks into its tent. Its current allies include the Awami National Party (ANP, which represents mostly Pashtuns in Karachi and areas bordering Afghanistan), the MQM (which enjoys support from Pakistan's Urdu-speaking peoples), and the Pakistan Muslim League-Quaid-e-Azam (PML-Q). Both the ANP and the MQM tend to be more secularly inclined and ethnic-based. Had Bhutto not been assassinated, Musharraf and his American supporters hoped that the PPP would win the 2007 elections with the support of the PML-Q.

Based on the *Herald*/SDPI survey, the current PPP alliance could secure 38 percent of the vote. Arrayed against the PPP are the PML-N and virtually all of the religious parties, as well as the PTI. If these parties were to form a coalition against the PPP, they could win the election. Such a grand coalition is unlikely but not impossible. With the race between the PPP and the PML-N so close, the PTI could choose to be a kingmaker by throwing in its lot with one of the big parties, or it could join forces in a grand coalition of all opposition parties.

Whatever political permutation wins out, it is clear that the opposition will be fierce and that the new government will have conflicted aims, with no clear mandate about which domestic and foreign policies it should pursue and prioritize.

Dangerous Gamble?

The army may have secured its near-term interests by weakening the PPP and ensuring that whatever government emerges from the elections will be feeble, divided, and vulnerable to inter-party and intra-party disagreements. In contrast to near-certain political deadlock, the army will seem comparatively mature, capable of steering Pakistan through dangerous waters.

But stability will remain elusive, because the army has yet to fully recuperate its image among Pakistanis. The bin Laden affair, raging internal insecurity, questions surrounding the military's cooperation with the US drone program, and the disappearances of citizens in Balochistan and elsewhere have left many Pakistanis suspicious about what their army is doing and with whose support.

With a weakened army and a shaky coalition, the new government will be unlikely to muster the political will to undertake the serious reforms needed to end Pakistan's long and dangerous descent. Pakistan's obstacles are enduring and massive. Among numerous difficult tasks, it needs fiscal reform, police reform, and an overhaul of the legal code to help the state deal with the many criminal and terrorist threats facing the nation.

To meet these policy challenges, Pakistan needs a strong democratic government that enjoys a broad base of support across provinces, ethnicities, ages, and genders. But this is an unlikely outcome. Whoever wins in the upcoming election, the loser will most certainly be the Pakistani voter, who can expect little improvement in governance or accountability.

Critical Thinking

1. Why is it difficult for Pakistan to make the transition to democracy?
2. Why does the author think that the election in 2013 will not improve governance in Pakistan?
3. Why did the U.S. support President Musharrat?

Create Central

www.mhhe.com/createcentral

Internet References

Ministry of Foreign Affairs-Islamabad
www.mofa.gov.pk/

Embassy of the Islamic Republic of Pakistan, Washington, DC.
www.embassyofpakistanusa.org/

National Democratic Institute
www.ndi.org/

C. CHRISTINE FAIR *is an assistant professor at Georgetown University's Security Studies Program in the Edmund A. Walsh School of Foreign Service and a* Current History *contributing editor.*

Prepared by: Robert Weiner, *University of Massachusetts/Boston*

Article

One Step Forward, Two Steps Back

Democracy is in retreat. And there's a surprising culprit

JOSHUA KURLANTZICK

Learning Outcomes

After reading this article, you will be able to:

- Explain the relationship between the global middle class and democracy.
- Explain what factors contribute to the consolidation of democracy.

OVER THE PAST TWO YEARS, the world's attention has been captured by previously unimaginable—even rapturous—changes throughout parts of the Arab world, Africa, and Asia, where political openings have been born in some of the most repressive and unlikely societies on Earth. In Burma, where only six years ago a thuggish junta ordered the shooting of red-robed monks in the streets, the past two years have seen a formal, and seemingly real, transition to a civilian democratic government. In Tunisia, Egypt, and Libya, longtime autocrats were toppled by popular revolutions, and citizens in these states seemed at last to be enjoying the trappings of freedom.

"The Arab Spring is the triumph of democracy," Tunisian President Moncef Marzouki, a former human rights activist, told the *Guardian* in 2012. The Arab peoples "have come up with their own answer to violent extremism and the abusive regimes we've been propping up. It's called democracy," wrote *New York Times* columnist Thomas Friedman.

Don't believe the hype. In reality, democracy is going into reverse. While some countries in Africa, the Arab world, and Asia have opened slightly in the past two years, in other countries once held up as examples of political change democratic meltdowns have become depressingly common. In fact, Freedom House found that global freedom dropped in 2012 for the seventh year in a row, a record number of years of consistent decline.

The Arab Spring has not only led to dictators like Syria's Bashar al-Assad and Bahrain's ruling Al Khalifa family digging in across the region, but it has also pushed autocrats around the world to take a harder line with their populations—whether it's China censoring even vague code words for protest or Russia passing broad new treason laws and harassing human rights NGOs. As Arch Puddington, Freedom House's vice president for research, put it, "Our findings point to the growing sophistication of modern authoritarians. . . . Especially since :he Arab Spring, they are nervous, which accounts for their intensified persecution of popular movements for change."

But it's not the Arab Spring alone that's to blame. According to Freedom House, democracy's "forward march" actually peaked around the beginning of the 2000s. A mountain of evidence supports that gloomy conclusion. One of the most comprehensive studies of global democracy, the Bertelsmann Foundation's Transformation Index, has declared that "the overall quality of democracy has deteriorated" throughout the developing world. The index found that the number of "defective" and "highly defective democracies"—those with institutions, elections, and political culture so flawed that they hardly resemble real democracies—was up to 52 in 2012.

In another major survey, by the Economist Intelligence Unit, democracy deteriorated in 48 of 167 countries surveyed in 2011. "The dominant pattern globally over the past five years has been backsliding," the report says. We're not just talking about the likes of Pakistan and Zimbabwe here. Thirteen countries on the Transformation Index qualified as "highly defective democracies," countries with such a lack of opportunity for opposition voices, such problems with the rule of law, and such unrepresentative political structures that they are now little better than autocracies.

Even countries often held up as new democratic models have regressed over the past decade. When they entered the European Union in 2004, the Czech Republic, Hungary, Poland, and

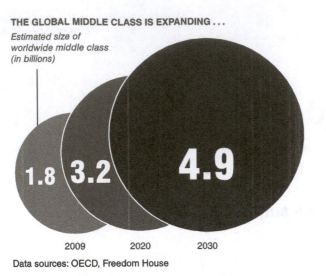

THE GLOBAL MIDDLE CLASS IS EXPANDING . . .

Estimated size of worldwide middle class (in billions)

1.8   3.2   4.9

2009   2020   2030

Data sources: OECD, Freedom House

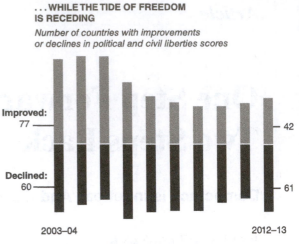

. . . WHILE THE TIDE OF FREEDOM IS RECEDING

Number of countries with improvements or declines in political and civil liberties scores

Improved:
77   42

Declined:
60   61

2003–04   2012–13

Slovakia were considered success stories. After nearly a decade as EU members, however, all of these bright lights have dimmed. Populist and far-right parties with little commitment to democratic norms gained steady popularity; public distaste for democracy increased; and governments showed more willingness to crack down on activists. Hungary has deteriorated so badly that its press freedoms rate barely better than they were under the communists.

Meanwhile, as European democracy falters, old-fashioned coups are returning elsewhere. In Africa, Asia, and Latin America, coups had become rare by the late 1990s. But between 2006 and 2012, militaries grabbed power in Bangladesh, Fiji, Guinea, Guinea-Bissau, Honduras, Madagascar, Mauritania, Mali, Niger, and Thailand, among others. In places like Ecuador, Mexico, Pakistan, and the Philippines, where the military did not launch an outright coup, it still managed to restore its power as a central actor in political life.

This is also true across the Middle East, where the Arab uprisings appear to be entrenching the power of militaries, sparking massive unrest, scaring middle-class liberals into exodus, and empowering Islamist majorities. Protesters may have bravely challenged leaders from Yemen to Egypt, but it's the loyalty of the military that has determined whether these rulers stay in power.

So WHAT WENT wrong? Let's start by blaming an unlikely culprit: the middle class. Contrary to the modernization theories of Samuel Huntington, Seymour Martin Lipset, and most Western world leaders, who have long argued that the growth of the middle class in developing countries is a boon to democratization, it hasn't worked out that way.

In theory, as the middle class expands, men and women should become more educated and more-demanding of greater economic, social, and ultimately political freedoms. And once a country reaches the per capita GDP of a middle-income country, it should rarely if ever return to authoritarian rule. "In virtually every country [that has democratized] the most

active supporters of democratization came from the urban middle class," Huntington wrote. Or consider the words of Russian economist Sergei Guriev, who declared just this past January that his country's booming middle class has become "too well-educated and too determined to enjoy increases in their quality of life" not to force an end to President Vladimir Putin's creeping authoritarianism. "They will demand that the Russian government is less corrupt and more accessible," Guriev said.

But they're not succeeding. Sure, it's true that the middle class globally is exploding; the World Bank estimates that between 1990 and 2005, the middle class tripled in size in developing countries in Asia, and in Africa it grew by a third over the past decade, according to the African Development Bank Group. Today, roughly 70 million people worldwide each year begin to earn enough to join the middle class.

It seems, however, that this new global middle is choosing stability over all else. From Algeria to Zimbabwe, the rising middle class has often supported the military as a bulwark against popular democracy, fearing that it might empower the poor, the religious, and the less-educated. In research for my book, I studied every coup attempt in the past 10 years in the developing world and then analyzed a comprehensive range of local polls and media. I found that in 50 percent of cases, middle-class men and women either agitated in advance for coups or subsequently expressed their wholehearted support for the army takeover. This is a shockingly high percentage, given that in many of these countries, such as Pakistan and Thailand, the middle class had originally been at the forefront of trying to get the army out of politics.

And in many countries, middle classes have increasingly come to disdain norms of democratic culture such as using elections, not violent demonstrations, to change leaders. From Bolivia to Venezuela to the Philippines, middle classes have turned to street protests or appeals to the judiciary to try to remove elected leaders.

And the trend is only growing stronger. Opinion polling from many developing countries shows that not only is the quality of democracy declining, but public views of democracy are deteriorating. The respected Globalbarometer series uses extensive questionnaires to ask people about their views on democracy. It has found declining levels of support for democracy throughout much of sub-Saharan Africa. In Central Asia and the former Soviet Union, the story is the same. In Kyrgyzstan, which despite its flaws remains the most democratic state in Central Asia, a majority of the population did not think that a political opposition is very or somewhat important. And recent polls show that only 16 percent of Russians surveyed said that it was "very important" that their country be governed democratically. Likewise, in Colombia, Ecuador, Honduras, Guatemala, Nicaragua, Paraguay, and Peru, either a minority or only a tiny majority of people thinks democracy is preferable to any other type of government.

Global economic stagnation since the 2008 crash has only weakened public support for democracy. New middle classes have been hit hard by the malaise, particularly in Eastern Europe. A comprehensive study of Central and Eastern Europe by the European Bank for Reconstruction and Development released in 2011 found that the crisis severely lowered support for democracy in all 10 of the new EU countries. "Those who enjoyed more freedoms wanted less democracy and markets when they were hurt by the crisis," the report noted.

Even in Asia, one of the world's most economically vibrant and globalized regions, polls show rising dissatisfaction with democracy—what some researchers have termed "authoritarian nostalgia." Indonesia, for example, is considered by many to be the democratic success story of the 2000s. Yet vote-buying and corruption among elected politicians have begun to wear. In a 2011 study, only 13 percent of respondents thought that the current group of democratic politicians was doing a better job than leaders during the era of Suharto's authoritarianism.

Even where democracy has deeper roots, disillusionment with the political process has exploded in recent years. From hundreds of thousands of Indians demonstrating against corruption to Israelis camping in the streets of Tel Aviv to protest their leaders' lack of interest in basic economic issues to the French pushing back against government austerity measures, middle classes are increasingly turning to street protests to make their points. "Our parents are grateful because they're voting," one young woman told a reporter in Spain, where unemployment now tops 50 percent for young people. "We're the first generation to say that voting is worthless."

In his second inaugural address, U.S. President Barack Obama, like every U.S. president for decades, spoke of America's role in helping promote democracy around the globe. "We will support democracy from Asia to Africa, from the Americas to the Middle East, because our interests and our conscience compel us to act on behalf of those who long for freedom," he declared. Obama may have the best of intentions, but in reality there is little he can do. The sad, troubling regression of democracy in developing countries isn't something that America can fix—because it has to be fixed at home too.

Critical Thinking

1. Why is the spread of democracy declining?
2. Why has democracy deteriorated throughout the developing world?
3. What is preventing democratization in Central and Eastern Europe?

Create Central

www.mhhe.com/createcentral

Internet References

Freedom House
   Freedomhouse.org
International Foundation for Election Systems
   www.ifes.org

JOSHUA KURLANTZICK, *fellow for Southeast Asia at the Council on Foreign Relations, is author of* Democracy in Retreat: The Revolt of the Middle Class and the Worldwide Decline of Representative Government.

Kurlantzick, Joshua. Reprinted in entirety by McGraw-Hill Education with permission from *Foreign Policy*, March/April 2013, pp. 20–22. www.foreignpolicy.com. © 2013 Washingtonpost.Newsweek Interactive, LLC.

Unit 3

UNIT

Prepared by: Robert Weiner, *University of Massachusetts/Boston*

Foreign Policy

In 2013, the foreign policy of the Obama administration remained focused on the idea of the American century in the face of the resurgence of the Declinist school of thought, which argued that the United States was losing its hegemonic position in the international system. By 2013, U.S. combat forces had been withdrawn from Iraq, although post-occupation Iraq continued to be plagued by sectarian conflict and suicide bombings that took a heavy toll among civilians. In the case of Afghanistan, the Obama administration reached an agreement with the Karzai regime that called for a withdrawal of U.S. forces by 2014, but with a commitment to help maintain stability in Afghanistan. A major development in Obama's second term was the announcement in June 2014 that the Taliban had opened an office in Qatar and that negotiations would take place between its representatives of the Taliban and the U.S. However, once the negotiations got underway, they were not going to include the representatives of the Karzai government at that time.

The winding down of the wars in Iraq and Afghanistan were accompanied by a major geopolitical shift in U.S. interests from Europe to the Asian–Pacific region as the Obama administration emphasized that it was also a Pacific power. As Secretary of State Hillary Clinton pointed out in an article in the November 2011 issue of *Foreign Policy,* the termination of the wars placed the United States at a pivot point in which the United States would emphasize the importance of the Asian–Pacific region. This would mean that more of the military assets of the United States would be shifted to the Pacific area partly to counter the rising naval power of China. However, a pivot toward the Asian–Pacific region enhances Chinese fear of encirclement by the United States and fits into the Chinese view that the United States is trying to block its rise. The United States fears the creation of a "Sino-centric bloc" in the Western Pacific; it is not surprising that Chinese economic power is being translated into military power. All realists emphasize the important relationship that exists between economic power and military power. One could compare the Chinese fear of encirclement by the United States to Kaiser Wilhelm's fear of the encirclement of Germany by the United Kingdom prior to the outbreak of the First World War. The Chinese fear a U.S. encirclement that could threaten its territorial integrity and also result in interference in its internal affairs. On the other hand, the United States is concerned that Chinese naval strategy is designed to push it out of the Western Pacific.

In 2013, U.S. standing in the international system suffered further damage, when Edward Snowden, a contractor working for the National Security Agency, revealed that U.S. intelligence efforts to gather metadata involved far more extensive eavesdropping on electronic communications around the world than imagined. The Russian decision to grant Snowden temporary asylum also had the effect off harming relations between Moscow and Washington, resulting in President Obama canceling a scheduled meeting with President Putin in St. Petersburg. The Snowden affair also provided Beijing with the opportunity to accuse the U.S. of hypocrisy in criticizing Chinese cyberwar capabilities while acting as a cyberbully itself. Moreover, the Snowden revelations caused further damage to relations between the U.S. and its European allies when it was revealed that the cell phone communications of Chancellor Merkl of Germany, among other leaders, had been monitored by the U.S. for a decade.

The Middle East continued to be a major foreign policy problem in 2013 as the Syrian civil conflict continued to rage, with the use of chemical weapons on a large scale. U.S. Secretary of State Kerry also expended a considerable amount of effort to restart Israel–Palestinian negotiations, which were taking place in the shadow of the Syrian conflict. Even though U.S. policy has been marked by a pivot or rebalancing toward Asia, the U.S. is not able to extricate itself from the conflicts in the Middle East.

Consequently, realism seems to be the dominant approach of the United States to the area, also given the opposition of Russia and China to any Western military intervention aimed at promoting regime change in Syria. 2013, however, witnessed the opening of negotiations between Iran and the United States, as newly elected Iranian President Rouhani seemed to take a more moderate and flexible stand in dealing with the United States, perhaps as a result of the UN's economic sanctions. The United States' decision to open talks with Iran in 2013 was somewhat controversial, especially in Saudi Arabia and Israel.

The growth of Chinese naval power has also resulted in a more aggressive policy of claiming sovereignty over disputed islands in the South China Sea, such as the Spratly Islands as well as the East China Sea, with a number of other states in the region. The disputes, fed by nationalism have resulted in several incidents between China and the other states claiming sovereignty over various islands and rocks. These threaten to escalate into a more serious conflict and can draw in the United States in support of its allies in the region. The United States has taken the position that a code of conduct should be adopted in dealing with these disputes. The conflict over the sovereignty of islands and rocks in the South China Sea and elsewhere are driven by the international law of the sea, which would allow the state whose sovereignty is recognized to exploit adjacent resources such as oil out to 200 miles. The Chinese are also concerned with maintaining control of the vital sea lanes in the region to ensure their access to the natural resources and the raw materials that they need.

Article

Prepared by: Robert Weiner, *University of Massachusetts/Boston*

The Future of United States–Chinese Relations: Conflict Is a Choice, Not a Necessity

HENRY A. KISSINGER

Learning Outcomes

After reading this article, you will be able to:

- Explain why the United States should not pursue a policy of confrontation with China.
- Analyze the basis for cooperation between China and the United States.

On January 19, 2011, U.S. President Barack Obama and Chinese President Hu Jintao issued a joint statement at the end of Hu's visit to Washington. It proclaimed their shared commitment to a "positive, cooperative, and comprehensive U.S.-China relationship." Each party reassured the other regarding his principal concern, announcing, "The United States reiterated that it welcomes a strong, prosperous, and successful China that plays a greater role in world affairs. China welcomes the United States as an Asia-Pacific nation that contributes to peace, stability and prosperity in the region."

Since then, the two governments have set about implementing the stated objectives. Top American and Chinese officials have exchanged visits and institutionalized their exchanges on major strategic and economic issues. Military-to-military contacts have been restarted, opening an important channel of communication. And at the unofficial level, so-called track-two groups have explored possible evolutions of the U.S.-Chinese relationship.

Yet as cooperation has increased, so has controversy. Significant groups in both countries claim that a contest for supremacy between China and the United States is inevitable and perhaps already under way. In this perspective, appeals for U.S.-Chinese cooperation appear outmoded and even naive.

The mutual recriminations emerge from distinct yet parallel analyses in each country. Some American strategic thinkers argue that Chinese policy pursues two long-term objectives: displacing the United States as the preeminent power in the western Pacific and consolidating Asia into an exclusionary bloc deferring to Chinese economic and foreign policy interests. In this conception, even though China's absolute military capacities are not formally equal to those of the United States, Beijing possesses the ability to pose unacceptable risks in a conflict with Washington and is developing increasingly sophisticated means to negate traditional U.S. advantages. Its invulnerable second-strike nuclear capability will eventually be paired with an expanding range of antiship ballistic missiles and asymmetric capabilities in new domains such as cyberspace and space. China could secure a dominant naval position through a series of island chains on its periphery, some fear, and once such a screen exists, China's neighbors, dependent as they are on Chinese trade and uncertain of the United States' ability to react, might adjust their policies according to Chinese preferences. Eventually, this could lead to the creation of a Sinocentric Asian bloc dominating the western Pacific. The most recent U.S. defense strategy report reflects, at least implicitly, some of these apprehensions.

No Chinese government officials have proclaimed such a strategy as China's actual policy. Indeed, they stress the opposite. However, enough material exists in China's quasi-official press and research institutes to lend some support to the theory that relations are heading for confrontation rather than cooperation.

U.S. strategic concerns are magnified by ideological predispositions to battle with the entire nondemocratic world. Authoritarian regimes, some argue, are inherently brittle, impelled to rally domestic support by nationalist and expansionist rhetoric and practice. In these theories- versions of which are embraced in segments of both the American left and the American right-tension and conflict with China grow out of China's domestic structure. Universal peace will come, it is asserted, from the global triumph of democracy rather than from appeals for cooperation. The political scientist Aaron Friedberg writes, for example, that "a liberal democratic China will have little cause to fear its democratic counterparts, still less to use force against them." Therefore, "stripped of diplomatic niceties, the ultimate aim of the American strategy [should be] to hasten a revolution, albeit a peaceful one, that will sweep away China's one-party authoritarian state and leave a liberal democracy in its place."

On the Chinese side, the confrontational interpretations follow an inverse logic. They see the United States as a wounded superpower determined to thwart the rise of any challenger, of which China is the most credible. No matter how intensely China pursues cooperation, some Chinese argue, Washington's fixed objective will be to hem in a growing China by military deployment and treaty commitments, thus preventing it from playing its historic role as the Middle Kingdom. In this perspective, any sustained cooperation with the United States is self-defeating, since it will only serve the overriding U.S. objective of neutralizing China. Systematic hostility is occasionally considered to inhere even in American cultural and technological influences, which are sometimes cast as a form of deliberate pressure designed to corrode China's domestic consensus and traditional values. The most assertive voices argue that China has been unduly passive in the face of hostile trends and that (for example, in the case of territorial issues in the South China Sea) China should confront those of its neighbors with which it has disputed claims and then, in the words of the strategic analyst Long Tao, "reason, think ahead and strike first before things gradually run out of hand . . . launch[ing] some tiny-scale battles that could deter provocateurs from going further."

The Past Need not be Prologue

Is there, then, a point in the quest for a cooperative U.S.-Chinese relationship and in policies designed to achieve it? To be sure, the rise of powers has historically often led to conflict with established countries. But conditions have changed. It is doubtful that the leaders who went so blithely into a world war in 1914 would have done so had they known what the world would be like at its end. Contemporary leaders can have no such illusions. A major war between developed nuclear countries must bring casualties and upheavals impossible to relate to calculable objectives Preemption is all but excluded, especially for a pluralistic democracy such as the United States.

If challenged, the United States will do what it must to preserve its security. But it should not adopt confrontation as a strategy of choice. In China, the United States would encounter an adversary skilled over the centuries in using prolonged conflict as a strategy and whose doctrine emphasizes the psychological exhaustion of the opponent. In an actual conflict, both sides possess the capabilities and the ingenuity to inflict catastrophic damage on each other. By the time any such hypothetical conflagration drew to a close, all participants would be left exhausted and debilitated. They would then be obliged to face anew the very task that confronts them today: the construction of an international order in which both countries are significant components.

The blueprints for containment drawn from Cold War strategies used by both sides against an expansionist Soviet Union do not apply to current conditions. The economy of the Soviet Union was weak (except for military production) and did not affect the global economy. Once China broke off ties and ejected Soviet advisers, few countries except those forcibly absorbed into the Soviet orbit had a major stake in their economic relationship with Moscow. Contemporary China, by contrast, is a dynamic factor in the world economy. It is a principal trading partner of all its neighbors and most of the Western industrial powers, including the United States. A prolonged confrontation between China and the United States would alter the world economy with unsettling consequences for all.

Nor would China find that the strategy it pursued in its own conflict with the Soviet Union fits a confrontation with the United States. Only a few countries-and no Asian ones-would treat an American presence in Asia as "fingers" to be "chopped off" (in Deng Xiaoping's graphic phrase about Soviet forward positions). Even those Asian states that are not members of alliances with the United States seek the reassurance of an American political presence in the region and of American forces in nearby seas as the guarantor of the world to which they have become accustomed. Their approach was expressed by a senior Indonesian official to an American counterpart: "Don't leave us, but don't make us choose."

China's recent military buildup is not in itself an exceptional phenomenon: the more unusual outcome would be if the world's second-largest economy and largest importer of natural resources did not translate its economic power into some increased military capacity. The issue is whether that buildup is open ended and to what purposes it is put. If the United States treats every advance in Chinese military capabilities as a hostile act, it will quickly find itself enmeshed in an endless series of disputes on behalf of esoteric aims. But China must be aware, from its own history, of the tenuous dividing line between defensive and offensive capabilities and of the consequences of an unrestrained arms race.

China's leaders will have their own powerful reasons for rejecting domestic appeals for an adversarial approach-as indeed they have publicly proclaimed. China's imperial expansion has historically been achieved by osmosis rather than conquest, or by the conversion to Chinese culture of conquerors who then added their own territories to the Chinese domain. Dominating Asia militarily would be a formidable undertaking. The Soviet Union, during the Cold War, bordered on a string of weak countries drained by war and occupation and dependent on American troop commitments for their defense. China today faces Russia in the north; Japan and South Korea, with American military alliances, to the east; Vietnam and India to the south; and Indonesia and Malaysia not far away. This is not a constellation conducive to conquest. It is more likely to raise fears of encirclement. Each of these countries has a long military tradition and would pose a formidable obstacle if its territory or its ability to conduct an independent policy were threatened. A militant Chinese foreign policy would enhance cooperation among all or at least some of these nations, evoking China's historic nightmare, as happened in the period 2009–10.

Dealing With the New China

Another reason for Chinese restraint in at least the medium term is the domestic adaptation the country faces. The gap in Chinese society between the largely developed coastal regions and the undeveloped western regions has made Hu's objective of a "harmonious society" both compelling and elusive. Cultural changes compound the challenge. The next decades will witness, for the first time, the full impact of one-child families on adult Chinese society. This is bound to modify cultural patterns in a society in which large families have traditionally taken care of the aged and the handicapped. When four grandparents compete for the attention of one child and invest him with the aspirations heretofore spread across many offspring, a new pattern of insistent achievement and vast, perhaps unfulfillable, expectations may arise.

All these developments will further complicate the challenges of China's governmental transition starting in 2012, in which the presidency; the vice-presidency; the considerable majority of the positions in China's Politburo, State Council, and Central Military Commission; and thousands of other key national and provincial posts will be staffed with new appointees. The new leadership group will consist, for the most part, of members of the first Chinese generation in a century and a half to have lived all their lives in a country at peace. Its primary challenge will be finding a way to deal with a society revolutionized by changing economic conditions, unprecedented and rapidly expanding technologies of communication, a tenuous global economy, and the migration of hundreds of millions of people from China's countryside to its cities. The model of government that emerges will likely be a synthesis of modern ideas and traditional Chinese political and cultural concepts, and the quest for that synthesis will provide the ongoing drama of China's evolution.

These social and political transformations are bound to be followed with interest and hope in the United States. Direct American intervention would be neither wise nor productive. The United States will, as it should, continue to make its views known on human rights issues and individual cases. And its day-to-day conduct will express its national preference for democratic principles. But a systematic project to transform China's institutions by diplomatic pressure and economic sanctions is likely to backfire and isolate the very liberals it is intended to assist. In China, it would be interpreted by a considerable majority through the lens of nationalism, recalling earlier eras of foreign intervention.

What this situation calls for is not an abandonment of American values but a distinction between the realizable and the absolute. The U.S.-Chinese relationship should not be considered as a zero-sum game, nor can the emergence of a prosperous and powerful China be assumed in itself to be an American strategic defeat.

A cooperative approach challenges preconceptions on both sides. The United States has few precedents in its national experience of relating to a country of comparable size, self-confidence, economic achievement, and international scope and yet with such a different culture and political system. Nor does history supply China with precedents for how to relate to a fellow great power with a permanent presence in Asia, a vision of universal ideals not geared toward Chinese conceptions, and alliances with several of China's neighbors. Prior to the United States, all countries establishing such a position did so as a prelude to an attempt to dominate China.

The simplest approach to strategy is to insist on overwhelming potential adversaries with superior resources and materiel. But in the contemporary world, this is only rarely feasible. China and the United States will inevitably continue as enduring realities for each other. Neither can entrust its security to the other-no great power does, for long-and each will continue to pursue its own interests, sometimes at the relative expense of the other. But both have the responsibility to take into account the other's nightmares, and both would do well to recognize that their rhetoric, as much as their actual policies, can feed into the other's suspicions.

China's greatest strategic fear is that an outside power or powers will establish military deployments around China's periphery capable of encroaching on China's territory or meddling in its domestic institutions. When China deemed that it faced such a threat in the past, it went to war rather than risk the outcome of what it saw as gathering trendsin Korea in 1950, against India in 1962, along the northern border with the Soviet Union in 1969, and against Vietnam in 1979.

The United States' fear, sometimes only indirectly expressed, is of being pushed out of Asia by an exclusionary bloc. The United States fought a world war against Germany and Japan to prevent such an outcome and exercised some of its most forceful Cold War diplomacy under administrations of both political parties to this end against the Soviet Union. In both enterprises, it is worth noting, substantial joint U.S.-Chinese efforts were directed against the perceived threat of hegemony.

Other Asian countries will insist on their prerogatives to develop their capacities for their own national reasons, not as part of a contest between outside powers. They will not willingly consign themselves to a revived tributary order. Nor do they regard themselves as elements in an American containment policy or an American project to alter China's domestic institutions. They aspire to good relations with both China and the United States and will resist any pressure to choose between the two.

Can the fear of hegemony and the nightmare of military encirclement be reconciled? Is it possible to find a space in which both sides can achieve their ultimate objectives without militarizing their strategies? For great nations with global capabilities and divergent, even partly conflicting aspirations, what is the margin between conflict and abdication?

That China will have a major influence in the regions surrounding it is inherent in its geography, values, and history. The limits of that influence, however, will be shaped by circumstance and policy decisions. These will determine whether an inevitable quest for influence turns into a drive to negate or exclude other independent sources of power.

For nearly two generations, American strategy relied on local regional defense by American ground forces-largely to avoid the catastrophic consequences of a general nuclear war. In recent decades, congressional and public opinion have impelled an end to such commitments in Vietnam, Iraq, and Afghanistan Now, fiscal considerations further limit the range of such an approach. American strategy has been redirected from defending territory to threatening unacceptable punishment against potential aggressors. This requires forces capable of rapid intervention and global reach, but not bases ringing China's frontiers. What Washington must not do is combine a defense policy based on budgetary restraints with a diplomacy based on unlimited ideological aims.

Just as Chinese influence in surrounding countries may spur fears of dominance, so efforts to pursue traditional American national interests can be perceived as a form of military encirclement. Both sides must understand the nuances by which apparently traditional and apparently reasonable courses can evoke the deepest worries of the other. They should seek together to define the sphere in which their peaceful competition is circumscribed. If that is managed wisely, both military confrontation and domination can be avoided; if not, escalating tension is inevitable. It is the task of diplomacy to discover this space, to expand it if possible, and to prevent the relationship from being overwhelmed by tactical and domestic imperatives.

Community or Conflict

The current world order was built largely without Chinese participation, and hence China sometimes feels less bound than others by its rules. Where the order does not suit Chinese preferences, Beijing has set up alternative arrangements, such as in the separate currency channels being established with Brazil and Japan and other countries. If the pattern becomes routine and spreads into many spheres of activity, competing world orders could evolve. Absent common goals coupled with agreed rules of restraint, institutionalized rivalry is likely to escalate beyond the calculations and intentions of its advocates. In an era in which unprecedented offensive capabilities and intrusive technologies multiply, the penalties of such a course could be drastic and perhaps irrevocable.

Crisis management will not be enough to sustain a relationship so global and beset by so many differing pressures within and between both countries, which is why I have argued for the concept of a Pacific Community and expressed the hope that China and the United States can generate a sense of common purpose on at least some issues of general concern. But the goal of such a community cannot be reached if either side conceives of the enterprise as primarily a more effective way to defeat or undermine the other. Neither China nor the United States can be systematically challenged without its noticing, and if such a challenge is noted, it will be resisted. Both need to commit themselves to genuine cooperation and find a way to communicate and relate their visions to each other and to the world.

Some tentative steps in that direction have already been undertaken. For example, the United States has joined several other countries in beginning negotiations on the Trans-Pacific Partnership (TPP), a free-trade pact linking the Americas with Asia. Such an arrangement could be a step toward a Pacific Community because it would lower trade barriers among the world's most productive, dynamic, and resource-rich economies and link the two sides of the ocean in shared projects.

Obama has invited China to join the tpp. However, the terms of accession as presented by American briefers and commentators have sometimes seemed to require fundamental changes in China's domestic structure. To the extent that is the case, the tpp could be regarded in Beijing as part of a strategy to isolate China. For its part, China has put forward comparable alternative arrangements. It has negotiated a trade pact with the Association of Southeast Asian Nations and has broached a Northeast Asian trade pact with Japan and South Korea.

Important domestic political considerations are involved for all parties. But if China and the United States come to regard each other's trade-pact efforts as elements in a strategy of isolation, the Asia-Pacific region could devolve into competing adversarial power blocs. Ironically, this would be a particular challenge if China meets frequent American calls to shift from an export-led to a consumption-driven economy, as its most recent five-year plan contemplates. Such a development could reduce China's stake in the United States as an export market even as it encourages other Asian countries to further orient their economies toward China.

The key decision facing both Beijing and Washington is whether to move toward a genuine effort at cooperation or fall into a new version of historic patterns of international rivalry. Both countries have adopted the rhetoric of community. They have even established a high-level forum for it, the Strategic and Economic Dialogue, which meets twice a year. It has been productive on immediate issues, but it is still in the foothills of its ultimate assignment to produce a truly global economic and political order. And if a global order does not emerge in the economic field, barriers to progress on more emotional and less positive-sum issues, such as territory and security, may grow insurmountable.

The Risks of Rhetoric

As they pursue this process, both sides need to recognize the impact of rhetoric on perceptions and calculations. American leaders occasionally launch broadsides against China, including specific proposals for adversarial policies, as domestic political necessities. This occurs even-perhaps especially-when a moderate policy is the ultimate intention. The issue is not specific complaints, which should be dealt with on the merits of the issue, but attacks on the basic motivations of Chinese policy, such as declaring China a strategic adversary. The target of these attacks is bound to ask whether domestic imperatives requiring affirmations of hostility will sooner or later require hostile actions. By the same token, threatening Chinese statements, including those in the semiofficial press, are likely to be interpreted in terms of the actions they imply, whatever the domestic pressures or the intent that generated them.

The American debate, on both sides of the political divide, often describes China as a "rising power" that will need to "mature" and learn how to exercise responsibility on the world stage. China, however, sees itself not as a rising power but as a returning one, predominant in its region for two millennia and temporarily displaced by colonial exploiters taking advantage of Chinese domestic strife and decay. It views the prospect of a strong China exercising influence in economic, cultural, political, and military affairs not as an unnatural challenge to world order but rather as a return to normality. Americans need not agree with every aspect of the Chinese analysis to understand that lecturing a country with a history of millennia about its need to "grow up" and behave "responsibly" can be needlessly grating.

On the Chinese side, proclamations at the governmental and the informal level that China intends to "revive the Chinese nation" to its traditional eminence carry different implications inside China and abroad. China is rightly proud of its recent strides in restoring its sense of national purpose following what it sees as a century of humiliation. Yet few other countries in Asia are nostalgic for an era when they were subject to Chinese suzerainty. As recent veterans of anti-colonial struggles, most Asian countries are extremely sensitive to maintaining their independence and freedom of action vis-à-vis any outside power, whether Western or Asian. They seek to be involved in as many overlapping spheres of economic and political activity as possible; they invite an American role in the region but seek equilibrium, not a crusade or confrontation.

The rise of China is less the result of its increased military strength than of the United States' own declining competitive position, driven by factors such as obsolescent infrastructure, inadequate attention to research and development, and a seemingly dysfunctional governmental process. The United States should address these issues with ingenuity and determination instead of blaming a putative adversary. It must take care not to repeat in its China policy the pattern of conflicts entered with vast public support and broad goals but ended when the American political process insisted on a strategy of extrication that amounted to an abandonment, if not a complete reversal, of the country's proclaimed objectives.

China can find reassurance in its own record of endurance and in the fact that no U.S. administration has ever sought to alter the reality of China as one of the world's major states, economies, and civilizations. Americans would do well to remember that even when China's gdp is equal to that of the United States, it will need to be distributed over a population that is four times as large, aging, and engaged in complex domestic transformations occasioned by China's growth and urbanization. The practical consequence is that a great deal of China's energy will still be devoted to domestic needs.

Both sides should be open to conceiving of each other's activities as a normal part of international life and not in themselves as a cause for alarm. The inevitable tendency to impinge on each other should not be equated with a conscious drive to contain or dominate, so long as both can maintain the distinction and calibrate their actions accordingly. China and the United States will not necessarily transcend the ordinary operation of great-power rivalry. But they owe it to themselves, and the world, to make an effort to do so.

Critical Thinking

1. Why dos China fear the United States?
2. How can China and the United States develop a more cooperative relationship?
3. Why does Kissinger contend that China is not rising, but is returning to its position of predominance in the international system?

Create Central

www.mhhe.com/createcentral

Internet References

The Association of Southeast Asian Nations
www.asean.sec.org
Institute of Southeast Asian Studies
www.iseas.edu.sg

Prepared by: Robert Weiner, *University of
Massachusetts/Boston*

Article

Do Presidents Matter?

**Where foreign policy is concerned, the most-valuable traits
are not always the ones we value most highly.**

Joseph S. Nye Jr.

Learning Outcomes

After reading this article, you will be able to:

- Explain what is meant by the contextual intelligence of
  presidents.
- Explain the importance of presidential leadership.

THE 21ST CENTURY began with an extraordinary imbalance in world power. The United States was the only country able to project military force globally; it represented more than a quarter of the world economy, and had the world's leading soft-power resources in its universities and entertainment industry. America's primacy appeared well established.

Americans seemed to like this situation. In the 2012 presidential campaign, both major-party candidates insisted that American power was not in decline, and vowed that they would maintain American primacy. But how much are such promises within the ability of presidents to keep? Was presidential leadership ever essential to the establishment of American primacy, or was that primacy an accident of history that would have occurred regardless of who occupied the Oval Office?

Leadership experts and the public alike extol the virtues of transformational leaders—those who set out bold objectives and take risks to change the world. We tend to downplay "transactional" leaders, whose goals are more modest, as mere managers. But in looking closely at the leaders who presided over key periods of expanding American primacy in the past century, I found that while transformational presidents such as Woodrow Wilson and Ronald Reagan changed how Americans viewed their nation's role in the world, some transactional presidents, such as Dwight D. Eisenhower and George H. W. Bush, were more effective in executing their policies.

Transformation involves large gambles, the outcomes of which are not always immediately evident. One of history's great strategists, Otto von Bismarck, successfully bet in 1870 that a manufactured war with France would lead to Prussian unification of Germany. But he also bet that he could annex Alsace-Lorraine, a move with enormous costs that became clear only in 1914.

Franklin D. Roosevelt and Harry Truman made transformational bets on, respectively, the nation's entry into World War II and the subsequent containment of the Soviet Union, but each did so only after cautious initial approaches (and in Roosevelt's case, only after the Japanese bombed Pearl Harbor). John F. Kennedy and Lyndon Johnson mistakenly bet that Vietnam would prove to be a game of dominoes, whereas Eisenhower—who, ironically, had coined the domino metaphor—wisely avoided combat intervention. And Richard Nixon, who successfully bet on an opening to China in 1971, lost a nearly simultaneous bet in severing the dollar's tie to gold, thus contributing to rampant inflation over the subsequent decade.

Compare Woodrow Wilson, a failed transformational president, with the first George Bush, a successful transactional one. Wilson made a costly and mistaken bet on the Treaty of Versailles at the conclusion of the First World War. His noble vision of an American-led League of Nations was partially vindicated in the long term. But he lacked the leadership skills to implement this vision in his own time, and this shortcoming contributed to America's retreat into isolationism in the 1930s. In the case of Bush 41, the president's lack of what he called "the vision thing" limited his ability to sway Americans' perceptions of the nation and its role in the world. But his execution and management of policy was first-rate.

Consider, too, the contrast between the elder Bush's presidency and that of his son, George W. Bush, who has been described as having been obsessed with being a transformational president. Members of the younger Bush's administration often compared him to Ronald Reagan or Harry Truman, but the 20th-century president he most resembled was Wilson. Both were highly religious and moralistic men who initially focused on domestic issues without an eye toward foreign policy. Both projected self-confidence, and both responded to a crisis boldly and resolutely. As Secretary of State Robert Lansing described Wilson's mind-set in 1917: "Even established facts were ignored if they did not fit in with his intuitive sense, this semi-divine power to select the right." Similarly, Tony Blair observed in 2010 that Bush "had great intuition. But his intuition was less . . . about politics and more about what he thought was right and wrong." Like Wilson, Bush placed a large, transformative bet on foreign policy—the invasion of Iraq—and, like Wilson, he lacked the skill to implement his plan successfully.

This is not an argument against transformational leaders in general. In turbulent situations, leaders such as Gandhi, Mandela, and King can play crucial roles in redefining a people's identity and aspirations. Nor is it an argument against transformational leaders in American foreign policy in particular. FDR and Truman made indelible contributions to the creation of the American era; others, such as Nixon, with his opening to China, or Carter, with his emphasis on human rights and nuclear nonproliferation, reoriented important aspects of foreign policy. But in judging leaders, we need to pay attention both to acts of commission and to acts of omission—dogs that barked and those that did not. For example, Ike refused to follow numerous recommendations by the military to use nuclear weapons during the Korean, Dien Bien Phu, and Quemoy-Matsu crises, at one point telling an adviser, "You boys must be crazy. We can't use those awful things against Asians for the second time in less than 10 years." In 1954, he explained his broader thinking to the Joint Chiefs of Staff. Suppose it would be possible to destroy Russia, he said. "Here would be a great area from the Elbe to Vladivostok . . . torn up and destroyed, without government, without its communications, just an area of starvation and disaster. I ask you, what would the civilized world do about it?" George H. W. Bush likewise largely eschewed transformational objectives, with one important exception: the reunification of Germany. But even here, he acted with caution. When the Berlin Wall was opened in November 1989, partly because of a mistake by East Germany, Bush was criticized for his low-key response. But his deliberate choice not to gloat or to humiliate the Soviets helped set the stage for the successful Malta summit with Mikhail Gorbachev a month later.

Transformational leaders are important because they make choices that most other leaders would not. But a key question is how much risk a democratic public wants its leaders to take in foreign policy. The answer very much depends on the context, and that context is enormously complex, involving not only potential international effects, but the intricacies of domestic politics in multiple societies. This complexity gives special relevance to the Aristotelian virtue of prudence. We live in a world of diverse cultures, and we know very little about social engineering and how to "build nations." And when we cannot be sure how to improve the world, hubristic visions pose a grave danger. For these reasons, the virtues of transactional leaders with good contextual intelligence are also very important. Good leadership in this century may or may not be transformational, but it will almost certainly require a careful understanding of the context of change.

Decline, for example, is a misleading description of the current state of American power—one that President Obama has thankfully rejected. American influence is not in absolute decline, and in relative terms, there is a reasonable probability that the country will remain more powerful than any other single state in the coming decades. We do not live in a "post-American world," but neither do we live any longer in the American era of the late 20th century. No one has a crystal ball, but the National Intelligence Council may be correct in its 2012 projection that although the unipolar moment is over, the U.S. most likely will remain *primus inter pares* at least until 2030 because of the multifaceted nature of its power and the legacies of its leadership.

The U.S. will certainly face a rise in the power of many others—both states and nonstate actors. Presidents will increasingly need to exert power *with* others as much as *over* others; our leaders' capacity to maintain alliances and create networks will be an important dimension of our hard and soft power. The problem of America's role in the 21st century is not the country's supposed decline, but its need to develop the contextual intelligence to understand that even the most powerful nation cannot achieve the outcomes it wants without the help of others. Educating the public to both understand the global information age and operate successfully in it will be the real task for presidential leadership.

All of which suggests that President Obama and his successors should beware of thinking that transformational proclamations are the key to successful adaptation amid these rapidly changing times. American power and leadership will remain crucial to stability and prosperity at home and abroad. But presidents will be better served by remembering their transactional predecessors' observance of the credo "Above all, do no harm" than by issuing stirring calls for transformational change.

Critical Thinking

1. Why is the U.S. unipolar moment over?
2. Why will the U.S. remain first among equals until 2030?
3. What is meant by the multifaceted nature of U.S. power?
4. Why will transactional presidents be more effective than transformational presidents?

Create Central

www.mhhe.com/createcentral

Internet References

National Security Agency
www.nsa.gov
Executive Office of the President
www.whitehouse.gov/administration/eop

JOSEPH S. NYE JR. is a University Distinguished Service Professor at Harvard. This article and the accompanying sidebar are adapted from his upcoming book, Presidential Leadership and the Creation of the American Era.

Prepared by: Robert Weiner, *University of Massachusetts/Boston*

Article

The Irony of American Strategy

Putting the Middle East in Proper Perspective

RICHARD N. HAASS

Learning Outcomes

After reading this article, you will be able to:

- Explain the effect that the pivot to Asia has on U.S. policy in the Middle East.
- Explain some of the difficulties associated with reaching a Palestinian–Israeli peace agreement.

The United States emerged from the Cold War with unprecedented absolute and relative power. It was truly first among unequals. Not surprisingly, its leaders were uncertain about what to do with such advantages, and for more than a decade following the dismantlement of the Berlin Wall, U.S. foreign policy was conducted without much in the way of an overarching strategy.

The 9/11 attacks changed all this, giving Washington a surfeit of purpose to go along with its preponderant power. Within weeks, in the opening act of what became known as the "global war on terrorism," the United States moved to oust the Taliban-led government of Afghanistan in order to prevent future attacks by al Qaeda and to send the message that governments that tolerated or abetted terrorism would not be secure.

Association with terrorism, however, was not the reason the United States attacked Iraq 17 months later. Nor was the reason preempting the use of weapons of mass destruction, for Iraq represented at most a gathering threat in that realm, not an imminent one. (Now, we know it did not represent even that, but at the time, it was widely believed that it did.) Rather, the principal rationale for attacking Iraq was to signal to the world that even after 9/11, the United States was not, in Richard Nixon's words, a "pitiful, helpless giant." Many of the war's proponents also believed that Iraq would quickly become a thriving democracy that would set an example for the rest of the Middle East.

The decision to attack Iraq in March 2003 was discretionary; it was a war of choice. There was no vital American interest in imminent danger, and there were alternatives to using military force, such as strengthening the existing sanctions. The war in Afghanistan, in contrast, started as a war of necessity.

Vital interests were at stake, and no other policy could have protected them in a timely fashion. But toward the end of the Bush administration, that conflict started to morph into something else, and it crossed a line in March 2009, when President Barack Obama decided to sharply increase American troop levels and declared that it was U.S. policy to "take the fight to the Taliban in the south and the east" of the country. With these escalations, Afghanistan, too, became a war of choice.

Around the middle of his first term, however, Obama seemed to conclude that the effort in Afghanistan, like the one in Iraq, made little sense, at least on the scale that the United States was conducting it, accomplishing little but costing a great deal and with no end in sight. Local realities trumped American ambitions. And so just as he had moved to end the U.S. military presence in Iraq (even though it might have been possible to arrange for a small, residual force to stay there), so he also moved to wind down U.S. military involvement in Afghanistan.

The drawdowns in Iraq and Afghanistan were part of a larger military distancing from the greater Middle East. When domestic upheavals rocked the region in the spring of 2011, the Obama administration tried to avoid getting deeply involved. Its participation in the operation to oust Libyan dictator Muammar al-Qaddafi was mostly limited to providing air and missile strikes and assisting the United States' NATO partners with intelligence and command and control—"leading from behind," in the words of an anonymous administration official—and the United States showed no appetite for participating in an effort to stabilize, much less rebuild, Libya in the aftermath of Qaddafi's fall. U.S. policy toward the civil war in Syria has been even more cautious. The United States has resisted not just direct military participation on behalf of the Syrian opposition (rejecting, for example, calls to establish no-fly zones and the like) but also supplying weapons. Instead, Washington has helped coordinate economic and political sanctions designed to weaken the regime while providing "nonlethal" political, intelligence, communications, and economic support to the opposition.

Washington's diplomatic involvement in the Middle East during these years was uneven. Efforts to promote peace between Israel and the Palestinians more or less came to an end after Israel's government rebuffed Obama's pressure to rein in settlement construction, and by 2012, the administration appeared more anxious to block UN consideration of the Palestinian issue than promote progress in other venues. In early 2011, the Obama team unceremoniously pushed Hosni Mubarak to give up power in Egypt but appeared reluctant to demand changes from his successors and said little about the resistance to reform in friendly monarchies.

Policies on individual issues or cases can be debated, but the thrust of the administration's approach has mostly been sensible. The greater Middle East had come to dominate and distort American foreign and defense policy, and a course correction was called for. The Obama administration's vehicle for this correction was the announcement of a "pivot," or "rebalancing," toward Asia, a region home to many of the world's largest and fastest-growing economies and one likely to be more central than the Middle East in shaping the world's future. A recognition that China was not just rising but becoming more assertive and even bullying gave the pivot some urgency, as did the sense that without it, other countries in the region, including some U.S. friends and allies, might soon start accommodating themselves to growing Chinese dominance or mobilizing their own, possibly destabilizing efforts to resist it.

By now, as the administration begins its second term, its argument for paying more attention to Asia has been widely accepted, and correctly so. The administration also faces major pressure to reduce federal spending, including on defense. But events in the greater Middle East are making it difficult for the United States to limit its involvement there. The irony is inescapable: a decade ago, Washington chose to immerse itself in the region when it did not have to, carrying out two decadelong wars of choice that involved a total of more than two million American servicemen and servicewomen and ended up costing more than 6,000 American dead, 40,000 wounded, $1.5 trillion, and enormous time and energy on the part of policymakers; but now that most Americans want little to do with the region, U.S. officials are finding it difficult to turn away. It is easy to imagine the president echoing Michael Corleone's lament in The Godfather, Part III: "Just when I thought I was out, they pull me back in."

The New New Middle East

Six and a half years ago, I wrote an essay for this magazine titled "The New Middle East." The piece argued that the era of American domination of the region was coming to an end and that the Middle East's future would be characterized by considerable but reduced U.S. influence, an imperial Iran with growing regional sway, a messy post-Saddam Iraq, a stagnant peace process, and the further spread of political Islam. This has largely come to pass, although the reality is even grimmer than the prediction. Egypt is now led by a Muslim Brotherhood president seeking to consolidate power. There is a deadly civil war going on in Syria, unrest and polarization in Bahrain, and growing signs of disquiet in Jordan. Saudi Arabia is suffering from a prolonged succession crisis, Iran is closer to possessing

nuclear weapons, and the prospects for a comprehensive and durable Middle East peace have deteriorated further.

The most immediate and difficult policy choice facing the United States in the region concerns Syria. The death toll in the civil war there has risen beyond 70,000; the conflict now involves several of Syria's neighbors and threatens to spread to them. At the same time, the ouster of the Assad regime would be a major blow to its Iranian patron. But the government remains in place, the Syrian opposition is divided and its future agenda unknowable, and the sectarian nature of the conflict guarantees that any armed intervention to end the fighting would have to be large scale, long lasting, and skillfully managed to have even a chance of success. It would be hard to justify so potentially costly and difficult an undertaking for less than vital interests. This does not mean that Washington should turn its back on the human suffering in Syria. But humanitarian intervention should not be equated with or limited to direct military action, particularly with ground forces. Washington should use other tools—such as tightening economic and political sanctions against the government, providing discrete military aid to opposition groups whose views it can accept, and diplomatic initiatives—to help remove the current leadership. It should also help refugees and internally displaced persons, working with Syria's neighbors, especially Jordan and Turkey.

Other challenges will arise from political turbulence in countries such as Bahrain, Jordan, and possibly even Saudi Arabia. In such situations, U.S. officials can encourage political and economic reform, but there is no guarantee that such advice will be accepted or that modest reforms would ensure stability. That said, the United States ought to give strong support to friendly governments that act responsibly and should think twice before distancing itself from regimes that fail to pursue desired reforms, given the important economic and security interests at stake and the strong possibility that successor regimes will be unfriendly and even more flawed.

Once political change has occurred—as in Egypt, Libya, and Tunisia—the United States should establish a more conditional relationship with the new government. Obama had it right when he described the new Egyptian government as neither an ally nor an adversary. In such circumstances, U.S. economic and military support (and U.S. backing for support from international financial institutions) should be tightly linked to how the new government treats American interests, its neighbors, and its citizens. U.S. officials should be willing to take their criticism public when such a course is warranted and might be effective.

As for Iran, the United States has many good reasons for trying to stop it from acquiring nuclear weapons. An Iran with nuclear weapons, or the capacity to acquire them quickly and easily, would be more able and willing to shape the Middle East in its anti-American image. It might be tempted to transfer nuclear weapons or material to a group such as Hezbollah, could threaten Israel, or could motivate other countries in the vicinity to develop or acquire nuclear weapons, creating a situation of enormous instability and potential destructiveness. Repeated U.S. statements that it would be "unacceptable" for

Iran to go nuclear, meanwhile, mean that American credibility is now at stake as well.

At the same time, Washington should try hard to avoid another costly war of choice. Before launching or supporting a preventive strike on Iranian nuclear assets, the United States should consider what the chances are that the strike would destroy much of Iran's relevant capacity, the costs of likely retaliation, the implications for other American interests in the region, the prospects that a nuclear Iran could be confidently deterred, the possibility that the proliferation aspirations of other regional states could be managed though alternative policies, and the impact of an attack on domestic Iranian political development. It is conceivable that when all these considerations are taken into account, a strike might make sense, but this would be a high bar to clear. And if such a course were to be embarked on, it should be only after clear offers of negotiation have been made and rejected, demonstrating Iran's unwillingness to accept a reasonable compromise.

As for the Israeli-Palestinian divide, the prospects for advancing reconciliation and peace are poor. But this is not an argument for standing pat; bad situations can and do get worse. Ideally, the Israeli government or the Palestinian Authority would put forward a comprehensive peace proposal that would generate real excitement and support both at home and across the divide; failing that, the United States should articulate principles for establishing a sustainable peace settlement that would leave all parties better off. Hopefully, a political process and negotiations would then ensue. Hamas, which controls Gaza, should be able to participate in negotiations only if it eschews violence and demonstrates a willingness to coexist with Israel. Washington should do what it can to bolster moderate forces in the Palestinian community and discourage Israel from engaging in activities—including, but not limited to, settlement construction—that will further undermine what few prospects remain to create a viable Palestinian state.

The United States retains important and in some cases vital interests in the Middle East, including a deep commitment to Israel's security, opposition to terrorism and the spread of nuclear weapons, and a commitment to safeguarding access to the region's energy resources. But today, the region is not an arena of decisive great-power competition, as it was at times during the Cold War, nor is it home to any major power. In addition, it is a part of the world where local realities can and often do limit the utility of military force, economic sanctions, and diplomacy. The fact that the United States is moving toward energy self-sufficiency gives it some added cushion (although not independence) from the consequences of the region's turbulence.

For more than a decade now, American foreign policy has been both distracted and distorted by the greater Middle East. But myriad policy choices lie between preoccupation and disengagement. Military interventions to overthrow hostile regimes or prop up friendly ones are becoming increasingly untenable and should almost always be avoided, given their high costs and uncertain payoffs, along with the existence of competing priorities at home and abroad. More discrete armed action—whether to help maintain the free movement of oil and gas, destroy or

set back programs to develop weapons of mass destruction, or attack terrorists—should be prepared for and carried out on a case-by-case basis. Where potential partners exist, Washington should also work to build up local government capacities to maintain order and combat terrorism. U.S. officials should push governments led by the Muslim Brotherhood and other Islamist movements to follow democratic norms and procedures; failing that, Washington should do what it can to make it difficult for such groups to consolidate their power. The staple of American involvement in this part of the world should be the provision or withholding of various forms of diplomatic, economic, intelligence, and military support, to influence a country's foreign policy and, in select cases, its domestic trajectory.

What Rebalancing Means

In contrast to the Middle East, Asia is a locus of great-power competition, where U.S. military presence and action may prove extremely useful in heading off or handling many potential problems. The Obama administration was wise to place a greater emphasis on this part of the world in 2011, although it could (and should) have done better in articulating and implementing its new course. "Pivot" implied too sharp a turn, both by suggesting too dramatic a pullback from the greater Middle East and by overlooking all that the United States has already done over the decades in East Asia. "Rebalancing," the administration's second label for its policy, better characterizes both the substance and the rationale of the new approach. The military dimensions of the new policy were also overemphasized at first. Maintaining and perhaps even selectively increasing the U.S. military presence in the region has been important, but more significant than the deployment of 2,500 marines in Australia is the direction of U.S. diplomacy vis-a-vis China and its neighbors, the availability of economic assistance to promote political and economic development in the region's poorer countries, and the ability to negotiate a new trade agreement (specifically, the Trans-Pacific Partnership) as quickly and inclusively as possible.

The best way to ensure Asia's stability is for the United States to stay active, be a reliable strategic partner, and be present in every sense and sphere, lest other countries in the region begin to accommodate their stronger neighbors or become more nationalist and aggressive themselves. Thus, it continues to make sense for a sizable American force (now 28,000 troops) to be stationed in South Korea, even though six decades have passed since the end of the Korean War and the South itself has grown rich and strong. Deterring a renewal of the conflict is a high priority, and on-the-scene American troops help achieve that. Making clear that any future conflict would end with the reunification of the entire peninsula under the Souths authority should increase the North's restraint, as well as reinforce China's efforts to rein in its obstreperous ally The United States could also try to reassure China that any reunified Korea would be nonnuclear and home to only a small number of U.S. troops, if any. Such reassurance might influence China's policy at a time when the new leadership in Beijing is showing signs of weariness with the antics of its longtime North Korean ally.

The rationale for defending South Korea is relatively straightforward given U.S. treaty commitments, but what Washington should do in several other potential scenarios in the region is less clear. The United States has obligations to Taiwan, as well as to Japan, the Philippines, and Australia, but it does not want to be drawn into a regional conflict without excellent reasons. So American foreign policy faces a delicate balancing act: it must communicate enough resolve so as to discourage aggression against its friends and allies, but it must avoid signaling unconditional support (the diplomatic version of moral hazard) lest it encourage those friends and allies to behave provocatively or recklessly. In practice, this means continuing to provide limited military support for Taiwan while dissuading it from unilateral efforts to alter the political status quo. It also means consulting closely with Japan, the Philippines, and other regional friends in order to see to it that Chinese assertiveness does not go unmet and that crises are avoided or, failing that, dampened rather than escalated.

Managing U.S.-Chinese relations in such a context will be far from easy, but doing so successfully is essential. There will be no more important challenge for U.S. diplomacy over the next generation than working to integrate China into regional and global arrangements, be it managing the economy, limiting climate change, or combating the proliferation of weapons of mass destruction and their delivery vehicles. China's help is needed to reunify Korea peacefully, prevent Iran from gaining nuclear weapons, and get Pakistan to change its ways. The original rationale for the rapprochement between Washington and Beijing (opposition to the Soviet Union) is no longer relevant, and the successor rationale (cooperating for mutual economic benefit) is too narrow to sustain harmony between the two countries by itself. Close cooperation on solving major regional and global challenges should be a crucial element in the mix.

Talk of a U.S.-Chinese "G-2" or of a "global condominium," however, is unrealistic. Chinese leaders remain focused on China's perceived internal needs and are busy raising the country's profile throughout the region. The wisest policy for the United States and others is thus to hedge, cooperating where possible with the People's Republic while maintaining a strong diplomatic, economic, and military presence in the region and a thick web of local ties. Such a stance will discourage China from acting aggressively, give local states confidence to stand up to their bigger neighbor, and provide the foundations for a robust response should China choose not to integrate into regional and global institutions and instead embark on a fundamentally nationalist and aggressive course. Still, given the high costs of containment, adopting it as policy before it is definitely required would be a mistake and might even help bring about an adversarial relationship that would serve the interests of no one.

Beyond Irony

The United States chose to immerse itself in the greater Middle East when it had little reason to dive in, and now that it has good reasons for limiting its involvement there, doing so has turned out to be difficult. But difficult is not impossible. There is not much to be gained by Washington's doing more

in the region right now, especially if more is defined in terms of large, prolonged military interventions designed to remake societies decidedly unripe for democracy. Where the interests at stake are less than vital and the likely risks and costs of acting on their behalf outweigh the likely benefits, the United States should learn to live with outcomes that are less than optimal. In Afghanistan, for example, it would be regrettable were the Taliban to stage a recovery, but it would not necessarily be intolerable, especially if al Qaeda or its offshoots were not allowed to operate from Afghan territory. Elsewhere in the region, it would be unfortunate if the Muslim Brotherhood and its allies came to power in several more countries, but that would not normally be grounds for armed (as opposed to other kinds of) intervention. This is not isolationism but strategy.

The old order in the Middle East is disappearing. The transition is still in its early phases, and what will follow (and when) is uncertain. But political Islam is sure to play a large part. Some borders are likely to be redrawn, and some new states may even emerge. Intra- and interstate conflict are likely to be commonplace. The United States can and should try to influence the course of events, but it is unlikely to have much control, and doing more will not necessarily give it more say. Some of the wisest actions the country could take, therefore, would include insulating itself as much as possible from regional events, continuing its development of energy sources outside the Middle East, and making itself more resilient to terrorism.

Asia, however, does call for more U.S. military, diplomatic, and economic involvement. Other regions can also stake a claim to a share of Washington's attention. Negotiating a transatlantic free-trade agreement (ideally, one involving both Canada and Mexico) would be a major economic and strategic accomplishment; so, too, would be negotiating and implementing a NAFTA 2.0 that would more closely link the United States with its immediate neighbors so as to better manage shared interests related to trade and investment, security, energy, infrastructure, and the flow of people. Narrowing the gap between global challenges and the current institutional arrangements for dealing with them is also an important issue, particularly in the case of climate change. Here and elsewhere, though, global accords with broad participation may not be possible, and it may be more realistic and rewarding to focus on agreements with narrower aims, less participation, or both.

Any U.S. rebalancing among regions and issues, finally, needs to be complemented by another sort of rebalancing, between the internal and the external, the domestic and the foreign. The United States needs to restore the foundations of American economic power so that it will once again have the resources to act freely and lead in the world, so that it can compete, so that it can discourage threats from emerging and contend with them if need be, so that it is less vulnerable to international developments it cannot control, and so that it can set an example others will want to emulate. The vast sums spent on the wars in Afghanistan and Iraq did not cause the nation's current budgetary or economic predicament, but they did contribute to it. Spending more on national security now would only make it more difficult to set things right. The goal at home must be to restore historical levels of domestic economic growth, reduce the ratio of debt to GDP, and improve the quality of the nation's infrastructure and human capital. During the next several years, facing no rival great power or existential threat, the United States is likely to enjoy something of a strategic respite. The question is whether the United States will take advantage of that respite to renew the sources of its strength or squander it through continued overreaching in the Middle East, not attending to Asia, and underinvesting at home.

Now that most Americans want little to do with the greater Middle East, U.S. officials are finding it difficult to turn away.

Today, the Middle East is not an arena of decisive great-power competition, nor is it home to any major power.

The wisest policy for the United States in Asia is to hedge, cooperating where possible with China while maintaining a strong presence in the region.

Closing time: a U.S. soldier leaving Baghdad, December 2011

Critical Thinking

1. What does Haass mean by the "new new" Middle East?

2. What policy should the U.S. follow toward the new new Middle East?

3. Why should the U.S. avoid a war of choice with Iran?

Create Central

www.mhhe.com/createcentral

Internet References

Middle East Online
   www.middle-east-online.com/english.

Arab net
   www.arab.net

Al Jazeera
   english.alijazeera.net

The National Security Archive
   www.gwu.edu/nsarchiv

RICHARD N. HAASS is president of the Council on Foreign Relations and the author, most recently, of *Foreign Policy Begins at Home.*

The Currency of Power: Want to Understand America's Place in the World? Write Economics Back into the Plan by Robert Zoellick

71

Article

Prepared by: Robert Weiner, *University of Massachusetts/Boston*

The Currency of Power

Want to Understand America's Place in the World? Write Economics Back into the Plan

ROBERT ZOELLICK

Learning Outcomes

After reading this article, you will be able to:

- Explain the relationship between the U.S. economic problems and the global economy.

- Explain the relationship between a U.S. default on its national debt and the U.S. position of leadership in the world.

Earlier this year, Bob Carr, Australia's foreign minister and a longtime friend of the United States, observed with Aussie clarity: "The United States is one budget deal away from restoring its global preeminence." He added a caution: "There are powers in the Asia-Pacific that are whispering that this time the United States will not get its act together, so others had best attend to them."

Carr's insight—that the connection between economics and security will determine America's future—is sound and persuasive. Yet ever since the rise of "national security" as a concept at the start of the Cold War, economics has become the unappreciated subordinate of U.S. foreign policy. Today, the power of deficits, debt, and economic trend lines to shape security is staring the United States in the face. Others see it, even if America does not.

Carr, a student of U.S. history, would probably not be surprised to learn that his warning echoes words drafted by Alexander Hamilton, America's first Treasury secretary, for President George Washington's farewell address: The new nation, Hamilton urged, must "cherish credit as a means of strength and security." Ironically, it took an admiral—Mike Mullen, then chairman of the U.S. Joint Chiefs of Staff—to recall Hamilton's warning about the link between credit and security. Mullen seized attention not by pointing out a danger to the fleet, but by telling CNN, "The most significant threat to our national security is our debt."

Mullen's observation should not come as a surprise, because strategists in uniform often look to history as their laboratory.

They also have to match means and capabilities to achieve ends. Officers at staff colleges may be inspired by the exciting chapters on Napoleon Bonaparte's bold campaigns, but the astute also discover that the key to Britain's victory in the Napoleonic Wars is found in the dry accounts of the budgets of William Pitt the Younger, the chancellor of the Exchequer and prime minister. By restoring Britain's credit after its costly imbroglio with the American colonies, Pitt enabled his country to fight a long war—and even repeatedly finance coalition partners—without choking Britain's economy.

In contrast, consider the foreign-policy debates of this U.S. election year. Journalists and commentators expound about wars and rumors of wars, political leaders and upheavals, human rights and duties to intervene, missiles and their defense. All serious and important topics. But how about a question on the eurozone crisis that threatens the integration of Europe, one of the 20th century's greatest security-policy achievements and America's closest ally and partner? What about America's connections to growth in East Asia, where economics is the coin of the realm? The reply is that these topics concern economics, not foreign policy!

America's security strategists seem to have lost the ability to integrate the two. Their perspectives on economics do not extend much beyond sanctions policies and paying for defense budgets. At best, the role of economics is assumed, not analyzed. We scarcely understand its effects on power, influence, diplomacy, ideas, and human rights. At worst, economic problems have become a justification for a "come home, America" isolationism. And economists—absorbed with mathematical models and debates about quantitative easing and stimulus policies—are content to operate in their separate universe.

Some, on the left and the right, disparage the role of economics in foreign policy as crass commercialism, narrow business interests, or, worse, affording undue influence to bankers. Others view international economics and trade policy as narrow specialties involving technical negotiations that just aggravate domestic constituencies.

Yet this separation of economics from U.S. foreign policy and security policy reflects a shift from earlier American experience. For its first 150 years, the American foreign-policy tradition was deeply infused with economic logic. Unfortunately, thinking about international political economy has become a lost art in the United States. How did this happen?

In 1773, a tribe of Bostonians threw 342 chests of tea into the harbor—without damaging other property, I should add—to protest taxes imposed to bail out the nearly bankrupt British East India Company. Their protest still inspires a political movement in our time. The incident was the most dramatic of waves of colonial "nonimportation" policies dating to the 1760s, early American efforts to employ trade as a tool of policy.

The new American republic was born amid a world of mercantilist empires. Navigating around the trading monopolies of the seasoned, established powers—and later blockades and bullying—the former colonies fought continually for what historian Philip Zelikow has called "freedom to trade." This principle was not "free trade" as we understand it today, but a challenge to the old order nevertheless.

The young United States, under President Thomas Jefferson, tried to exert its own leverage with nonimportation acts and even a disastrous embargo on foreign commerce in 1807. Ironically, it took the failure of Jefferson's trade sanctions, as well as the War of 1812, for the United States to start developing the manufacturing base that Hamilton sought and Jefferson opposed.

Britain was not the only object of U.S. economic security policy. From 1801 to 1805, in the face of the Barbary pirates' attacks on U.S. ships, Jefferson rejected demands for tribute and instead sent the U.S. Navy to the shores of Tripoli. As the U.S. Marine Corps' hymn has memorialized, this Libya expedition was not "led from behind."

In an age when power arose from the expansion of territory, resources, people, and commerce, America's implicit strategy understandably concentrated on the North American continent and open immigration. Land and settlement provided security, especially when buffered by two vast oceans.

Wielding a tool of diplomacy lost today, the United States resolved disputes by buying lands: Louisiana; Florida; old New Mexico, California, and the Gadsden Purchase; Alaska; and even the Virgin Islands at the start of the 20th century. (Admittedly, in some cases use of force led to price discounts.) In another touch of irony, Jefferson needed Hamilton's Bank of the United States and credit system, which Jefferson had opposed, for his greatest achievement, the Louisiana Purchase.

The theme of Western Hemispheric integration—a partnership of young democracies, not an empire—was advanced by Secretary of State Henry Clay in the 1820s, revived in the 1880s and 1890s, and found first fruits a century later in the North American Free Trade Agreement (NAFTA) and then five more U.S. free trade agreements with Latin America. Today, the partners in those free trade agreements account for more than half of the hemisphere's non-U.S. GDP. In the 21st century, comprehensive free trade agreements could turn out to be the ties that bind, like the alliances of old.

The *Federalist Papers,* the touchstone of American constitutionalism, are replete with references to the need for a strong federal government to secure the United States' place among foreign countries, including through healthy commerce and credit. The founders understood the link between economics and security. In a prescient example, John Jay, in Federalist No. 4, cautioned in 1787 that trade with China and India could one day draw the United States into conflict with competitors. It was no coincidence that the first U.S. forays into international relations were called treaties of "Amity and Commerce."

The oceans that were barriers to armies became highways for the U.S. Navy and American mariners seeking markets. In 1854, Commodore Matthew Perry "opened" Japan to trade. By 1899, Secretary of State John Hay was resisting carving up China, as Africa had been, in favor of an "Open Door" policy to secure equal commercial opportunity.

This race through U.S. foreign economic policy is not intended to suggest that the American system was all about peaceful commerce. To the contrary, even if the connection was driven by interests and not explicit planning, the economic and security policies worked hand in hand. These interests were infused with a healthy dose of what those generations called spreading "civilization," and what we call "values." With trade and the flag came missionaries and their schools. After the defeat of the Boxer Rebellion in 1900, the United States pragmatically used its share of the indemnity imposed on China—which the United States had opposed—to found Tsinghua University in Beijing and fund scholarships for Chinese students to attend universities in the United States.

As the United States settled its home continent around the opening of the 20th century, a debate arose about expansion to territories beyond U.S. shores. Some wanted markets or coaling stations, and others sought to carry "civilization" to foreign peoples. Some simply wanted to keep strategic places out of the hands of others. But "imperialism" did not sit well with many Americans, who proudly recalled that their new nation had freed itself from old empires. The U.S. war with Spain in 1898, precipitated by conflicts over Cuba, led the United States to acquire the Philippines (for $20 million) to keep the islands from being grabbed by others whose fleets were hovering—but the United States did not take Cuba. President Theodore Roosevelt stirred up a revolt in Panama so he could build a canal that linked the two great oceans, commerce, and fleets of the U.S. Navy.

America's foreign economic policy also helped spur early interest in international law—what we now call "rules-based systems"—to resolve disputes. The United States was an active participant in the 1899 Hague Conference and lent its support to a convention to resolve disputes peacefully through third-party mediation, international commissions, and a Permanent Court of Arbitration. Secretary of State Elihu Root negotiated arbitration treaties with 25 countries early in the 20th century.

The decades that followed continued the pattern of melding U.S. economic interests with foreign policy and security policy. "Dollar diplomacy," as historians have dubbed the strategy, sought to support U.S. enterprises in Latin America and East Asia through what we now call transnational actors—but in those days were railroad and mining engineers, bankers, and merchants. In World War I, Britain shrewdly played on the U.S. commitment to neutral rights on the seas to draw

President Woodrow Wilson to its side against Germany and its U-boats.

After the war, reacting against what the United States viewed as the old European politics of perpetuated hostilities, America withdrew from European military security. Yet even during the 1920s and 1930s, the United States relied on banker-statesmen to negotiate debt and reparations to revive broken economies. To secure peaceful seas, the United States even launched the idea of global naval arms control in the 1920s.

Reeling from the Great Depression, however, America withdrew from the world economy, enacting the Smoot-Hawley tariff wall to block imports and subverting a last-gasp effort for international economic cooperation at the 1933 World Economic Conference. Political-military isolationism followed.

Then came 1941, and the United States again learned, through harsh experience, that economics and security were linked. The United States had imposed embargoes on the sale of petroleum and scrap iron to Japan in response to Japan's invasion of China and its threats to Southeast Asia. Imperial Japan responded with a surprise attack, in part to secure its sources of oil and raw materials. The United States, caught unprepared, paid a terrible price.

World War II and the opening of the Cold War led to a sharp break in the American foreign-policy tradition. At least that is the impression left by the masters of mid-20th-century security studies.

In their narrative, the dawn of the nuclear age and the face-off between communism and the West required a new approach: a national security strategy. The traditional aims of amity and commerce seemed quaint and outdated in a world of superpower confrontation and containment. For the first time, the United States maintained a large conventional army, a significant part based in Europe, with hundreds of thousands of other troops fighting in Asia over decades.

Instead of Milton Friedman's idea that economic freedom is an end in itself and an indispensable means toward achieving political freedom, economics became a resource factor—and the handmaiden of the strategic policy process. The U.S. National Security Act of 1947 is full of references to new offices to mobilize people and resources for total war. Yet the act did not even make the Treasury secretary a statutory member of the new National Security Council. Ever since, the U.S. government has struggled to integrate economics into its national security strategies.

Because the United States has not faced up to its economic problems at home, its voice on international economic challenges does not carry, its power has waned, and its strategic designs drift.

The transformation of U.S. foreign-policy priorities signaled a change in the training of the stewards of American foreign policy. The new specialties were Soviet studies, political-military affairs, defense policy, and eventually Middle East policy. Short of homegrown talent on the central front, America even outsourced security strategy to immigrants from continental Europe—Henry Kissinger and Zbigniew Brzezinski—who had grown up in a world of threat and the complications of balancing power in Eurasia.

Now, we need to rewrite economics back into the narrative of the Cold War and all that follows. We need a fuller appreciation of the links between economics and security to match the times. The world continues to struggle through a global economic crisis that began in the United States. Fears, fragilities, and failures fuel tensions within and among countries. Leaders are under protectionist and nationalist pressures—in trade, but also regarding currencies, investments, resources, and the oceans. These frictions risk a downward economic spiral and even conflict. Because the United States has not faced up to its economic problems at home, its voice on international economics does not carry, its power has waned, and its strategic designs drift with the currents of the day's news. Without healthy economic growth, the United States will be unable to lead. Just as dangerously, it will lose its identity on the global stage if it loses its economic dynamism. America's unique strength is the ability to reinvent itself.

To better appreciate the political economy story—and its significance for security—it is helpful to consider three phases since the end of World War II: from the creation of the Bretton Woods system to its breakdown in the 1970s; then, a capitalist revival from the late 1970s through the end of the Cold War; and on to the rise of globalization in the 1990s, extending to the crash of 2008. We are now stumbling into a fourth phase that is vital for the United States to shape. To do so, America has to look "back to the future" to recognize the critical connections between economics and security.

PHASE 1: From Bretton Woods to the 1970s

Even as World War II raged on, the United States and Britain began creating new international economic institutions to address currency exchange rates, trade, reconstruction, and development. The United States and Europe then launched the Marshall Plan—and Europe created an economic community—to shore up the free world's economic foundations. The United States exported capital and imported goods to boost recoveries in Europe, Japan, and then South Korea and other developing countries.

The economic internationalists of the Bretton Woods system and the European Economic Community were not driven primarily by a plan for "containment" or to counter the Soviet Union. That came later. Indeed, a primary architect of Bretton Woods, Harry Dexter White, was later doomed by his sympathies for the Soviet Union. These strategists were trying to avoid a repeat of the economic causes of the political and security breakdown in the 1920s and 1930s. Only over time did the imperatives of the Cold War lead to a pragmatic convergence of the national security planners and the economic designers.

Still, the national security model treated the economy as a source of benefits to be exchanged to support security aims. Trade concessions. Foreign assistance. Military aid. The ends were not necessarily inclusive growth, good governance, and open, competitive markets. The national security logic assumed economics was about static sources of resources for the accounting and balancing of power.

The national security perspective of state power overlooked a vital reality: that sound economic policies are the underpinning of both individual freedom and national power—not only military power, but also the dynamism, innovation, and influence of the economy and society. The 20th-century concept of national security also overlooked how economic change can be a powerful force of its own in international relations. Economics is the study of a continual dynamic, so its concepts of "stability," "balance," and even "regime change" reflect very different perspectives from those of most security strategists.

President Dwight Eisenhower understood this distinction. He invested political capital in balanced budgets, low taxes, and sound monetary policies. He recognized the underlying strength generated by investments in national highways, education, and science. On the international stage, as British Prime Minister Anthony Eden learned to his sorrow in the Suez crisis, Eisenhower would even use the power of the U.S. dollar over the British pound to stop the use of force in Egypt.

In the 1970s, a new generation of international relations thinkers, led by Robert Keohane and Joseph Nye, questioned the realist power model of Hans Morgenthau that accompanied the Cold War national security concept. They did not dismiss power; indeed, they recognized the vital role of military force. But Keohane and Nye supplemented realism with a description of complex interdependence. Their framework reinserted economic and other considerations not involving force into foreign policy. Their attention to diverse regimes that govern relationships of interdependence and international organizations was timely, given the breakdown of the Bretton Woods system of fixed exchange rates and the rise of oil power through OPEC in the 1970s.

Keohane and Nye were also describing how the first phase of the post-World War II international economic system came to a messy close. As the 1970s limped to an end, the world economy stumbled toward a new reality of floating exchange rates, oil shocks, big bank loans of petrodollars to developing-world sovereigns, and stagflation.

It is intriguing to observe that Kissinger—the master strategist of classic realism—struggled to integrate these seminal economic events into his *Weltanschauung* during his time as U.S. national security advisor and secretary of state. Some critics even argued that Kissinger's lack of understanding of economics led to a balancing strategy based on the flawed assumption of a Spenglerian decline of the West. From the dominant national security perspective, however, Kissinger had shrewdly extricated the United States from military defeat in Vietnam while opening relations with China as a counterweight to his Soviet foe.

Keohane and Nye recognized that they had a blind spot too. As presenters of a "framework for using *international factors* to explain change and stability in world politics," they acknowledged that they did not examine the connection between domestic and foreign policies. They focused on international systems—rules, norms, regimes. Yet critical domestic choices, particularly economic ones, will infuse foreign policy and shape the principles, practices, and diplomacy of those international systems, for good or ill.

It took the rise of a singular statesman—and one tough stateswoman—to turn the intellectual connection between domestic strength and international influence into action.

PHASE 2: The Revival and Success of Capitalism

British Prime Minister Margaret Thatcher and U.S. President Ronald Reagan intuitively understood the connection between national economic revival and foreign policy. Their priority was to revive capitalism at home—and then extend it to the world. In doing so, they defined a second phase of the postwar international economy.

The promotion of global capitalism seemed to be disruptive to many, the antithesis of rebuilding an international economic system still reeling from the shocks of the 1970s. After all, Joseph Schumpeter had explained that capitalism is "creative destruction." Yet this very disruptive quality enables capitalism to respond flexibly and continually to technological and other changes. There are risks to overregulation—just as there are to weak supervision. Zealous social protections, though perhaps well intended, can add costly rigidities.

The reform of capitalism was not just an Anglo-American venture. West Germany's commitment to sound economic policies and export competitiveness demonstrated that social market economies can work. East Germans watching West German TV saw the stark contrast between their grim existence in a "workers' paradise" and the lifestyles of their wealthier cousins. Japanese manufacturers responded to the oil shocks with a huge increase in energy efficiency. Together, those domestic economic policies enabled the West to adjust to the 1970s breakdown of the post-World War II economic security system.

The Soviet Union could not adapt to its economic challenges. It could not cope with changing information technology, new drivers of productivity and competition, and eventually $15-a-barrel oil. The U.S. intelligence community, geared toward the Cold War calculations of national security, largely missed the story. Soviet leader Mikhail Gorbachev, facing the combination of the democracies' economic regeneration, the U.S. military buildup with advanced technologies, and transatlantic solidarity over euromissiles, concluded that he had to reform communism. But his perestroika didn't work.

The adaptation to markets on a truly global scale, integrating developed and developing countries alike, was bound to be complicated and disjointed. It was.

Reagan believed that the regimes and institutions of interdependence should be tested for effectiveness in boosting

The Currency of Power: Want to Understand America's Place in the World? Write Economics Back into the Plan by Robert Zoellick

75

international growth, opportunity, and human rights. After all, those were the standards of economic freedom he was applying at home. Moreover, at a time of economic flux, the international economic system needed to foster adaptation. Reagan did not want international rules to constrict domestic economic revival, and he stirred controversy by rejecting counterproductive international schemes such as the United Nations' New World Information Order and New International Economic Order, and the deep seabed mining regime in the Law of the Sea treaty.

Still, after a pause, the United States would lead, with pragmatism and compromise, in reshaping international economic relations. In Reagan's second term, the United States steered the International Monetary Fund to a new role in the Latin American debt crisis. It led a major recapitalization of the World Bank to support developing countries' economic reforms and debt reschedulings—until banks could write down losses. In 1985, Treasury Secretary James Baker launched a process of international economic coordination in the G-7. The United States pushed to expand global trade through the launch of the Uruguay Round of trade talks, completed much of that negotiation under President George H.W Bush, and closed the deal under President Bill Clinton to create the more influential and liberalized World Trade Organization (WTO) out of the old GAIT. Bush also initiated the Asia-Pacific Economic Cooperation (APEC) forum and negotiated NAFTA, which Clinton enacted.

The end of the Cold War and its immediate aftermath brought the second phase of the postwar international economy to a close with astounding success as renewed economic vitality led to a vast expansion of private markets and advances in international economic regimes based on market principles and openness. The national security aims of the Cold War in Europe were achieved with hardly a shot fired. It may not be coincidental that the principal U.S. secretaries of state in this period—George Shultz and Baker—had served as Treasury secretaries too.

PHASE 3: Globalization's Promise and Perils

The end of the Cold War reunited Europe. The European Community became a deeper, wider union and launched its own currency. Just as importantly, China, India, and other developing countries moved from planned socialism and import-substitution schemes to market competition. Over a decade, the number of people engaged in or actively affected by the world market economy surged from about 1 billion to four or five times that. Information technology swept ahead. Capital raced around the globe. Rather than the world economy of Bretton Woods, the earlier era of globalization before World War I—with its large movements of capital, trade, and people, spurred by new technologies in transport and communication—seemed to offer a closer parallel.

Yet the adaptation to markets on a truly global scale, integrating developed and developing countries alike, was bound to be complicated and disjointed. It was. In the late 1990s, countries in East Asia and Latin America faced harsh financial blows and painful restructurings. Almost all are now stronger for the experience.

But the recovery strategies of some developing countries planted the seeds of a new problem: "imbalances"—whether of savings, reserves, trade accounts, or other dimensions. Developing economies in East Asia saved and exported more, and the United States and some European countries increased borrowing, consumption, and imports. Interventions to lessen the value of Asian currencies constrained their imports and expanded exports. Some economists maintain that the low prices of goods available from new suppliers led central bankers to persist in easy monetary policies for too long, risking widespread asset-price inflation, especially in real estate markets. Then the bubbles burst.

The institutions of the international economic system adapted incrementally, often with difficulty. The economic firefighting of the IMF and World Bank made them principal targets of an anti-globalization movement in the 1990s. The continuing boom almost put the IMF out of business. Unfortunately, neither international nor domestic supervisors of financial markets kept up with the innovations—or the frauds and foolishness that inevitably come with long boom periods.

The WTO added many new members. The trading system even withstood terrorist attacks—and fears of more. But the travails of the WTO's Doha Round of trade negotiations, launched in 2001, signaled a new challenge. The traditional developed economies wanted the middle-income countries—China, Brazil, India, and others—to assume more responsibility for lowering barriers to trade, while all would offer special treatment for Africa and the poorest. The major developing economies, in turn, pointed to their large numbers of poor people and wanted to maintain the privileges of what WTO practice refers to as "special and differential treatment." This debate reverberates not only in trade, but in monetary affairs, investment, development, energy, and the environment.

The 9/11 attacks concentrated America's attention on terrorism, homeland security, and the long wars that followed. Yet the connections of economics to the new security threats are also strong. When al Qaeda targeted the United States, it aimed for the World Trade Center—its twin towers the symbols of American capitalism—as well as Washington. In addition to shock and destruction, the terrorists wanted to strangle economic and political freedom. As Osama bin Laden boasted in 2004, his aim was "bleeding America to the point of bankruptcy."

Even as America fought in Iraq and Afghanistan and against terrorist threats around the globe, other forces of history did not stand still. China, India, and other emerging economies began to change the landscape of power. The failed political and stunted economic systems of North Africa and the Middle East sparked upheavals that will shake the region for a generation.

PHASE 4: After the Crash

The crash of 2008 has ushered in a fourth phase of the post-World War II economic experience. Global financial capitalism now faces a new crisis—of credit, conduct, and even confidence.

After harsh blows, the advanced economies are struggling to reduce debt and revive jobs and productivity through structural reforms. Unemployment is up. Confidence is down. Protectionism is rising. Publics are anxious. Politicians are struggling. Conflicts and tensions within and between countries are building. Developing economies have been hit too, though many have fared relatively better. In a profound shift, the 60-year leadership of developed economies is in question.

Will the eurozone and the historic success of Europe's peaceful integration survive—and with it, Europe's influence in the world?

Will the high-growth developing countries overcome the so-called "middle-income trap" to become high-income countries and "responsible stakeholders" in an international system that has benefited them—but that they did not design?

Will the poorest—the "bottom billion"—have an opportunity to prosper too, or will they be breeding grounds for transnational insecurities?

Will the new political systems of the Middle East and North Africa lead to new economic policies for inclusive growth and peaceful integration into the world economy?

Will the United States show leadership—at home and internationally—in reviving its core economic strength while simultaneously leveraging those capabilities through an activist economic diplomacy? Will the United States connect its foreign economic policy with security interests in freedom of the seas, open skies, and protection of cyberspace?

Foreign Minister Carr's warning about America's need to resolve its budget mess is correct: The United States must restore its credit, both for its own health and to enable it to lead. But the United States does not need just any budget deal. It needs one that rebuilds the fundamentals of long-term growth. It needs to limit government spending. It needs to encourage private-sector innovation and productivity. It needs inclusive growth that empowers all its citizens to fulfill their potential. It needs to revive a free trade agenda that has stalled in recent years. It needs to favor makers over takers.

By restoring America's credit and reviving growth, the next president and Congress would add to the country's power and influence in reinforcing ways. Strong, sustainable growth would boost public and private resources, while disciplining the debt would halt the burdening of future generations to pay for current excesses. Greater public resources would pay—not borrow—for vital purposes, starting with national defense, but also including the public goods of education, research, infrastructure, and the environment. A comprehensive budget and growth deal would also remove the weight of costly uncertainty from the private sector. Success at home would strengthen America's standing around the world as a can-do country with the means, ideas, and willpower to reinvent capitalism yet again.

In his classic study, *The World in Depression, 1929–1939,* the economist Charles Kindleberger argued that it was critical for one major power to take the lead in shaping an international economic system. This power could not dictate, but instead needed to invest in encouraging a shared approach to trade, capital flows, currencies, and reliance on markets. Kindleberger described how during the Great Depression, the United States had the means but not the will to lead, while Britain had the will but no longer the means. If the United States does not lead now, who will?

Critical Thinking

1. Why does global financial capitalism face a new crisis?
2. Why is the link between economics and security important for the United States?
3. What is meant by complex interdependence?

Create Central

www.mhhe.com/createcentral

Internet References

International Monetary Fund
www.imf.org

World Bank
www.worldbank.org

ROBERT ZOELLICK, former World Bank president, is senior fellow at Harvard University's Belfer Center for Science and International Affairs and distinguished visiting fellow at the Peterson Institute for International Economics. This article is adapted from his Alastair Buchan Memorial Lecture at the International Institute for Strategic Studies.

Zoellick, Robert. Reprinted in entirety by McGraw-Hill Education with permission from *Foreign Policy*, November 2012, pp. 68–73. www.foreignpolicy.com. © 2012 Washingtonpost.Newsweek Interactive, LLC.

Article

Prepared by: Robert Weiner, *University of Massachusetts/Boston*

Beyond the Pivot: A New Road Map for U.S.–Chinese Relations

KEVIN RUDD

Learning Outcomes

After reading this article, you will be able to:

- Explain what is meant by rebalancing to Asia.
- Discuss why the goal is to avoid a major confrontation with China.

Debate about the future of U.S.-Chinese relations is currently being driven by a more assertive Chinese foreign and security policy over the last decade, the region's reaction to this, and Washington's response—the "pivot," or "rebalance," to Asia. The Obama administration's renewed focus on the strategic significance of Asia has been entirely appropriate. Without such a move, there was a danger that China, with its hard-line, realist view of international relations, would conclude that an economically exhausted United States was losing its staying power in the Pacific. But now that it is clear that the United States will remain in Asia for the long haul, the time has come for both Washington and Beijing to take stock, look ahead, and reach some long-term conclusions as to what sort of world they want to see beyond the barricades.

Asia's central tasks in the decades ahead are avoiding a major confrontation between the United States and China and preserving the strategic stability that has underpinned regional prosperity. These tasks are difficult but doable. They will require both parties to understand each other thoroughly, to act calmly despite multiple provocations, and to manage the domestic and regional forces that threaten to pull them apart. This, in turn, will require a deeper and more institutionalized relationship— one anchored in a strategic framework that accepts the reality of competition, the importance of cooperation, and the fact that these are not mutually exclusive propositions. Such a new approach, furthermore, should be given practical effect through a structured agenda driven by regular direct meetings between the two countries' leaders.

Hidden Dragon no Longer

The speed, scale, and reach of China's rise are without precedent in modern history. Within just 30 years, China's economy has grown from smaller than the Netherlands' to larger than those of all other countries except the United States. If China soon becomes the largest economy, as some predict, it will be the first time since George III that a non-English-speaking, non-Western, nondemocratic country has led the global economy. History teaches that where economic power goes, political and strategic power usually follow. China's rise will inevitably generate intersecting and sometimes conflicting interests, values, and worldviews. Preserving the peace will be critical not only for the three billion people who call Asia home but also for the future of the global order. Much of the history of the twenty-first century, for good or for ill, will be written in Asia, and this in turn will be shaped by whether China's rise can be managed peacefully and without any fundamental disruption to the order.

The postwar order in Asia has rested on the presence and predictability of U.S. power, anchored in a network of military alliances and partnerships. This was welcomed in most regional capitals, first to prevent the reemergence of Japanese militarism, then as a strategic counterweight to the Soviet Union, and then as a security guarantee to Tokyo and Seoul (to remove the need for local nuclear weapons programs) and as a damper on a number of other lesser regional tensions. In recent years, China's rise and the United States' fiscal and economic difficulties had begun to call the durability of this framework into question. A sense of strategic uncertainty and some degree of strategic hedging had begun to emerge in various capitals. The Obama administration's "rebalance" has served as a necessary corrective, reestablishing strategic fundamentals. But by itself, it will not be enough to preserve the peace—a challenge that will be increasingly complex and urgent as great-power politics interact with a growing array of subregional conflicts and intersecting territorial claims in the East China and South China seas.

China views these developments through the prism of its own domestic and international priorities. The Standing Committee of the Politburo, which comprises the Communist Party's top leaders, sees its core responsibilities as keeping the Communist Party in power, maintaining the territorial integrity of the country (including countering separatist movements and defending offshore maritime claims), sustaining robust economic growth by transforming the country's growth model, ensuring China's energy security, preserving global and regional stability so as not to derail the economic growth agenda, modernizing China's military and more robustly asserting China's foreign policy interests, and enhancing China's status as a great power.

China's global and regional priorities are shaped primarily by its domestic economic and political imperatives. In an age when Marxism has lost its ideological relevance, the continuing legitimacy of the party depends on a combination of economic performance, political nationalism, and corruption control. China also sees its rise in the context of its national history, as the final repudiation of a century of foreign humiliation (beginning with the Opium Wars and ending with the Japanese occupation) and as the country's return to its proper status as a great civilization with a respected place among the world's leading states. China points out that it has little history of invading other countries and none of maritime colonialism (unlike European countries) and has itself been the target of multiple foreign invasions. In China's view, therefore, the West and others have no reason to fear China's rise. In fact, they benefit from it because of the growth of the Chinese economy. Any alternative view is castigated as part of the "China threat" thesis, which in turn is seen as a stalking-horse for a de facto U.S. policy of containment.

What China overlooks, however, is the difference between "threat" and "uncertainty"—the reality of what international relations theorists call "the security dilemma"—that is, the way that Beijing's pursuit of legitimate interests can raise concerns for other parties. This raises the broader question of whether China has developed a grand strategy for the longer term. Beijing's public statements—insisting that China wants a "peaceful rise" or "peaceful development" and believes in "win-win" or a "harmonious world"—have done little to clarify matters, nor has the invocation of Deng Xiaoping's axiom "Hide your strength, bide your time." For foreigners, the core question is whether China will continue to work cooperatively within the current rules-based global order once it has acquired great-power status or instead seek to reshape that order more in its own image. This remains an open question.

XI Who must be Obeyed

Within the parameters of China's overall priorities, Xi Jinping, the newly appointed general secretary of the Communist Party and incoming president, will have a significant, and perhaps decisive, impact on national policy. Xi is comfortable with the mantle of leadership. He is confident of both his military and his reformist backgrounds, and having nothing to prove on these fronts gives him some freedom to maneuver. He is well read and has a historian's understanding of his responsibilities to his country. He is by instinct a leader and is unlikely to be satisfied with simply maintaining the policy status quo. Of all

his predecessors, he is the most likely Chinese official since Deng to become more than primus inter pares, albeit still within the confines of collective leadership.

Xi has already set an unprecedented pace. He has bluntly stated that unless corruption is dealt with, China will suffer chaos reminiscent of the Arab Spring, and he has issued new, transparent conflict-of-interest rules for the leadership. He has set out Politburo guidelines designed to cut down on pointless meetings and political speechifying, supported taking action against some of the country's more politically outspoken publications and websites, and praised China's military modernizers. Most particularly, Xi has explicitly borrowed from Deng's political handbook, stating that China now needs more economic reform. On foreign and security policy, however, Xi has been relatively quiet. But as a high-ranking member of the Central Military Commission, which controls the country's armed forces (Xi served as vice chair from 2010 to 2012 and was recently named chair), Xi has played an important role in the commission's "leading groups" on policy for the East China and South China seas, and Beijing's recent actions in those waterways have caused some analysts to conclude that he is an unapologetic hard-liner on national security policy. Others point to the foreign policy formulations he used during his visit to the United States in February 2012, when he referred to the need for "a new type of great-power relationship" with Washington and was apparently puzzled when there was little substantive response from the American side.

It is incorrect at present to see Xi as a potential Gorbachev and his reforms as the beginning of a Chinese glasnost. China is not the Soviet Union, nor is it about to become the Russian Federation. However, over the next decade, Xi is likely to take China in a new direction. The country's new leaders are economic reformers by instinct or intellectual training. Executing the massive transformation they envisage will take most of their political capital and will require continued firm political control, even as the reforms generate strong forces for social and political change. There is as yet no agreed-on script for longer-term political reform; there is only the immediate task of widening the franchise within the 82-million-member party. When it comes to foreign policy, the centrality of the domestic economic task means that the leadership has an even stronger interest in maintaining strategic stability for at least the next decade. This may conflict occasionally with Chinese offshore territorial claims, but when it does, China will prefer to resolve the conflicts rather than have them derail that stability. On balance, Xi is a leader the United States should seek to do business with, not just on the management of the tactical issues of the day but also on broader, longer-term strategic questions.

Obama's Turn to Take the Initiative

More than just a military statement, the Obama administration's rebalancing is part of a broader regional diplomatic and economic strategy that also includes the decision to become a member of the East Asia Summit and plans to develop the

Trans-Pacific Partnership, deepen the United States' strategic partnership with India, and open the door to Myanmar (also called Burma). Some have criticized Washington's renewed vigor as the cause of recent increased tensions across East Asia. But this does not stand up to scrutiny, given that the proliferation of significant regional security incidents began more than half a decade ago.

China, a nation of foreign and security policy realists where Clausewitz, Carr, and Morgenthau are mandatory reading in military academies, respects strategic strength and is contemptuous of vacillation and weakness. Beijing could not have been expected to welcome the pivot. But its opposition does not mean that the new U.S. policy is misguided. The rebalancing has been welcomed across the other capitals of Asia—not because China is perceived as a threat but because governments in Asia are uncertain what a China-dominated region would mean. So now that the rebalance is being implemented, the question for U.S. policymakers is where to take the China relationship next.

One possibility would be for the United States to accelerate the level of strategic competition with China, demonstrating that Beijing has no chance of outspending or outmaneuvering Washington and its allies. But this would be financially unsustainable and thus not credible. A second possibility would be to maintain the status quo as the rebalancing takes effect, accepting that no fundamental improvement in bilateral relations is possible and perpetually concentrating on issue and crisis management. But this would be too passive and would run the risk of being overwhelmed by the number and complexity of the regional crises to be managed; strategic drift could result, settling on an increasingly negative trajectory.

A third possibility would be to change gears in the relationship altogether by introducing a new framework for cooperation with China that recognizes the reality of the two countries' strategic competition, defines key areas of shared interests to work and act on, and thereby begins to narrow the yawning trust gap between the two countries. Executed properly, such a strategy would do no harm, run few risks, and deliver real results. It could reduce the regional temperature by several degrees, focus both countries' national security establishments on common agendas sanctioned at the highest levels, and help reduce the risk of negative strategic drift.

A crucial element of such a policy would have to be the commitment to regular summitry. There are currently more informal initiatives under way between the United States and China than there are ships on the South China Sea. But none of these can have a major impact on the relationship, since in dealing with China, there is no substitute for direct leader-to-leader engagement. In Beijing, as in Washington, the president is the critical decision-maker. Absent Xi's personal engagement, the natural dynamic in the Chinese system is toward gradualism at best and stasis at worst. The United States therefore has a profound interest in engaging Xi personally, with a summit in each capital each year, together with other working meetings of reasonable duration, held in conjunction with meetings of the G-20, the Asia-Pacific Economic Cooperation, and the East Asia Summit.

Both governments also need authoritative point people working on behalf of the national leaders, managing the agenda between summits and handling issues as the need arises. In other words, the United States needs someone to play the role that Henry Kissinger did in the early 1970s, and so does China.

Globally, the two governments need to identify one or more issues currently bogged down in the international system and work together to bring them to successful conclusions. This could include the Doha Round of international trade talks (which remains stalled despite approaching a final settlement in 2008), climate-change negotiations (on which China has come a considerable way since the 2009 UN Conference on Climate Change in Copenhagen), nuclear non-proliferation (the next review conference for the Nuclear Nonproliferation Treaty is coming up), or specific outstanding items on the G-20 agenda. Progress on any of these fronts would demonstrate that with sufficient political will all around, the existing global order can be made to work to everyone's advantage, including China's. Ensuring that China becomes an active stakeholder in the future of that order is crucial, and even modest successes would help. Regionally, the two countries need to use the East Asia Summit and the Association of Southeast Asian Nations' Defense Ministers' Meeting-Plus forum to develop a series of confidence- and security-building measures among the region's 18 militaries. At present, these venues run the risk of becoming permanently polarized over territorial disputes in the East China and South China seas, so the first item to be negotiated should be a protocol for handling incidents at sea, with other agreements following rapidly to reduce the risk of conflict through miscalculation.

At the bilateral level, Washington and Beijing should upgrade their regular military-to-military dialogues to the level of principals such as, on the U.S. side, the secretary of defense and the chairman of the Joint Chiefs of Staff. This should be insulated from the ebbs and flows of the relationship, with meetings focusing on regional security challenges, such as Afghanistan, Pakistan, and North Korea, or major new challenges, such as cybersecurity. And on the economic front, finally, Washington should consider extending the Trans-Pacific Partnership to include both China and Japan, and eventually India as well.

Toward a New Shanghai Communique

Should such efforts begin to yield fruit and reduce some of the mistrust currently separating the parties, U.S. and Chinese officials should think hard about grounding their less conflictual, more cooperative relationship in a new Shanghai Communique. Such a suggestion usually generates a toxic response in Washington, because communiques are seen as diplomatic dinosaurs and because such a process might threaten to reopen the contentious issue of Taiwan. The latter concern is legitimate, since Taiwan would have to be kept strictly off the table for such an exercise to succeed. But this should not be an insurmountable problem, because cross-strait relations are better now than at any time since 1949.

As for the charge that communiques are of little current value, this may be less true for China than it is for the United States. In China, symbols carry important messages, including for the military, so there could be significant utility within the Chinese system in using a new communique to reflect and lock in a fresh, forward-looking, cooperative strategic mindset—if one could be worked out. Such a move should follow the success of strategic cooperation, however, rather than be used to start a process that might promise much but deliver little.

Skeptics might argue that the United States and China must restore their trust in each other before any significant strategic cooperation can occur. In fact, the reverse logic applies: trust can be built only on the basis of real success in cooperative projects. Improving relations, moreover, is increasingly urgent, since the profound strategic changes unfolding across the region will only make life more complicated and throw up more potential flash points. Allowing events to take their own unguided course would mean running major risks, since across Asia, the jury is still out as to whether the positive forces of twenty-first-century globalization or the darker forces of more ancient nationalisms will ultimately prevail.

The start of Obama's second term and Xi's first presents a unique window of opportunity to put the U.S.—Chinese relationship on a better course. Doing that, however, will require sustained leadership from the highest levels of both governments and a common conceptual framework and institutional structure to guide the work of their respective bureaucracies, both civilian and military. History teaches that the rise of new great powers often triggers major global conflict. It lies within the power of Obama and Xi to prove that twenty-first-century Asia can be an exception to what has otherwise been a deeply depressing historical norm.

Critical Thinking

1. How can the U.S. avoid a confrontation with China?
2. Why is the pivot a necessary corrective to China's rise?
3. What is China's grand strategy?
4. What is meant by U.S.–Chinese strategic competition ?

Create Central

www.mhhe.com/createcentral

Internet References

The Foreign Ministry of China
www.mfa.gov.cn/eng
Asia-Pacific Economic Cooperation Forum
www.apec.org/

KEVIN RUDD is a Member of the Australian Parliament. He served as Prime Minister of Australia from 2007 to 2010 and Foreign Minister from 2010 to 2012.

Article

Prepared by: Robert Weiner, *University of Massachusetts/Boston*

The Cuban Missile Crisis at 50: Lessons for U.S. Foreign Policy Today

GRAHAM ALLISON

Learning Outcomes

After reading this article, you will be able to:

- Explain how the lessons learned from the Cuban missile crisis help in dealing with Iran, North Korea, and China.

- Explain why the Cuban missile crisis can serve as a guide on how to defuse crises.

Fifty years ago, the Cuban missile crisis brought the world to the brink of nuclear disaster. During the standoff, U.S. President John F. Kennedy thought the chance of escalation to war was "between 1 in 3 and even," and what we have learned in later decades has done nothing to lengthen those odds. We now know, for example, that in addition to nucleararmed ballistic missiles, the Soviet Union had deployed 100 tactical nuclear weapons to Cuba, and the local Soviet commander there could have launched these weapons without additional codes or commands from Moscow. The U.S. air strike and invasion that were scheduled for the third week of the confrontation would likely have triggered a nuclear response against American ships and troops, and perhaps even Miami. The resulting war might have led to the deaths of 100 million Americans and over 100 million Russians.

The main story line of the crisis is familiar. In October 1962, a U.S. spy plane caught the Soviet Union attempting to sneak nuclear-tipped missiles into Cuba, 90 miles off the United States' coast. Kennedy determined at the outset that this could not stand. After a week of secret deliberations with his most trusted advisers, he announced the discovery to the world and imposed a naval blockade on further shipments of armaments to Cuba. The blockade prevented additional materiel from coming in but did nothing to stop the Soviets from operationalizing the missiles already there. And a tense second week followed during which Kennedy and Soviet Premier Nikita Khrushchev stood "eyeball to eyeball," neither side backing down.

Saturday, October 27, was the day of decision. Thanks to secret tapes Kennedy made of the deliberations, we can be flies on the wall, listening to the members of the president's ad hoc Executive Committee of the National Security Council, or ExComm, debate choices they knew could lead to nuclear Armageddon. At the last minute, the crisis was resolved without war, as Khrushchev accepted a final U.S. offer pledging not to invade Cuba in exchange for the withdrawal of the Soviet missiles.

Every president since Kennedy has tried to learn from what happened in that confrontation. Ironically, half a century later, with the Soviet Union itself only a distant memory, the lessons of the crisis for current policy have never been greater. Today, it can help U.S. policymakers understand what to do-and what not to do-about Iran, North Korea, China, and presidential decision-making in general.

What Would Kennedy Do?

The current confrontation between the United States and Iran is like a Cuban missile crisis in slow motion. Events are moving, seemingly inexorably, toward a showdown in which the U.S. president will be forced to choose between ordering a military attack and acquiescing to a nuclear-armed Iran.

Those were, in essence, the two options Kennedy's advisers gave him on the final Saturday: attack or accept Soviet nuclear missiles in Cuba. But Kennedy rejected both. Instead of choosing between them, he crafted an imaginative alternative with three components: a public deal in which the United States pledged not to invade Cuba if the Soviet Union withdrew its missiles, a private ultimatum threatening to attack Cuba within 24 hours unless Khrushchev accepted that offer, and a secret sweetener that promised the withdrawal of U.S. missiles from Turkey within six months after the crisis was resolved. The sweetener was kept so secret that even most members of the ExComm deliberating with Kennedy on the final evening were in the dark, unaware that during the dinner break, the president had sent his brother Bobby to deliver this message to the Soviet ambassador.

Looking at the choice between acquiescence and air strikes today, both are unattractive. An Iranian bomb could trigger a cascade of proliferation, making more likely a devastating conflict in one of the world's most economically and strategically critical regions. A preventive air strike could delay Iran's nuclear progress at identified sites but could not erase the knowledge and skills ingrained in many Iranian heads. The truth is that any outcome that stops short of Iran having a nuclear bomb will still leave it with the ability to acquire one down the road, since Iran has already crossed the most significant "redline" of proliferation: mastering the art of enriching uranium and building a bomb covertly. The best hope for a Kennedyesque third option today is a combination of agreed-on constraints on Iran's nuclear activities that would lengthen the fuse on the development of a bomb, transparency measures that would maximize the likelihood of discovering any cheating, unambiguous (perhaps secretly communicated) threats of a regime changing attack should the agreement be violated, and a pledge not to attack otherwise. Such a combination would keep Iran as far away from a bomb as possible for as long as possible.

The Israeli factor makes the Iranian nuclear situation an even more complex challenge for American policymakers than the Cuban missile crisis was. In 1962, only two players were allowed at the main table. Cuban Prime Minister Fidel Castro sought to become the third, and had he succeeded, the crisis would have become significantly more dangerous. (When Khrushchev announced the withdrawal of the missiles, for example, Castro sent him a blistering message urging him to fire those already in Cuba.) But precisely because the White House recognized that the Cubans could become a wild card, it cut them out of the game. Kennedy informed the Kremlin that it would be held accountable for any attack against the United States emanating from Cuba, however it started. His first public announcement said, "It shall be the policy of this Nation to regard my nuclear missile launched from Cuba against any nation in the Western Hemisphere as an attack by the Soviet Union on the United States, requiring a full retaliatory response upon the Soviet Union."

Today, the threat of an Israeli air strike strengthens U.S. President Barack Obama's hand in squeezing Iran to persuade it to make concessions. But the possibility that Israel might actually carry out a unilateral air strike without U.S. approval must make Washington nervous, since it makes the crisis much harder to manage. Should the domestic situation in Israel reduce the likelihood of an independent Israeli attack, U.S. policymakers will not be unhappy.

Carrots Go Better with Sticks

Presented with intelligence showing Soviet missiles in Cuba, Kennedy confronted the Soviet Union publicly and demanded their withdrawal, recognizing that a confrontation risked war. Responding to North Korea's provocations over the years, in contrast, U.S. presidents have spoken loudly but carried a small stick. This is one reason the Cuban crisis was not repeated whereas the North Korean ones have been, repeatedly.

In confronting Khrushchev, Kennedy ordered actions that he knew would increase the risk not only of conventional war but also of nuclear war. He raised the U.S. nuclear alert status to defcon 2, aware that this would loosen control over the country's nuclear weapons and increase the likelihood that actions by other individuals could trigger a cascade beyond his control. For example, nato aircraft with Turkish pilots loaded active nuclear bombs and advanced to an alert status in which individual pilots could have chosen to take off, fly to Moscow, and

drop a bomb. Kennedy thought it necessary to increase the risks of war in the short run in order to decrease them over the longer term. He was thinking not only about Cuba but also about the next confrontation, which would most likely come over West Berlin, a free enclave inside the East German puppet state. Success in Cuba would embolden Khrushchev to resolve the Berlin situation on his own terms, forcing Kennedy to choose between accepting Soviet domination of the city and using nuclear weapons to try to save it.

During almost two dozen face-offs with North Korea over the past three decades, meanwhile, U.S. and South Korean policymakers have shied away from such risks, demonstrating that they are deterred by North Korea's threat to destroy Seoul in a second Korean war. North Korean leaders have taken advantage of this fear to develop an effective strategy for blackmail. It begins with an extreme provocation, blatantly crossing a redline that the United States has set out, along with a threat that any response will lead to a "sea of fire." After tensions have risen, a third party, usually China, steps in to propose that "all sides" step back and cool down. Soon thereafter, side payments to North Korea are made by South Korea or Japan or the United States, leading to a resumption of talks. After months of negotiations, Pyongyang agrees to accept still more payments in return for promises to abandon its nuclear program. Some months after that, North Korea violates the agreement, Washington and Seoul express shock, and they vow never to be duped again. And then, after a decent interval, the cycle starts once more.

If the worst consequence of this charade were simply the frustration of being bested by one of the poorest, most isolated states on earth, then the repeated Korean crises would be a sideshow. But for decades, U.S. presidents have declared a nuclear-armed North Korea to be "intolerable" and "unacceptable." They have repeatedly warned Pyongyang that it cannot export nuclear weapons or technology without facing the "gravest consequences." In 2006, for example, President George W. Bush stated that "the transfer of nuclear weapon or material by North Korea to state or nonstate entities would be considered a grave threat to the United States, and North Korea would be held fully accountable for the consequences." North Korea then proceeded to sell Syria a plutonium producing reactor that, had Israel not destroyed it, would by now have produced enough plutonium for Syria's first nuclear bomb. Washington's response was to ignore the incident and resume talks three weeks later.

One lesson of the Cuban missile crisis is that if you are not prepared to risk war, even nuclear war, an adroit adversary can get you to back down in successive confrontations. If you do have redlines that would lead to war if crossed, then you have to communicate them credibly to your adversary and back them up or risk having your threats dismissed. North Korea's sale of a nuclear bomb to terrorists who then used it against an American target would trigger a devastating American retaliation. But after so many previous redlines have been crossed with impunity, can one be confident that such a message has been received clearly and convincingly? Could North Korea's new leader, Kim Jong Un, and his advisers imagine that they could get away with it?

The Rules

A similar dynamic may have emerged in the U.S. economic relationship with China. The Republican presidential candidate Mitt Romney has announced that "on day one of my presidency I will designate [China] a currency manipulator and take appropriate counteraction." The response from the political and economic establishment has been a nearly unanimous rejection of such statements as reckless rhetoric that risks a catastrophic trade war. But if there are no circumstances in which Washington is willing to risk a trade confrontation with China, why would China's leaders not simply take a page from North Korea's playbook? Why should they not continue, in Romney's words, "playing the United States like a fiddle and smiling all the way to the bank" by undervaluing their currency, subsidizing domestic producers, protecting their own markets, and stealing intellectual property through cybertheft?

Economics and security are separate realms, but lessons learned in one can be carried over into the other. The defining geopolitical challenge of the next half century will be managing the relationship between the United States as a ruling superpower and China as a rising one. Analyzing the causes of the Peloponnesian War more than two millennia ago, the Greek historian Thucydides argued that "the growth of the power of Athens, and the alarm which this inspired in Sparta, made war inevitable." During the Cuban missile crisis, Kennedy judged that Khrushchev's adventurism violated what Kennedy called the "rules of the precarious status quo" in relations between two nuclear superpowers. These rules had evolved during previous crises, and the resolution of the standoff in Cuba helped restore and reinforce them, allowing the Cold War to end with a whimper rather than a bang.

The United States and China will have to develop their own rules of the road in order to escape Thucydides' trap. These will need to accommodate both parties' core interests, threading a path between conflict and appeasement. Overreacting to perceived threats would be a mistake, but so would ignoring or papering over unacceptable misbehavior in the hope that it will not recur. In 1996, after some steps by Taipei that Beijing considered provocative, China launched a series of missiles over Taiwan, prompting the United States to send two aircraft carrier battle groups into harm's way. The eventual result was a clearer understanding of both sides' redlines on the Taiwan issue and a calmer region. The relationship may need additional such clarifying moments in order to manage a precarious transition as China's continued economic rise and new status are reflected in expanded military capabilities and a more robust foreign posture.

Do Process

A final lesson the crisis teaches has to do not with policy but with process. Unless the commander in chief has sufficient time and privacy to understand a situation, examine the evidence, explore various options, and reflect before choosing among them, poor decisions are likely. In 1962, one of the first questions Kennedy asked on being told of the missile discovery was, How long until this leaks? McGeorge Bundy, his national security adviser, thought it would be a week at most. Acting on that advice, the president took six days in secret to deliberate, changing his mind more than once along the way. As he noted afterward, if he had been forced to make a decision in the first 48 hours, he would have chosen the air strike rather than the naval blockade-something that could have led to nuclear war.

In today's Washington, Kennedy's week of secret deliberations would be regarded as a relic of a bygone era. The half-life of a hot secret is measured not even in days but in hours. Obama learned this painfully during his first year in office, when he found the administration's deliberations over its Afghanistan policy playing out in public, removing much of his flexibility to select or even consider unconventional options. This experience led him to demand a new national security decision-making process led by a new national security adviser. One of the fruits of the revised approach was a much more tightly controlled flow of information, made possible by an unprecedented narrowing of the inner decision-making circle. This allowed discussions over how to handle the discovery of Osama bin Laden's whereabouts to play out slowly and sensibly, with the sexiest story in Washington kept entirely secret for five months, until the administration itself revealed it after the raid on bin Laden's Abbottabad compound.

It has been said that history does not repeat itself, but it does sometimes rhyme. Five decades later, the Cuban missile crisis stands not just as a pivotal moment in the history of the Cold War but also as a guide for how to defuse conflicts, manage great-power relationships, and make sound decisions about foreign policy in general.

Critical Thinking

1. What is the relationship between Thucydides' history of the Peloponnesian war and the Cuban missile crisis?
2. Should the U.S. risk war to prevent Iran from developing the bomb?
3. What should U.S. policy be in dealing with North Korea?

Create Central

www.mhhe.com/createcentral

Internet References

World at the Brink
    http:microsites.jfklibrary.org/cmc/
National Archives and JFK Library Mark 50th Anniversary of Cuban Missile Crisis in October
    www.archives.gov/press/press-releases/2012/nr12-146.html.

GRAHAM ALLISON is Professor of Government and Director of the Belfer Center for Science and International Affairs at Harvard University's Kennedy School of Government.

Allison, Graham. From *Foreign Affairs*, July/August 2012, pp. 11–16. Copyright © 2012 by Council on Foreign Relations, Inc. Reprinted by permission of Foreign Affairs. www.ForeignAffairs.com

Article

Prepared by: Robert Weiner, *University of Massachusetts/Boston*

Israel's New Politics and the Fate of Palestine

Akiva Eldar

Learning Outcomes

After reading this article, you will be able to:

- Discuss the Israeli position on peace negotiations.
- Explain why Israel has chosen to continue the status quo.

In my vision of peace, there are two free peoples living side by side in this small land, with good neighborly relations and mutual respect, each with its flag, anthem and government. . . . If we get a guarantee of demilitarization, and if the Palestinians recognize Israel as the Jewish state, we are ready to agree to a real peace agreement, a demilitarized Palestinian state side by side with the Jewish state.

—Benjamin Netanyahu, June 14, 2009

Seemingly, it was a historic moment. The prime minister of Israel and leader of the Likud Party publicly embraced the two-state solution. A short while into his second term in office, ten days after the newly inaugurated president of the United States promised in Cairo to "personally pursue this outcome," Netanyahu declared an about-face, shifting from the traditional course he and his political camp had once pursued.

Thus, more than ninety years after the Balfour Declaration of November 1917, it appeared the successors of the founders of Zionism were moving toward a historic compromise to resolve the conflict embedded in that intentionally vague statement. It is the conflict between "the establishment in Palestine of a national home for the Jewish people" and "nothing shall be done which may prejudice the civil and religious rights of existing non-Jewish communities in Palestine."

Now it appeared that this dispute, which for decades had split Israeli society into rival political camps, could be resolved. Forty-two years after the occupation of the West Bank and the Gaza Strip, formerly held by Jordan and Egypt, a right-wing prime minister declared his willingness to return these territories to the people living in them, as well as his consent for the establishment of a new, independent state of Palestine.

But almost immediately, other voices emerged questioning whether this solution—dividing the land into two independent, coexisting states—was still feasible; whether the "window of opportunity" that might have been available in the past had already closed for good; whether the Israeli settlement enterprise in the West Bank had reached a point of no return, creating a new situation that did not allow for any partition; and whether the division of political powers within Israeli society had changed, making the dramatic move impossible. As Robert Serry, UN special coordinator for the Middle East peace process, put it:

> If the parties do not grasp the current opportunity, they should realize the implication is not merely slowing progress toward a two-state solution. Instead, we could be moving down the path toward a one-state reality, which would also move us further away from regional peace.

This article focuses on the Israeli side of this equation in part because the Palestinian leadership, as far back as 1988, made a strategic decision favoring the two-state solution, presented in the Algiers declaration of the Palestinian National Council. The Arab League, for its part, voted in favor of a peace initiative that would recognize the state of Israel and set the terms for a comprehensive Middle East settlement. Meanwhile, various bodies of the international community reasserted partition of the land as their formal policy. But Israel, which signed the Oslo accords nearly two decades ago, has been moving in a different direction. And Netanyahu's stirring words of June 2009 now ring hollow.

Israel never overtly spurned a two-state solution involving land partition and a Palestinian state. But it never acknowledged that West Bank developments had rendered such a solution impossible. Facing a default reality in which a one-state solution seemed the only option, Israel chose a third way—the continuation of the status quo. This unspoken strategic decision has dictated its polices and tactics for the past decade, simultaneously safeguarding political negotiations as a framework for the future and tightening Israel's control over the West Bank. In essence, a "peace process" that allegedly is meant to bring the occupation to an end and achieve a two-state solution has become a mechanism to perpetuate the conflict and preserve the status quo.

This reality and its implications are best understood through a brief survey of the history that brought the Israelis and Palestinians to this impasse. The story is one of courage, sincere efforts, internal conflicts on both sides, persistent maneuvering and elements of folly.

In August 1993, the foreign ministers of Israel and the Palestine Liberation Organization (PLO), Shimon Peres and Mahmoud Abbas, signed a declaration of principles. In September of that year, Israeli prime minister Yitzhak Rabin and PLO chairman Yasir Arafat exchanged the "letters of recognition," which led to an impressive signing ceremony on the White House lawn. Words about historical compromise, reconciliation and peace filled the air. The world perceived a true, deep change sweeping the Middle East, with both sides resolved to divide the land into two states.

Nevertheless, the negotiating partners' starting points remained far apart. The Palestinians considered engaging in a process based on the acceptance of the 1967 borders to be a major compromise in itself. They believed their willingness to settle for territory representing 22 percent of mandatory Palestine was already an immense compromise foreclosing much further concession. Israel, in contrast, considered these borders the starting point for talks and never intended to withdraw fully from the occupied territory.

Prime Minister Rabin accentuated this position in seeking Knesset support for the interim agreement, or Oslo II:

> We would like this to be an entity which is less than a state, and which will independently run the lives of Palestinians under its authority. The borders of the State of Israel, during the permanent solution, will be beyond the lines which existed before the Six Day War. We will not return to the 4 June 1967 lines.

Rabin further referred to different areas of the West Bank that Israel would insist on keeping, including regions that no Palestinian negotiator could give up.

Because of these differences, the Oslo accords were originally labeled an interim agreement "for a transitional period not exceeding five years," meant to lay the foundations for "a permanent settlement based on Security Council Resolutions 242 (1967) and 338 (1973)." Yet, even though the final objective intentionally remained vague, the agreement itself listed detailed timetables for the implementation of interim phases, including, most remarkably, an Israeli withdrawal from the cities of Gaza and Jericho in three months. Already in this sensitive initial phase, cracks appeared. "No dates are sacred," said Rabin in December 1993, as the deadline for withdrawal was being postponed.

Nevertheless, despite the evident differences between both sides and the difficulties that were clear from the beginning,

two-state-solution supporters believed the dynamics of the process would generate their own power, which would force the parties to take brave steps and reach an ultimate resolution. Whatever actual force these developments could have set in motion, the effort suffered a fatal blow on November 4, 1995, when an opponent of the agreement killed the prime minister.

Six months later, Israel conducted elections between two candidates for prime minister—Shimon Peres, perceived as a progenitor of the Oslo plan, and Benjamin Netanyahu, a fierce opponent of the process throughout his time as head of the parliamentary opposition. Netanyahu won narrowly.

His election marked a new era in Israel's attitude toward the negotiations. Prior to Rabin's assassination, one could reasonably argue that the main motivation of the government was to conclude an agreement. But Netanyahu did everything possible to safeguard the negotiations as a framework while concurrently evading their declared objective. All of his successors as prime minister followed this pattern.

Netanyahu himself testified to this scheme and his way of handling it in a private conversation in 2001, when he was out of office. Unaware that he was being recorded, he bragged about the manipulative tactics he had used in his first tenure as prime minister to undermine the Oslo accords. He explained that he had insisted the Clinton administration provide him with a written commitment that Israel alone would be able to determine the borders of the "defined military sites" that would remain under its control. He went on to say that by defining the entire Jordan Valley as a military location, he "actually stopped the Oslo Accord." He was right. Without this large area, the Palestinians wouldn't have a viable state.

Indeed, under Netanyahu's first tenure as prime minister, which ended in 1999, little progress was made in implementing the agreed-upon phases or moving toward a final-status agreement. When the five years allocated for the transitional period passed, no Palestinian state seemed near.

No one publicly embraced this status quo choice as a policy, and yet it seemed to generate its own momentum as various players quietly understood that it served their purposes.

During his first term, Netanyahu came under attack from both sides. Those opposed to dividing the land were furious that he didn't spurn the peace process overtly. Supporters of the accords, meanwhile, protested against his foot-dragging in implementing the agreement's provisions. All condemned Netanyahu's indecision. But these critics failed to perceive that Israel's new status quo approach was actually a choice—and, indeed, a policy.

No one publicly embraced this decision, and yet it seemed to generate its own momentum as various players quietly understood that it served their purposes. A report published by the

International Crisis Group, tellingly titled "The Emperor Has No Clothes: Palestinians and the End of the Peace Process," lists benefits to the various partners in the so-called peace process, including the entities known collectively as the "quartet" (the UN, United States, European Union and Russia). The Europeans, said the report, wanted influence in the Middle East, and by funding the Palestinian Authority (PA) they found they could get a seat at some prestigious diplomatic tables. Russia and the UN harbored similar desires for diplomatic advancement.

Meanwhile, Washington knew its support for the ongoing peace process, however much it may be a sham, allowed it to maintain good relations with Arab countries even as it nurtured its "special relationship" with Israel. Thus, the United States saw in the status quo an opportunity to preserve its influence in the Middle East by maintaining a delicate balance in its ties with most major regional players. But this approach is far removed from the evenhanded policy championed by President Dwight Eisenhower in the early years of Israel's existence. Israel today shows immense confidence in the financial aid and large diplomatic umbrella it gets from America, as reflected in Netanyahu's oft-quoted comment:

> *I know what America is. America is something that can be easily moved. Moved to the right [direction]. . . . They won't get in our way. They won't get in our way. . . . So let's say they say something. So they said it! They said it! 80 percent of the Americans support us.*

Even the Palestinians get sucked into this status quo game, although they pay the highest price for the current stalemate and have demonstrated in recent years open hostility to continuing the barren peace talks. But in reality, under such extremely asymmetrical circumstances, they likely would suffer the most if the process were to collapse. Since the days of Yasir Arafat, and more intensely since the beginning of Mahmoud Abbas's presidency, the PA leadership has relied almost solely on the international community for generous financial aid and global attention. Thus, the PA is highly dependent on foreign support. Its leaders fear that if they take actions that upset the international community, and particularly the United States, they will lose their aid—and consequently face a possible collapse in their political standing within the Palestinian community.

So, lacking any better alternative, the existence of the PA allows for a kind of welfare for large portions of the West Bank's political and economic elite. This is true of Fatah, whose raison d'être has become maintaining the ongoing process. It also includes tens of thousands of families whose livelihoods depend on the PA. For these families, stopping the aid would be disastrous. Thus, if the peace process has become an addiction for many participants, as the International Crisis Group report notes, this addiction has become an absolute reliance for the people of the PA.

Whatever motivates most participants in the process, Israel's embrace of it is most intense, for good reasons—including religion, historical traumas, national security, territorial aspirations, control over natural resources, the threat of internal social division and political survival. Yet, to understand how deeply rooted this imperative is for Israel, one must examine the foundation on which Israeli society and the ethos of its collective identity are built.

If Israeli citizens were to create a collective identification card, most would probably embrace the words "Jewish and democratic." From the 1940s, when Israel was yet to be established, up until today, these two adjectives have been almost a binding code, the vision with which the different elements of the state were to act. This sensibility was embodied in the country's declaration of independence, the basis of Israel's establishment. A body of commentary, scholarship and civic documents emerged that sought to examine whether those two terms were contradictory. These studies included the "basic laws," the groundwork for a possible future Israeli constitution, restrictions imposed on the platforms of parties running for the Knesset, and many hundreds of news and academic articles.

Yet, since Israel is not merely an abstract idea but an actual political entity, these two concepts—one connected to a collective cultural and religious identity, the other a method for governing—must be merged with the realities of geography. The relationship between the three sides of this triangle—geography, demography and democracy—has influenced Israel's nature and policies from day one.

When the United Nations General Assembly voted on the partition plan in 1947, two-thirds of the inhabitants of mandatory Palestine were Arabs, while Jews constituted a third of the population. Of course, this situation did not allow for the existence of a state that would be both Jewish and democratic. But only a few months later, with the establishment of the state inside what would become the 1949 armistice line, 84 percent of the population of newly born Israel—spread over 78 percent of the land—were Jews. The formation of an almost absolute identity between the geographic partition and the demographic division over the different parts of what had been mandatory Palestine was anything but accidental. Israel's first prime minister, David Ben-Gurion, summarized the consequences of the 1948 war:

> *The IDF could have conquered the entire territory between the [Jordan] River and the Sea. But what kind of state would we have? . . . We would have a Knesset with an Arab majority. Having to choose between the wholeness of the land or a Jewish State, we chose the Jewish State.*

In other words, the demographic concern was the dominant factor in Israel's decisions on how to conduct its first war—initially, by encouraging more than seven hundred thousand Arab inhabitants to leave the territories over which it took control, then by refraining from conquering additional territory.

However, this consonance between geography and demography changed dramatically nineteen years later, with Israel's decisive victory in the Six-Day War of 1967. Israel's military took control over vast amounts of land, including the Gaza Strip and the West Bank, the latter encompassing a 30 percent increase in territory over what Israel had controlled before the war. But these territories were not empty. And although many Palestinians on those lands left their homes, some for the second time, a large number remained. Thus did Israel's ability to retain simultaneously a Jewish and a democratic identity become endangered. But this departure from the Ben-Gurion formula was not quickly perceived by Israeli leaders, even though the triangle of demography, geography and democracy became much more complex and explosive.

Israel's geographic expansion in the 1967 war—and the new demographic proportions between Jews and Arabs under its control—once again forced Israel to make a choice: Which sides of the triangle would strengthen, and which would weaken? Seemingly, the territorial conquests undermined the demographic edge, meaning the Jewish majority. However, no one intended to allow a weakening in this fundamental component of the state's identity.

"The key phrase in the Israeli experience is 'a Jewish majority.' Israelis will do anything—wage war or make peace—to maintain a Jewish majority and preserve the Israeli tribal bonfire." These were the words of Daniel Ben-Simon, former journalist and current Labor Party member of the Knesset. A senior member of the rival party has expressed a similar position. In a conference held in March 2002, at the peak of the suicide bombings that killed many Israelis, Dan Meridor, deputy prime minister and minister of intelligence and atomic energy in the Israeli cabinet, said: "Of all the various questions—security, the Middle East peace process, etc.—the demographic-democratic problem is the chief imminent threat that we simply cannot evade." More recently, the newly elected chairman of the Kadima Party, Shaul Mofaz, declared the so-called demographic threat the most dangerous of all to the existence of Israel.

This outlook, embraced by the most prominent figures of the mainstream political parties, is shared by the Jewish Israelis they represent. This is seen in public-opinion polls such as the Democracy Index, which found in 2010 that 86 percent of Israeli Jews believed decisive choices for the state must be taken on the basis of a Jewish majority.

Therefore, a careful analysis of the triangle model cannot focus on the strength of each side independently but must focus on possible two-side combinations. On the collective identity card, the definition of "Jewish and democratic" is being replaced with "Jewish and geographic." Whenever two of the edges are dominant, the third tends to weaken, and the third in this instance is the democratic component.

The move toward a "Jewish and geographic" state became even more prominent following changes undergone by Israeli society in recent decades. Settlers, although they composed a relatively small fraction of the population, became the vanguard that directed political thinking for most of the Jewish religious public. The ultra-Orthodox political parties, which previously had been considered the swing faction between dovish and hawkish political camps, accepted the settlers' doctrine that occupied territories represented Israeli land. They stood by the right-wing parties in opposing partition. This political drift took place at a time when religious groups in Israel became larger in both absolute and relative terms. A survey conducted by Israel's Central Bureau of Statistics in 2008 showed that only 40 percent of Israeli Jews between the ages of twenty and twenty-four identified themselves as nonreligious or secular. This trend has great influence on the direction Israeli society is taking nowadays.

In a survey conducted on the tenth anniversary of Prime Minister Rabin's assassination, Israeli Jews were asked to assess whether the decision to engage in the Oslo process had been correct. While 62 percent of the secular respondents answered affirmatively, the answer given by religious and ultra-Orthodox respondents was the complete opposite; among those respondents, representing a growing segment of Israeli society, more than 70 percent said it had been wrong. Placing "greater Israel" at the top of the value system meant that democracy and demography were undermined among the wider public, to the point where they believed the executive and the Knesset did not have the mandate to decide on territorial withdrawals. This is reflected in a recent statement by Benny Katzover, former chairman of the Shomron settlers' regional council and a settler leader: "The main role of Israeli democracy now is to disappear. Israeli democracy has finished its role, and it must disassemble and give way to Judaism."

Gabriel Sheffer, a prominent expert on the study of regime and societal relations in Israel, views the lack of separation between religion and state in Israel as the key factor in understanding the country's recent history. In a 2005 article, he stressed that the historical failure to separate ethnic-national identity and religious belief is the primary cause of events in Jewish society and in the relationship between Israelis, Arabs and Palestinians. He explained that this issue distorts Israeli democracy. More recently, he characterized Israel as a Jewish-national-religious state that naturally excludes many citizen groups from any serious influence on public policy.

Even so, it would be a mistake to explain Israeli society's right-wing drift only in terms of the growing power of religious groups. Another factor is the mass immigration of the early 1990s and the corollary collapse of the so-called Zionist Left.

In 1992, Israeli general elections ended with a change of government: the Labor and Meretz parties, which represented the Zionist Left in parliament, together won fifty-six of the Knesset's 120 seats. This outcome enabled Rabin to form a Center-Left coalition government that set in motion the historical recognition of the PLO as the representative of the Palestinian people and signed the declaration of principles. Seventeen years later, during the 2009 elections—the most recent in Israel—these two parties won only sixteen seats. Public-opinion surveys prior to the elections showed that 72 percent of Jewish respondents defined themselves as "right-wing." These results illustrate the rise of the Israeli political Right, which has been growing in force since 1967.

Washington knew its support for the peace process, however much it may be a sham, allowed it to maintain good relations with Arab countries even as it nurtured its "special relationship" with Israel.

During the 1990s, nearly a million immigrants arrived in Israel, about 85 percent from the former Soviet Union. This group's size and demographic characteristics had a crucial effect on the composition and nature of Israeli society. These newcomers found in Israel a refuge from a crumbling communist empire that had shaped much of their historical and cultural thinking. Natan Sharansky, a "refusenik" and an immigrant from the Soviet Union, explained to President Clinton, perhaps jocularly, why he was the only Israeli cabinet member who opposed the peace agreement the president was trying to promote at Camp David in 2000: "I can't vote for this, I'm Russian. . . . I come from one of the biggest countries in the world to one of the smallest. You want me to cut it in half. No, thank you."

The 2009 Democracy Index revealed that "in general, the immigrants' attitudes are less liberal and less tolerant in almost every realm and concerning every topic examined." For example, 77 percent of former Soviet immigrants in the survey supported policies to encourage Arab emigration from Israel. The right-wing sensibility of these people, who are largely secular, stems not from religious attitudes but from a perception of the Jewish society as "landlord" of Israel, with aspirations to exercise strong national sovereignty over a territory that should be as extensive and secure as possible.

Former Knesset member Mossi Raz of Meretz, in analyzing the rise of the immigrant Right and the dovish political camp's unprecedented decline in the latest elections, said that "these million and a half immigrants, who arrived in recent decades, constitute 20 percent of the voters, but Meretz and the Labor Party together received only 5 percent of their votes."

Even traditional supporters of the Zionist Left, such as secular people of the middle and upper classes, shifted toward the Right, in part due to Prime Minister Ariel Sharon and the Likud Party's success in creating a conceptual turnabout in Israeli political culture. The conservative Right successfully separated the notion of "prosperity" from the term "peace" and convinced many Israelis that economic growth would emerge if the government merely managed the Israeli-Palestinian conflict and practiced a neoliberal economic policy. This trend accelerated when the West declared a "war on terror" following the September 11, 2001, attacks, which gave Israeli enterprises new access to wide markets. As *Forbes* magazine noted, Israel became the destination for those seeking antiterrorism technology. The stability and prosperity of Israel's economy, even without conflict resolution, diminished the imperative of peace for many.

Israel's Palestinian citizens also have undergone significant political changes since Oslo. These shifts, seen in voting patterns, result from the deterioration in the relationship between the Jewish and Arab populations. These, in turn, reflect a growing sense of Israel's changing nature as a state; a mistrust between the two population groups; and a rise in the intensity of hostility and violence between Israel and the Palestinians in the West Bank and the Gaza Strip.

This process had a twofold impact on voting patterns: first, Arab voter participation declined; and second, more Arabs who did participate gave their votes to Arab rather than Zionist parties. In 1996, for instance, 79.3 percent of eligible Arab voters took part in the first elections after Rabin's assassination. In 2003, it was 63 percent; in 2009, only 53.6 percent. Yet, as more of these Arab participants voted for Arab parties, the number of parliamentary seats granted to the three Arab political parties rose to eleven, the highest ever. In 1992, only 47.7 percent of Arab voters supported these parties, but in the elections of 1996, after the assassination of Rabin, sectarian voting jumped to 67.3 percent. In the latest elections, 82.1 percent of Palestinians who are Israeli citizens voted for one of these three parties.

The balance of political power inside Israel is unsustainable, given the demographic facts between the Mediterranean Sea and the Jordan River. For the first time since the establishment of the state, the proportion of Jews and Arabs living under Israeli jurisdiction is approaching equilibrium. Sharon, who was aware of this, tried in 2005 to exclude a million and a half Palestinians from this calculation by withdrawing Israeli forces and settlers from the Gaza Strip. Yet, since Israel continued to exercise control over Gaza's airspace and sea—and to a very large extent over its land borders—Israel is still responsible for this territory and its inhabitants, according to a widely accepted interpretation of international law. Sergio della Pergola, an expert on demography, estimates that by Israel's hundredth anniversary, the demographic balance between the Mediterranean and the Jordan River will return to what it was before Israel's declaration of independence: two-thirds Arabs and non-Jews and one-third Jews. Demographers estimate that by 2030, the proportion of Jews in the population will decline to 46 percent. According to another estimate, a similar percentage will be reached by 2020, and some even suggest that by that time Jews will constitute only 40 percent of the population. Regardless, by the end of the present decade, Jews are expected to become a minority between the sea and the river.

The Oslo accords were intended to mark the beginning of a gradual end to the Israeli presence in the occupied territories. Instead, the accords opened a new era for the settlement enterprise, which continues its expansion in the so-called C areas, which encompass 60.2 percent of the West Bank territory and remain under full Israeli control. "This is one of the strangest maps of existing and potential autonomous territories ever agreed-upon by two conflicting parties," said Elisha Efrat, Israel Prize winner for geographical research. He referred to the way 176 "orange stains" (B areas), representing the Palestinian rural space, are spread throughout the map, with C areas separating them from one another and leaving Palestinians with mere isolated enclaves that preclude any national self-sustainment. Jeff Halper, a human-rights activist, compares this to the Japanese game of Go, in which "you win by immobilizing your opponent, by gaining control of key points of a matrix so that every time s/he moves s/he encounters an obstacle of some kind."

Since Israel refuses to undertake any commitment to freeze settlement, it uses the interim phases, whose purpose was to advance toward a two-state solution, to create obstacles that would impede a fair, agreed-upon partition of the territory. In the decade following the Oslo accords from 1993 to 2003, the number of West Bank settlers doubled, from 110,000 to 224,000 (not including East Jerusalem). Since then, the figure has risen to more than 340,000. Together with Israelis residing in Israeli-constructed neighborhoods in East Jerusalem, they now represent more than six hundred thousand people. The number of existing settlements authorized by Israel is 124, to which one should add twelve East Jerusalem neighborhoods and more than a hundred "outposts" built by settlers without formal approval by the state (though with the help of public authorities and branches of the government). Many of those outposts were located carefully to prevent any territorial contiguity in a future Palestinian state. It is in these strategic areas of the mountain strip and across the separation wall that the Jewish West Bank population grew the most during 2011.

At the same time, and more formally, Israeli governments worked to increase the settler population in "block settlements" in order to eventually annex these areas, as was openly declared. Some of these blocks are close to the 1967 borders, and, in informal negotiations (such as the Geneva Initiative), Palestinians agreed in principle to the idea that they would be annexed, as long as the Palestinian state would be compensated with separate territory equivalent in size. However, they strongly rejected Israeli annexation of areas such as the Ariel and Karnei Shomron blocks, necessary for any viable Palestinian state with territorial contiguity.

To exercise control over the land without giving up its Jewish identity, Israel has embraced various policies of "separation." It has separate legal systems for traditional Israeli territory and for the territory it occupies; it divides those who reside in occupied lands based on ethnic identity; it has retained control over occupied lands but evaded responsibility for the people living there; and it has created a conceptual distinction between its democratic principles and its actual practices in the occupied territories. These separations have allowed Israel to manage the occupation for forty-five years while maintaining its identity and international status. No other state in the twenty-first century has been able to get away with this, but it works for Israel, which has little incentive to change it.

This article was written shortly after a coalition government controlling ninety-four seats out of 120 was formed in Israel. The coalition agreement between the two largest parties, Likud and Kadima, does not leave room for hope regarding a future breakthrough toward a two-state solution. The sides talked only in general terms about the resumption of the political process and instead emphasized the importance of maintaining Israel as a Jewish and democratic state. For this reason, they added a clarification regarding "the importance of maintaining defensible borders," a phrase implying that any compromise contemplated by the coalition government centers on gaps between the positions of Likud and Kadima more than on those between Israelis and Palestinians.

At present, only fourteen Knesset members (a little more than 10 percent) constitute the opposition, which supports dividing the land into two states on the basis of the 1967 borders. Eleven of them are Palestinians who are Israeli citizens,

and three are members of Meretz representing the Zionist Left. Even if we include the Labor Party, which facilitated the formation of Rabin's cabinet some two decades ago with a majority of sixty-one seats, this faction's presence has now been reduced to twenty-two Knesset members. Recent opinion polls indicate that, if the elections were held today, this political bloc would win thirty-two seats, a little more than a quarter of the parliament. Thus, the formation of the new unity government represents the monolithic nature of Israeli society. For decades, the boundaries of Israeli Jewish society, based on the Jews' relationship with the Palestinians and the question of dividing the land, were the focus of disputes that at times split Israelis into separate groups. But now a consensual answer has emerged. While the hawkish political camp has adopted some rhetoric that used to characterize the dovish bloc, the latter has been forced to accept the political reality of being dominated. Thus does Israel's grip on the occupied territories tighten, even as the issue wanes on Israelis' public agenda.

No doubt, the present unity government can promote almost anything it wishes. That means it is unlikely to use its power to promote a new partition of the land into two states in any way acceptable to Palestinians. History teaches us that Israelis are only willing to take brave and honest steps toward peace when they know the cost of failing to do so will be even greater. Unfortunately, given the realities of the current situation, there is little reason to think that the Israelis will take these steps anytime soon.

Critical Thinking

1. How does geography, demography, and democracy affect the Israeli–Palestinian peace processes?
2. What should the U.S. policy toward Palestinian–Israeli peace negotiations be?
3. What are some of the problems associated with the two-state solution?

Create Central
www.mhhe.com/createcentral

Internet References

Central Intelligence Agency
www.cia.gov

Israeli Ministry of Foreign Affairs
www.mfa.gov.il

Permanent Observer Mission of the State of Palestine to the United Nations
http://palestineun.org/

AKIVA ELDAR is the chief political columnist and an editorial writer for *Haaretz*. He wishes to thank his researcher, Eyal Raz, for assistance with this article.

Unit 4

UNIT

Prepared by: Robert Weiner, *University of Massachusetts/Boston*

War, Arms Control, and Disarmament

War is a method of conflict resolution that has been institutionalized over the centuries, according to the classical realist view of international relations. Political scientists also argue that there is a relationship between the internal regime of a state and war, so that liberal democracies do not wage war on other liberal democracies. Neorealists, such as Kenneth Waltz, argue that it is the international political structure within which states function that explains the phenomenon of war. War can also result from the miscalculations and misperceptions of the opposing sides. For example, Robert Jervis writes that "Misperception . . . includes inaccurate inferences, miscalculations of consequences, and misjudgments about how one will react to one's policies." Leaders may underestimate or overestimate the intentions and threats of their rivals. In a crisis situation, foreign policy decision-making elites may be overloaded with information, have difficulty screening it, and also may be subject to the phenomenon of cognitive dissonance. Cognitive dissonance occurs when an individual is so overloaded with information that he or she reverts to stereotypes that reinforce preexisting beliefs.

One of the major concerns of the international community is the problem of the proliferation of nuclear weapons. The acquisition of nuclear weapons by "rogue" states such as Iran and North Korea is viewed by the international community not only as a threat to regional stability, but also a threat to international stability. In 2013, it was clear that Iran was continuing to work on the development of the capacity to produce enriched uranium, which would bring it closer to a nuclear weapon capacity. Diplomacy, efforts at dialogue, heavy economic sanctions, the assassination of Iranian nuclear scientists, and the sabotaging of Iranian computers working on uranium enrichment have not worked. However, in 2013, a breakthrough occurred when the moderate president of Iran opened negotiations with the United States. Yet Israel was not convinced of the sincerity of the Iranian and continued believing that it was facing what it dubbed an existential threat. Israel could launch a preemptive strike against Iran to destroy its nuclear infrastructure.

One of the major problems that the United States faces in Afghanistan is the difficult relationship with Pakistan. Some elements in the Pakistani military and intelligence services are playing a double-edged game by supporting the Afghan Taliban to the extent that they can exercise control over it, ostensibly to counter an enhancement of Indian influence in Afghanistan. At the same time, Pakistan is considered to be a U.S. ally in the war against the Taliban and Al Qaeda. Relations between the United States and Pakistan have also been exacerbated by "collateral damage," which results in Pakistani civilian casualties due to the excessive use of drone strikes and commando raids in Pakistan by U.S. forces, which the Pakistani government considers to be a violation of its sovereignty.

President Obama has made an agreement with President Karzai to withdraw U.S. troops from Afghanistan by 2014, but with a commitment of U.S. forces to help Pakistan after 2014. The question is whether in view of the U.S. draw-down of forces, the Afghan security forces will be ready to deal with the Taliban.

While the international community has placed a great deal of emphasis on the arms control and disarmament of weapons of mass destruction, the trade in conventional weapons, which runs the gamut from revolvers to tanks to jet aircraft, continues to pose a serious problem to human security. In 2011, the global arms trade added up to more than $85 billion, with the United States as the world's leading arms supplier, according to a recent report by the Congressional Research Service. Millions of innocent civilians have been killed by conventional weapons since the end of World War II. They have perished at the hands of repressive governments, revolutionaries, criminal gangs, and paramilitaries and militias, with many of the weapons supplied by private arms dealers or "merchants of death." Although there are a number of treaties and agreements dealing with the trade in conventional weapons, there is no universal, comprehensive, and binding treaty that sets international standards to regulate the trade. An international conference met at the United Nations from July 2–27, 2012, but was unable to arrive at an agreement on an Arms Trade Treaty (ATT). The seizure of northern Mali by Al Qaeda in the Islamic Maghreb also showed how easy it was for a terrorist group to seize part of a country, resulting in hundreds of thousands of refugees and displaced persons. Asymmetrical war, such as the Mallian war in 2013,or Shabab's attack on Westgate Mall in Kenya, underscored the need for the international community to control the trade in conventional weapons, as millions of innocent civilians have been killed. Finally, in April 2013, the UN General Assembly adopted the Arms Trade Treaty by a majority vote. Nonetheless, major arms exporters like the US and Russia, still have not ratified the ATT.

Chemical weapons are considered to be one of the most reprehensible weapons of mass destruction. Chemical weapons, such as mustard gas, had been used extensively during the First World War, resulting in an estimated 1.3 million casualties. In 1925, the Protocol for the Prohibition of the Use in War of Asphyxiating Poisonous or Other Gases and Bacteriological Methods of Warfare was adopted. Syria was a signatory of the protocol. The protocol banned the use of chemical weapons, but did not ban their production and stockpiling. Chemical weapons were never used by the belligerents against each other during the Second World War, although the Nazis used poisonous gases to exterminate their victims in the concentration camps. After the Second World War, however, a number of states developed a chemical weapons capacity. Chemical weapons were used by Iraq in the war against Iran (1980–1988). Chemical weapons were also used by the regime of Saddam Hussein against Kurdish villagers in the conflict against the Kurds seeking an independent Kurdistan. In 1997, the international community adopted the chemical weapons convention. It is formally known as the Convention on the Prohibition of the Development, Production, Stockpiling and Use of Chemical Weapons and on Their Destruction. The Organization for the Prohibition of Chemical Weapons (OPCW) was established to enforce the convention. According to the OPCW, 81% of the world's chemical weapons stockpiles have been destroyed, with the U.S. claiming to have destroyed 90% of its stockpile, and Russia 70%. North Korea, which is reputed to have the second–largest stockpile of chemical weapons in the world, is not a signatory to the convention Israel has signed, but not ratified the convention.

Prepared by: Robert Weiner, *University of Massachusetts/Boston*

Article

Why Iran Should Get the Bomb

Nuclear Balancing Would Mean Stability

KENNETH N. WALTZ

Learning Outcomes

After reading this article, you will be able to:

- Understand the relationship between neo-realism and military balance.

- Analyze the logic of deterrence as it applies to Iran.

The past several months have witnessed a heated debate over the best way for the United States and Israel to respond to Iran's nuclear activities. As the argument has raged, the United States has tightened its already robust sanctions regime against the Islamic Republic, and the European Union announced in January that it will begin an embargo on Iranian oil on July 1. Although the United States, the EU, and Iran have recently returned to the negotiating table, a palpable sense of crisis still looms.

It should not. Most United States, European, and Israeli commentators and policymakers warn that a nuclear-armed Iran would be the worst possible outcome of the current standoff. In fact, it would probably be the best possible result: the one most likely to restore stability to the Middle East.

Power Begs to Be Balanced

The crisis over Iran's nuclear program could end in three different ways. First, diplomacy coupled with serious sanctions could convince Iran to abandon its pursuit of a nuclear weapon. But this outcome is unlikely: the historical record indicates that a country bent on acquiring nuclear weapons can rarely be dissuaded from doing so. Punishing a state through economic sanctions does not inexorably derail its nuclear program. Take North Korea, which succeeded in building its weapons despite countless rounds of sanctions and UN Security Council resolutions. If Tehran determines that its security depends on possessing nuclear weapons, sanctions are unlikely to change its mind. In fact, adding still more sanctions now could make Iran feel even more vulnerable, giving it still more reason to seek the protection of the ultimate deterrent.

The second possible outcome is that Iran stops short of testing a nuclear weapon but develops a breakout capability, the capacity to build and test one quite quickly. Iran would not be the first country to acquire a sophisticated nuclear program without building an actual bomb. Japan, for instance, maintains a vast civilian nuclear infrastructure. Experts believe that it could produce a nuclear weapon on short notice.

Such a breakout capability might satisfy the domestic political needs of Iran's rulers by assuring hard-liners that they can enjoy all the benefits of having a bomb (such as greater security) without the downsides (such as international isolation and condemnation). The problem is that a breakout capability might not work as intended.

The United States and its European allies are primarily concerned with weaponization, so they might accept a scenario in which Iran stops short of a nuclear weapon. Israel, however, has made it clear that it views a significant Iranian enrichment capacity alone as an unacceptable threat. It is possible, then, that a verifiable commitment from Iran to stop short of a weapon could appease major Western powers but leave the Israelis unsatisfied. Israel would be less intimidated by a virtual nuclear weapon than it would be by an actual one and therefore would likely continue its risky efforts at subverting Iran's nuclear program through sabotage and assassination—which could lead Iran to conclude that a breakout capability is an insufficient deterrent, after all, and that only weaponization can provide it with the security it seeks.

The third possible outcome of the standoff is that Iran continues its current course and publicly goes nuclear by testing a weapon. United States and Israeli officials have declared that outcome unacceptable, arguing that a nuclear Iran is a uniquely terrifying prospect, even an existential threat. Such language is typical of major powers, which have historically gotten riled up whenever another country has begun to develop a nuclear weapon of its own. Yet so far, every time another country has managed to shoulder its way into the nuclear club, the other members have always changed tack and decided to live with it. In fact, by reducing imbalances in military power, new nuclear states generally produce more regional and international stability, not less.

Israel's regional nuclear monopoly, which has proved remarkably durable for the past four decades, has long fueled instability in the Middle East. In no other region of the world does a lone, unchecked nuclear state exist. It is Israel's nuclear arsenal, not Iran's desire for one, that has contributed most to the current crisis. Power, after all, begs to be balanced. What is surprising about the Israeli case is that it has taken so long for a potential balancer to emerge.

Of course, it is easy to understand why Israel wants to remain the sole nuclear power in the region and why it is willing to use force to secure that status. In 1981, Israel bombed Iraq to prevent a challenge to its nuclear monopoly. It did the same to Syria in 2007 and is now considering similar action against Iran. But the very acts that have allowed Israel to maintain its nuclear edge in the short term have prolonged an imbalance that is unsustainable in the long term. Israel's proven ability to strike potential nuclear rivals with impunity has inevitably made its enemies anxious to develop the means to prevent Israel from doing so again. In this way, the current tensions are best viewed not as the early stages of a relatively recent Iranian nuclear crisis but rather as the final stages of a decades-long Middle East nuclear crisis that will end only when a balance of military power is restored.

Unfounded Fears

One reason the danger of a nuclear Iran has been grossly exaggerated is that the debate surrounding it has been distorted by misplaced worries and fundamental misunderstandings of how states generally behave in the international system. The first prominent concern, which undergirds many others, is that the Iranian regime is innately irrational. Despite a widespread belief to the contrary, Iranian policy is made not by "mad mullahs" but by perfectly sane ayatollahs who want to survive just like any other leaders. Although Iran's leaders indulge in inflammatory and hateful rhetoric, they show no propensity for self-destruction. It would be a grave error for policymakers in the United States and Israel to assume otherwise.

Yet that is precisely what many United States and Israeli officials and analysts have done. Portraying Iran as irrational has allowed them to argue that the logic of nuclear deterrence does not apply to the Islamic Republic. If Iran acquired a nuclear weapon, they warn, it would not hesitate to use it in a first strike against Israel, even though doing so would invite massive retaliation and risk destroying everything the Iranian regime holds dear.

Although it is impossible to be certain of Iranian intentions, it is far more likely that if Iran desires nuclear weapons, it is for the purpose of providing for its own security, not to improve its offensive capabilities (or destroy itself). Iran may be intransigent at the negotiating table and defiant in the face of sanctions, but it still acts to secure its own preservation. Iran's leaders did not, for example, attempt to close the Strait of Hormuz despite issuing blustery warnings that they might do so

after the EU announced its planned oil embargo in January. The Iranian regime clearly concluded that it did not want to provoke what would surely have been a swift and devastating American response to such a move.

Nevertheless, even some observers and policymakers who accept that the Iranian regime is rational still worry that a nuclear weapon would embolden it, providing Tehran with a shield that would allow it to act more aggressively and increase its support for terrorism. Some analysts even fear that Iran would directly provide terrorists with nuclear arms. The problem with these concerns is that they contradict the record of every other nuclear weapons state going back to 1945. History shows that when countries acquire the bomb, they feel increasingly vulnerable and become acutely aware that their nuclear weapons make them a potential target in the eyes of major powers. This awareness discourages nuclear states from bold and aggressive action. Maoist China, for example, became much less bellicose after acquiring nuclear weapons in 1964, and India and Pakistan have both become more cautious since going nuclear. There is little reason to believe Iran would break this mold.

As for the risk of a handoff to terrorists, no country could transfer nuclear weapons without running a high risk of being found out. United States surveillance capabilities would pose a serious obstacle, as would the United States' impressive and growing ability to identify the source of fissile material. Moreover, countries can never entirely control or even predict the behavior of the terrorist groups they sponsor. Once a country such as Iran acquires a nuclear capability, it will have every reason to maintain full control over its arsenal. After all, building a bomb is costly and dangerous. It would make little sense to transfer the product of that investment to parties that cannot be trusted or managed.

Another oft-touted worry is that if Iran obtains the bomb, other states in the region will follow suit, leading to a nuclear arms race in the Middle East. But the nuclear age is now almost 70 years old, and so far, fears of proliferation have proved to be unfounded. Properly defined, the term "proliferation" means a rapid and uncontrolled spread. Nothing like that has occurred; in fact, since 1970, there has been a marked slowdown in the emergence of nuclear states. There is no reason to expect that this pattern will change now. Should Iran become the second Middle Eastern nuclear power since 1945, it would hardly signal the start of a landslide. When Israel acquired the bomb in the 1960s, it was at war with many of its neighbors. Its nuclear arms were a much bigger threat to the Arab world than Iran's program is today. If an atomic Israel did not trigger an arms race then, there is no reason a nuclear Iran should now.

Rest Assured

In 1991, the historical rivals India and Pakistan signed a treaty agreeing not to target each other's nuclear facilities. They realized that far more worrisome than their adversary's nuclear deterrent was the instability produced by challenges to it. Since then, even in the face of high tensions and risky provocations, the two countries have kept the peace. Israel and Iran would do well to consider this precedent. If Iran goes nuclear, Israel and Iran will deter each other, as nuclear powers always have. There has never been a full-scale war between two nuclear-armed states. Once Iran crosses the nuclear threshold, deterrence will apply, even if the Iranian arsenal is relatively small. No other country in the region will have an incentive to acquire its own nuclear capability, and the current crisis will finally dissipate, leading to a Middle East that is more stable than it is today.

For that reason, the United States and its allies need not take such pains to prevent the Iranians from developing a nuclear weapon. Diplomacy between Iran and the major powers should continue, because open lines of communication will make the Western countries feel better able to live with a nuclear Iran. But the current sanctions on Iran can be dropped: they primarily harm ordinary Iranians, with little purpose.

Most important, policymakers and citizens in the Arab world, Europe, Israel, and the United States should take comfort from the fact that history has shown that where nuclear capabilities emerge, so, too, does stability. When it comes to nuclear weapons, now as ever, more may be better.

Critical Thinking

1. Why does Waltz believe that a nuclear armed Iran does not constitute a threat to stability in the region?
2. Should Israel engage in a preemeptive strike against Iran? Why?
3. Should the international community continue to use diplomacy and economic sanctions to prevent Iran from going nuclear? Why?

Create Central

www.mhhe.com/createcentral

Internet References

International Atomic Energy Agency(IAEA)
www.iaea.org
Non-Proliferation Treaty
www.un.org/disarmament /WMD/Nuclear/NPT.shtml.

KENNETH N. WALTZ is Senior Research Scholar at the Saltzman Institute of War and Peace Studies and Adjunct Professor of Political Science at Columbia University.

Waltz, Kenneth N. From *Foreign Affairs*, vol. 91, no. 4, July/August 2012, pp. 2–5. Copyright © 2012 by Council on Foreign Relations, Inc. Reprinted by permission of Foreign Affairs. www.ForeignAffairs.com

Article

Prepared by: Robert Weiner, *University of Massachusetts/Boston*

Talking Tough to Pakistan

How to End Islamabad's Defiance

STEPHEN D. KRASNER

Learning Outcomes

After reading this article, you will be able to:

- Explain why Pakistan is not a very good U.S. ally.
- Determine the policy that the U.S. should follow toward Pakistan.

On September 22, 2011, Admiral Mike Mullen, then chairman of the United States Joint Chiefs of Staff, made his last official appearance before the Senate Armed Services Committee. In his speech, he bluntly criticized Pakistan, telling the committee that "extremist organizations serving as proxies for the government of Pakistan are attacking Afghan troops and civilians as well as United States soldiers." The Haqqani network, he said, "is, in many ways, a strategic arm of Pakistan's Inter-Services Intelligence Agency [ISI]." In 2011 alone, Mullen continued, the network had been responsible for a June attack on the Intercontinental Hotel in Kabul, a September truck-bomb attack in Wardak Province that wounded 77 United States soldiers, and a September attack on the United States embassy in Kabul.

These observations did not, however, lead Mullen to the obvious conclusion: Pakistan should be treated as a hostile power. And within days, military officials began walking back his remarks, claiming that Mullen had meant to say only that Islamabad gives broad support to the Haqqani network, not that it gives specific direction. Meanwhile, unnamed United States government officials asserted that he had overstated the case. Mullen's testimony, for all the attention it received, did not signify a new United States strategy toward Pakistan.

Yet such a shift is badly needed. For decades, the United States has sought to buy Pakistani cooperation with aid: $20 billion worth since 9/11 alone. This money has been matched with plenty of praise. At his first press conference in Islamabad following his 2007 appointment as chairman of the Joint Chiefs, Mullen called Pakistan "a steadfast and historically." In 2008, then Secretary of State Condoleezza Rice even said that she "fully believed" that Pakistan "does not in any way want to be associated with terrorist elements and is indeed fighting to root them out wherever [Pakistani officials] find them." Meanwhile, United States leaders have spent an outsized amount of face time with their Pakistani counterparts. As secretary of state, Hillary Clinton has made four trips to Pakistan, compared with two to India and three to Japan. Mullen made more than 20 visits to Pakistan.

To be sure, Mullen was not the first United States official to publicly point the finger at Islamabad, nor will he be the last. In 2008, the CIA blamed Pakistan's ISI for aiding the bombing of the Indian embassy in Kabul. In July 2011, two months after United States Navy SEALS raided Osama bin Laden's compound near the prestigious Pakistan Military Academy, Admiral James Winnefeld, vice chair of the Joint Chiefs, told the Senate Armed Services Committee, "Pakistan is a very, very difficult partner, and we all know that." And in an October press conference with Afghan President Hamid Karzai, Clinton noted that the Obama administration intended to "push the Pakistanis very hard," adding, "they can either be helping or hindering."

Washington's tactic—criticism coupled with continued assistance—has not been effectual. Threats and censure go unheeded in Pakistan because Islamabad's leaders do not fear the United States. This is because the United States has so often demonstrated a fear of Pakistan, believing that although Pakistan's policies have been unhelpful, they could get much worse. Washington seems to have concluded that if it actually disengaged and as a result Islamabad halted all its cooperation in Afghanistan, then United States counterinsurgency efforts there would be doomed. Even more problematic, the thinking goes, without external support, the already shaky Pakistani state would falter. A total collapse could precipitate a radical Islamist takeover, worsening Pakistani relations with the United States-backed Karzai regime in Afghanistan and escalating tensions, perhaps even precipitating a nuclear war, between Pakistan and India.

Weighing of Deeds

The United States-Pakistani relationship has produced a few modest successes. Pakistan has generally allowed NATO to transport supplies through its territory to Afghanistan. It has helped capture some senior al Qaeda officials, including Khalid Sheik Mohammed, the 9/11 mastermind. It has permitted the United States to launch drone strikes from bases in Baluchistan.

Yet these accomplishments pale in comparison to the ways in which Pakistan has proved uncooperative. The country is the world's worst nuclear proliferator, having sold technology to Iran, Libya, and North Korea through the A. Q. Khan network. Although Islamabad has attacked those terrorist groups, such as al Qaeda and the Pakistani Taliban, that target its institutions, it actively supports others, such as the Haqqani network, the Afghan Taliban, and Hezb-i-Islami, that attack coalition troops and Afghan officials or conspire against India. Pakistan also hampers United States efforts to deal with those groups; although many Pakistani officials privately support the drone program, for example, they publicly exaggerate the resulting civilian deaths. Meanwhile, they refuse to give the United States permission to conduct commando raids in Pakistan, swearing that they will defend Pakistani sovereignty at all costs.

A case in point was the raid that killed bin Laden. Rather than embrace the move, Pakistani officials reacted with fury. The police arrested a group of Pakistani citizens who were suspected of having helped the United States collect intelligence prior to the operation and delayed United States interrogations of bin Laden's three wives for more than a week. Lieutenant General Ahmed Shuja Pasha, head of the ISI, condemned the United States raid before a special session of parliament, and the government passed a resolution pledging to revisit its relationship with the United States. Of course, the operation was embarrassing for the Pakistani military, since it showed the armed forces to be either complicit in harboring bin Laden or so incompetent that they could not find him under their own noses. But Pakistan could easily have saved face by publicly depicting the operation as a cooperative venture.

The fact that Pakistan distanced itself from the raid speaks to another major problem in the relationship: despite the billions of dollars the United States has given Pakistan, public opinion there remains adamantly anti-American. In a 2010 Pew survey of 21 countries, those Pakistanis polled had among the lowest favorability ratings of the United States: 17 percent. The next year, another Pew survey found that 63 percent of the population disapproved of the raid that killed bin Laden, and 55 percent thought it was a bad thing that he had died.

Washington's current strategy toward Islamabad, in short, is not working. Any gains the United States has bought with its aid and engagement have come at an extremely high price and have been more than offset by Pakistan's nuclear proliferation and its support for the groups that attack Americans, Afghans, Indians, and others.

Rational Choice

It is tempting to believe that Pakistan's lack of cooperation results from its weakness as a state. One version of this argument is that much of Pakistan's civilian and military leadership might actually want to be more aligned with the United States but is prevented from being so by powerful hard-line Islamist factions. Its advocates point to the fact that pubic officials

shrank from condemning the bodyguard who in January 2011 shot Salman Taseer, the governor of Punjab, who had spoken out against Pakistan's blasphemy law. Similar silence followed the March assassination of Shahbaz Bhatti, the minorities minister and only Christian in the cabinet, who had also urged reforming the law. Presumably, the politicians held their tongues out of fear of reprisal. Another explanation of the weakness of the Pakistani state is that the extremists in the government and the military who support militants offer that support despite their superiors' objections. For example, the May 2011 terrorist attack on Pakistan's naval air base Mehran, which the top military brass condemned, was later suspected to have been conducted with help from someone on the inside.

Still, there is a much more straightforward explanation for Pakistan's behavior. Its policies are a fully rational response to the conception of the country's national interest held by its leaders, especially those in the military. Pakistan's fundamental goal is to defend itself against its rival, India. Islamabad deliberately uses nuclear proliferation and deterrence, terrorism, and its prickly relationship with the United States to achieve this objective.

Pakistan's nuclear strategy is to project a credible threat of first use against India. The country has a growing nuclear arsenal, a stockpile of short-range missiles to carry warheads, and plans for rapid weapons dispersion should India invade. So far, the strategy has worked; although Pakistan has supported numerous attacks on Indian soil, India has not retaliated.

Transnational terrorism, Pakistanis believe, has also served to constrain and humiliate India. As early as the 1960s, Pakistani strategists concluded that terrorism could help offset India's superior conventional military strength. They were right. Pakistani militant activity in Kashmir has led India to send hundreds of thousands of troops into the province—as many as 500,000 during a particularly tense moment with Pakistan in 2002. Better that India sends its troops to battle terrorists on its own territory, the Pakistani thinking goes, than march them across the border. Further, the 2008 Mumbai attack, which penetrated the heart of India, was a particularly embarrassing episode; the failure to prevent it, and the feeble response to it, demonstrated the ineffectuality of India's security forces.

Pakistan's double game with the United States has been effective, too. After 9/11, Pakistan's leaders could hardly resist pressure from Washington to cooperate. But they were also loath to lose influence with the insurgents in Afghanistan, which they believed gave Pakistan strategic depth against India. So Islamabad decided to have things both ways: cooperating with Washington enough to make itself useful but obstructing the coalition's plans enough to make it nearly impossible to end the Afghan insurgency. This has been an impressive accomplishment.

Caring By Neglecting

As Mullen's comments attest, United States officials do recognize the flaws in their country's current approach to Pakistan. Yet instead of making radical changes to that policy, Washington continues to muddle through, working with Pakistan where possible, attempting to convince its leaders that they should focus on internal, rather than external, threats, and hoping for the best. For their part, commentators mostly call for marginal changes, such as engaging the Pakistani military more closely on the drone program and making the program more transparent, opening United States textile markets to Pakistani trade, helping Pakistan address its energy deficit, focusing on a peaceful resolution of the Kashmir dispute, and developing closer ties to civilian officials. Many of these suggestions seem to be based on the idea that if millions of dollars in United States aid has not been enough to buy Pakistani support, perhaps extra deal sweeteners will be.

The one significant policy change since 2008 has been the retargeting of aid to civilians. Under the Obama administration, total assistance has increased by 48 percent, and a much higher percentage of it is economic rather than security related: 45 percent in 2010 as opposed to 24 percent in 2008. The Enhanced Partnership With Pakistan Act of 2009, which committed $7.5 billion to Pakistan over five years, conditioned disbursements on Pakistan's behavior, including cooperation on counterterrorism and the holding of democratic elections.

Despite Pakistan's ongoing problematic behavior, however, aid has continued to flow. Clinton even certified in March 2011 that Pakistan had made a "sustained commitment" to combating terrorist groups. Actions such as this have undermined American credibility when it comes to pressuring Pakistan to live up to its side of the bargain. The United States has shown that the sticks that come with its carrots are hollow.

The only way the United States can actually get what it wants out of Pakistan is to make credible threats to retaliate if Pakistan does not comply with United States demands and offer rewards only in return for cooperative actions taken. United States officials should tell their Pakistani counterparts in no uncertain terms that they must start playing ball or face malign neglect at best and, if necessary, active isolation. Malign neglect would mean ending all United States assistance, military and civilian; severing intelligence cooperation; continuing and possibly escalating United States drone strikes; initiating cross-border special operations raids; and strengthening United States ties with India. Active isolation would include, in addition, declaring Pakistan a state sponsor of terrorism, imposing sanctions, and pressuring China and Saudi Arabia to cut off their support, as well.

Of course, the United States' new "redlines" would be believable only if it is clear to Pakistan that the United States would be better off acting on them than backing down. (And the more believable they are, the less likely the United States will have to carry them out.) So what would make the threats credible?

First, the United States must make clear that if it ended its assistance to Pakistan, Pakistan would not be able to retaliate. The United States could continue its drone strikes, perhaps using the stealth versions of them that it is currently developing. It could suppress Pakistani air defenses, possibly with electronic jammers, so as to limit military deaths and collateral damage. And even if Pakistan shot down some drones, it could not destroy them all. The United States might even be able to conduct some Special Forces raids, which would be of such short duration and against such specific targets that Pakistan would not be able to retaliate with conventional forces. Pakistan might attempt to launch strikes against NATO and Afghan forces in Afghanistan, but its military would risk embarrassing defeat if those campaigns did not go well. Pakistan might threaten to cut off its intelligence cooperation, but that cooperation has never really extended to sharing information on the Afghan Taliban, one of the United States' main concerns in Afghanistan. Moreover, if Pakistan started tolerating or abetting al Qaeda on its own soil, the country would be even more at risk. Al Qaeda could turn against the state and attempt to unseat the government. And the United States would surely begin striking Pakistan even more aggressively if al Qaeda found haven there.

Second, the United States must show that it can neutralize one of Pakistan's trump cards: its role in the war in Afghanistan. Washington must therefore develop a strategy for Afghanistan that works without Pakistan's help. That means a plan that does not require transporting personnel or materiel through Pakistan. Nearly 60 percent of the NATO supplies sent into Afghanistan are already routed through the north, through Russia and Central Asia. The United States military is hoping to increase that number to 75 percent. Without Pakistan, therefore, the coalition could still support a substantial force in Afghanistan, but not one as big as the current one of 131,000 troops. The basic objective of that force would necessarily be counterterrorism, not counterinsurgency. Counterterrorism is less personnel- and resource-intensive because it aims only to prevent the country from becoming a haven for Islamist extremists, not to transform it into a well-functioning democracy. Given the Obama administration's current plans to withdraw 24,000 United States troops by the summer of 2012, with many more to follow, such a strategy is already inescapable.

Finally, Washington must shed its fear that its withdrawal of aid or open antagonism could lead to the Pakistani state's collapse, a radical Islamist takeover, or nuclear war. Pakistanis, not Americans, have always determined their political future. Even substantial United States investments in the civilian state and the economy, for example, have not led to their improvement or to gains in stability. With or without United States aid to Pakistan, the Pakistani military will remain the most respected institution in the country. In a 2011 Pew poll of Pakistanis, 79 percent of respondents said that the military was having a good influence on the country's direction, compared with 20 percent who said that the national government was.

As for the possibility of an Islamist takeover, the country's current power centers have a strong interest in maintaining control and so will do whatever they can to keep it—whatever Washington's policy is. It is worth remembering that Pakistan has already proved itself able to take out the terrorist networks that threaten its own institutions, as it did in the Swat Valley and the district of Buner in 2009. Moreover, government by radical Islamists has not proved to be a popular choice among Pakistanis. In the last general election, the Muttahida Majlis-e-Amal, a coalition of Islamist parties, won only seven out of 340 seats in the National Assembly.

The possibility that nuclear weapons could wind up in the hands of terrorists is a serious risk, of course, but not one that the United States could easily mitigate whatever its policy in the region. Pakistan's nuclear posture, which involves rapid dispersion, a first-strike capability, and the use of tactical weapons, increases the chances of the central government's losing control. Even so, Pakistan will not alter that posture because it is so effective in deterring India. Meanwhile, previous United States efforts to help tighten Pakistan's command-and-control systems have been hampered by mutual distrust. Any new such efforts would be, too. Finally, since India has both a first- and a second-strike capability, Pakistan would not likely strike India first in the event of a crisis. In any case, even if things did escalate, there is not much that the United States, or anyone else, could do—good relations or not.

From a United States perspective, then, there is no reason to think that malign neglect or active isolation would make Pakistan's behavior or problems any worse.

Heads I Win, Tails You Lose

Even as the United States threatens disengagement, it should emphasize that it would still prefer a productive relationship. But it should also make clear that the choice is Pakistan's: if the country ends its support for terrorism; works in earnest with the United States to degrade al Qaeda, the Taliban, and the Haqqani network; and stops its subversion in Afghanistan, the United States will offer generous rewards. It could provide larger assistance programs, both civilian and military; open United States markets to Pakistani exports; and support political arrangements in Kabul that would reduce Islamabad's fear of India's influence. In other words, it is only after Pakistan complies with its demands that the United States should offer many of the policy proposals now on the table. And even then, these rewards should not necessarily be targeted toward changing Pakistan's regional calculus; they should be offered purely as payment for Pakistan's cooperation on the United States' most important policies in the region.

A combination of credible threats and future promises offers the best hope of convincing Islamabad that it would be better off cooperating with the United States. In essence, Pakistan would be offered a choice between the situation of Iran and that of Indonesia, two large Islamic states that have chosen very different paths. It could be either a pariah state surrounded by hostile neighbors and with dim economic prospects or a country with access to international markets, support from the United States and Europe, and some possibility of détente with its neighbors. The Indonesian path would lead to increased economic growth, an empowered middle class, strengthened civil-society groups, and a stronger economic and social foundation for a more robust democracy at some point in the future. Since it would not directly threaten the military's position, the Indonesian model should appeal to both pillars of the Pakistani state. And even if Islamabad's cooperation is not forthcoming, the United States is better off treating Pakistan as a hostile power than continuing to spend and get nothing in return.

Implicit in the remarks Mullen made to the Senate was the argument that Washington must get tough with Pakistan. He was right. A whole variety of gentle forms of persuasion have been tried and failed. The only option left is a drastic one. The irony is that this approach won't benefit just the United States: the whole region, including Pakistan, could quickly find itself better off.

Critical Thinking

1. What are the issues which have created a rift in United States-Pakistani relations?

2. Why is Pakistan supporting the Taliban?

3. What can the United States do to encourage more Pakistani cooperation in the war against the Taliban and Al Qaeda?

Create Central

www.mhhe.com/createcentral

Internet References

International Security Assistance Force
www.nato.int/ISAF

National Defense University
www.ndu.edu

STEPHEN D. KRASNER is Professor of International Relations at Stanford University and a Senior Fellow at Stanford's Freeman Spogli Institute for International Studies and at the Hoover Institution. He was Director of Policy Planning at the United States State Department in 2005–07.

Prepared by: Robert Weiner, *University of Massachusetts/Boston*

Article

Syria's Long Civil War

"Without an acknowledgment of possible defeat, neither the regime nor the opposition will accept a grand bargain in which compromise is central."

GLENN E. ROBINSON

Learning Outcomes

After reading this article, you will be able to:

- Explain the difficulties of holding another Syrian peace conference in Geneva.
- Explain the relationship between the Syrian Civil War and Syrian national identity.
- Discuss the possible two outcomes of the war.

The first year and a half in the current round of Syria's long civil war took as many lives as the three-week orgy of violence in the city of Hama that ended the last round in 1982. In both cases, some 25,000 to 30,000 people were killed, and in both cases, the root issues and the competing sides have been the same: a minority-based regime, allied with other minorities along with privileged elements from the majority population, ruling over a poor and often dysfunctional state that does not tolerate dissenters.

The last round in the Syrian civil war began after the regime of Hafez al-Assad in 1976 intervened in Lebanon's civil war. While that intervention had broad regional and international support, it was far more controversial at home. For Syrian forces to come to the aid of Lebanon's Christians—who were on the verge of defeat at the hands of Muslim forces—was seen by pious Sunni Muslims in Syria as proof positive of the heretical nature of the Assad regime, a regime dominated by Alawites, an offshoot of Islamic Shiism.

The resulting low-intensity civil war, instigated by the Muslim Brotherhood and fueled by forces that had given rise to the rapid growth of Islamist politics throughout the Middle East in the 1970s, continued in Syria for six years. Assassinations, attacks on Alawite military cadets, the mass murder of Muslim Brotherhood prisoners, and ultimately a crippling commercial strike brought the Assad government to the brink of collapse.

To ensure the survival of his regime, Hafez al-Assad cut a political deal with his bitter rivals, the Sunni bourgeoisie, heirs of the notable class that had dominated Syrian politics for centuries. This alliance between Alawite military power and Sunni (and Christian) economic muscle gave Assad the political cover he needed to launch an assault on Hama, the stronghold of Muslim Brotherhood power in Syria. By leveling much of the city with a relentless artillery barrage, Assad drove the Muslim Brotherhood underground, thereby winning the first round of Syria's long civil war.

A Bigger Lebanon

Syria's troubles go well beyond warring ethnic and confessional groups, to the fact that Syria as a political entity—as a nation—hardly exists. To be sure, the country's two major cities, Damascus and Aleppo, have very long histories and strong localized identities. However, until the twentieth century, Syria was never a country unto itself. During the half millennium when it was part of the Ottoman Empire, Syria was not even constituted as a single administrative district within the empire, but was split among several districts.

The invention of modern Syria following the First World War was based largely on agreements between the French and the British. Syria was not unique in this. Indeed, the modern borders of scores of countries in the developing world were based more on the interests of the colonial powers than on any historical or geographic reality.

What was different about Syria was that both the French colonial power and the ruling Arabs in Damascus worked to deny the construction of a modern Syrian national identity. France went beyond its usual divide-and-conquer strategy, and actually tried to split the Mandate of Syria into a half-dozen nominally independent states. Given strong local opposition in the 1920s, this effort never fully materialized, but two of those proposed states ultimately went their own way: an independent Lebanon, and the Hatay (Alexandretta) province of Turkey, ceded by France on the eve of the Second World War.

Syria's independence period, 1946–63, was marked by nearly constant turmoil. Political actors, both internal and international, sought to wield power and influence in the weak new state, often by using friendly surrogate forces inside the country. Egypt may have been the most successful at this, briefly merging with Syria to form the United Arab Republic from 1958 to 1961.

When the Arab nationalist Baath party seized power in a military coup in 1963, it replicated one part of French colonial rule by denying Syrian nationalism and national identity. Syria's new rulers rejected national identity as a European colonial construct. Instead, they regarded Syria as a "regional command" of the Baath party and of the Arab nation, not as a separate country in its own right. It was not until the waning years of Hafez al-Assad's rule, and then a bit more strongly under his son Bashar, that the Syrian state began to use its powers of political socialization (in school textbooks, on state television, and in other media) to tentatively promote a separate Syrian national identity.

This failure to construct a strong Syrian national identity has left in place a broad Arab nationalism among a dying cadre of true believers, and very strong parochial identities along ethnic and religious lines. There is still little sense of a shared national Syrian community. In this regard, Syria is just a bigger version of Lebanon. Neither would be accused of having a "melting pot" political culture.

Syria encompasses five primary ethno-confessional parochial identities, and many more fragmentary ones. Because a proper census has not been conducted since French colonial days, one can only estimate percentages. Roughly two-thirds of Syrians are both Arab by ethnicity and Sunni Muslim by religion. The bulk of this population appears to oppose the regime in Damascus.

Alawites, Christians, and Kurds each make up about 10 percent of Syria's population. However, significant minorities exist within each of these minorities. For example, the Christian community, split among Greek Orthodox, Maronite Catholics, Roman Catholics, Armenians, and others, hardly speaks with a single voice. The Druze, about 4 percent of the population, are an offshoot Islamic sect that incorporates strong folk traditions into religious practice.

Heretics in Charge

The Alawite community is the most politically powerful of Syria's minorities. Although it originated in Shiite Islam, the Alawite sect evolved in isolation over the past millennium in northwest Syria. Viewed as a "heretical sect" by orthodox Sunni Muslims, the Alawites have a long history of persecution by the Sunnis, resulting in a well-honed sense of grievance. They kept apart, scratching out an existence as subsistence farmers in the dusty hills overlooking the eastern Mediterranean. Historically, the Alawites have been among the poorest of Syrians, a position that has improved—but not markedly—in recent decades.

French colonialism paved the way for Alawite upward mobility. France's creation of a Syrian national army gave poor young Syrian minorities an opportunity for social advancement. Joining the army meant education, a chance to travel, and a steady job, all of which were in short supply at the time, and Alawites, along with Druze, signed up in large numbers.

Although the French army of Syria was disbanded in 1946 with Syrian independence, the tradition of military service among the Alawites was established. They enjoyed a good reputation at it; in fact, the French considered the Alawites a "martial race." As newly independent Syria rebuilt its army, the Alawites, eager to serve and to improve their lot in life, joined the army in droves. Their prevalence in the officer corps in particular helps explain why they were in such a good position to dominate Syrian politics beginning in the 1960s.

The Baathist military coup of 1963 was engineered by true believers in Arab nationalism, people who rejected on ideological grounds the importance of parochial identities. But the sociological changes outlined above guaranteed that many leading members of the Baathist regime came from minority groups, especially from the Alawite and Druze communities. By 1966, Syria had its first Alawite president, while many Sunni Baathists were purged from leading ranks. Druze Baathists were purged next, so when Hafez al-Assad came to power in 1970, virtually all the top power brokers in Syria were Alawites. In a volatile country, the politics of trust trumped ideological purity.

The dominant Alawite narrative today is "kill or be killed." Memories of persecution at the hands of vengeful Sunnis remain strong. Most Alawites appear to believe that, were the regime to collapse and the opposition come to power, some version of ethnic cleansing would be undertaken against the Alawite community as a whole. They are right to be concerned.

It is the same fear of a hard-line, vengeful Sunni Islamist regime that keeps other minorities tacitly supporting the regime, even after a year and a half of bloody repression. Christians, an important part of the business community and professional classes, do not necessarily fear a bloodletting so much as relegation to second-class status as a persecuted minority under an Islamist regime. They appear to prefer a flawed but secular regime, sharing the concerns of many Coptic Christians in Egypt under a post–Hosni Mubarak regime.

Historically, the Kurdish community in northeast Syria has had troubled relations with the Baathist regime. Since they are not Arabs, Kurds could not be expected to cheer on pan-Arab nationalism in Syria. However, the Kurds realize that a hardline Islamist regime would not likely be supportive of Kurdish rights, either. Furthermore, in mid-2012, the regime negotiated a withdrawal from Kurdish population centers. While this was likely done as retaliation against Turkey, which fears independence demands from its own Kurdish population, the reality is that today the Kurds of Syria have an autonomous homeland. Kurds from northern Iraq, including the well-trained Peshmerga militia, are widely reported to be crossing into Syria to support their Kurdish brethren. It is doubtful that any future regime in Damascus will ever be able to extend the writ of government into Kurdish Syria.

When one adds up the active support for the regime from the Alawite community with the tacit support of most Christians, many other minorities, and a small slice of the Sunni bourgeoisie, this government can count on the backing of about one-quarter of the Syrian population. That is far from a majority, to be sure, but it likely would be enough to hold onto power for a long time, barring a change in the logic of civil war.

The Second Round

The second round of Syria's long civil war began iconically (rarely are periodizations so precise) in March 2011, when regime thugs brutalized boys who were writing anti-regime graffiti on walls in Dara'a. This dusty border town just north of Jordan had previously crossed the Western imagination only as the place where Ottoman soldiers briefly imprisoned and sexually tortured Lawrence of Arabia.

The daily stories from Syria since the outbreak of protests and violence have been well covered elsewhere, but several issues should be underlined. One involves the influence of hard-line Islamists and jihadists within the opposition, and whether Christians and other minorities have a reasonable fear for their future in an opposition-led Syria. The short answer is yes, but not for the reason that media accounts usually proffer, that of an influx of jihadists from other countries. Foreign jihadists have entered Syria, but their numbers are still very small, and their influence is limited.

Syria has plenty of its own homegrown jihadists, from two primary sources. The first is Syria's underground Muslim Brotherhood. Unlike the situation in North Africa and Jordan, where the Muslim Brotherhood has been part of the political process for decades and has been "tamed" by having to negotiate with various groups that do not share its ideology, Syria's Muslim Brotherhood never left its militant phase. After the Syrian group was defeated militarily in 1982, a few leaders went into exile in London and Paris and reformed their goals, but the actual fighters just went home, nursing their wounds and their grudges. They and their sons have returned for round two, determined to exact revenge.

On top of this, the US invasion of Iraq bred a new generation of Syrian jihadists who went east to help fight the Americans and their Shiite allies. From 2003 to about 2007, the Assad regime clearly encouraged these Sunnis to fight, and perhaps die, in Iraq. But those who returned to Syria had learned valuable urban warfare skills, and had likely grown even more attuned to the "heretical" nature of the Alawite regime in Damascus.

Perhaps the best evidence of the early presence of jihadists in the current round of Syria's civil war is the body count. While the earliest protests were nonviolent, as elsewhere in the Arab Spring, the regime immediately succeeded in militarizing the conflict for its own interests. But the opposition was not so ill-prepared for a militarized conflict. The estimated body count ratio from the earliest months of the conflict last year ran about 4 to 1: For every four opponents killed, one regime security member was killed.

This ratio is quite good for an unorganized opposition, and suggests a certain skill set present from the beginning. By contrast, in the long Israeli-Palestinian conflict, the kill ratio is typically about 20 to 1 in Israel's favor. In the 2009 Gaza conflict, it was over 100 to 1. During 2012, the kill ratio in Syria narrowed to 2 to 1. Then the regime adopted the tactics of Russia's second Chechen war.

Outsiders in the Mix

The role of outside powers is another issue worth underlining. Given the number of Russians who are in Syria helping the government, it is no surprise that the Assad regime has implemented a Chechen war strategy. The Russian military, humiliated by Chechen secessionists in the 1994–96 war, seized the opportunity for payback in 1999. Vladimir Putin gave the military a free hand. The dominant new tactic was to use air power against urban targets without regard for civilian casualties, something considered since World War II to be beyond the pale of civilized warfare. As of this writing, the results in Syria have not yet been as bloody as those in Grozny. Still, it is clear that the use of air power against a largely civilian insurgency comes straight from the Russian experience in Chechnya.

Iran is playing an equally direct role in Syria. The linchpin of Iran's regional Arab strategy, Syria is the only Arab country that Iran can count as an ally. The alliance dates to 1980, when their mutual rival, Iraq, invaded Iran; it has withstood the test of time and runs considerably deeper than most analysts thought. Ideas of "flipping Syria" out of the Iranian orbit seem superficial in hindsight. By its own admission, Iran has sent Revolutionary Guards to help the regime win the civil war, an admission prompted when the opposition captured a bus full of Iranian fighters. Outside analysts believe thousands of Iranian Revolutionary Guard members are in Syria.

A final point to be made about the conflict today: In the years before the second round of the long civil war, Bashar al-Assad significantly narrowed the base of his regime, antagonizing the broader Alawite community in the process. Whereas Hafez al-Assad made sure that all major segments of the Alawite community benefited from regime jobs and largesse, Bashar al-Assad cut out the broader community in favor of his immediate family. Political and economic power was concentrated in the hands of Bashar, his brother Maher, his brother-in-law Asef Shawkat, his cousin Rami Makhluf, and a handful of others.

Now the opposition has assassinated Shawkat and (according to Russian reports) seriously wounded Maher al-Assad. With regime and perhaps community survival at stake, the Alawites have rallied around Bashar al-Assad. However, given the discontent within the Alawite community, it is not inconceivable that there would be support for abandoning the Assads as part of any grand political bargain that ends the conflict.

Which Outcome?

Logically, the current round of Syria's civil war must end in one of four ways: regime victory, opposition victory, stalemate with no end, or stalemate leading to a political resolution. The first two outcomes are the worst for all parties, and the last is the best plausible outcome. But for the better options to be plausible, both sides must believe they can actually lose the civil war. This is key. Without an acknowledgment of possible defeat, neither the regime nor the opposition will accept a grand bargain in which compromise is central.

Although the Arab Spring has caused a number of governments to fall, a victory for the Syrian regime is a very real possibility. It is also one of the two worst outcomes. International sanctions likely would remain in place for years to come, leaving Syria isolated and poor. Russia, Iran, China, and a few others could be expected to ignore such sanctions and provide an economic lifeline. But there can be no doubt that Syria would be a grim place for a long time.

A regime victory, moreover, likely would guarantee a third round of civil war sometime in the future, since the country's basic structural problems would remain. The regime will never allow the majority population to essentially vote it out of power. A government victory would also prompt greater emigration among those able to leave: typically, that part of the population most essential for rebuilding the economy.

While most Syrians "lose" under this scenario, the winners would be the quarter of the population that backs the regime—and the Kurds. An autonomous Kurdistan in Syria would almost certainly remain, creating a serious political problem for Turkey, among others. Russia, Iran, and Hezbollah (the Iranian- and Syrian-backed Islamic militant group in Lebanon) would also enjoy a major strategic victory in the region, while US allies Turkey, Saudi Arabia, and Qatar would suffer a significant foreign policy defeat after their support for regime change in Damascus.

Regime collapse and an outright opposition victory (never mind that there is not a unified opposition) represent an equally unappealing outcome. To the degree that any one group would be able to consolidate power in Damascus after a complete collapse of the government, it would more likely than not be an Islamist regime hardened by many months of violent insurgency and repression. The democrats of Syria—and there are many—are not strategically positioned to come to power or even to wield significant influence over a victorious opposition.

Such an outcome would almost certainly lead to significant levels of revenge killings against Alawites, and perhaps even to ethnic cleansing of Alawites in areas outside of their historic Latakia homeland. An opposition victory would also certainly lead to the emigration of hundreds of thousands of Syrian Christians. Tens of thousands already have taken up residence in nearby Lebanon, waiting out the civil war.

Many thousands of (mostly Sunni) Muslims are already refugees in Lebanon, Turkey, and Jordan; they would likely return to Syria if the regime were defeated. Christians and perhaps Alawites would likely take their place as refugees, while those who are able to would leave for Europe and the Americas. As with any plausible scenario, the now autonomous Kurdish region would maintain its newly won status. From the perspective of the United States and its allies in the region, the one positive result under the scenario of an opposition victory—besides the downfall of a brutal regime—is the strategic black eye it would give the Iranians, Russians, and Hezbollah.

More of the Same?

A third logical outcome is continuity, with the current civil war settling in for the long haul, though probably at a somewhat lower level of violence. The analogy for such an outcome is Lebanon from 1975 to 1989, albeit without an external occupying force. Central authority would officially remain in place, but be honored mostly in the breach. Zones with no formal governmental authority would be sprinkled widely throughout the country, with various militias having their own checkpoints. The external patrons of the opposing forces would make sure "their" side does not lose, keeping munitions and other resources flowing and extending the life of the civil war.

Such an outcome carries obvious downsides: most of all, the suffering of ordinary Syrians. From a strategic perspective, the likelihood of regional spillover is high. Since all of Syria's neighbors have friendly relations with the United States, including NATO member Turkey, the possibility of regional destabilization cannot be dismissed.

At the strategic level, however, a continuing civil war may be preferable, from America's perspective, to a victory for either the regime or the opposition. The rough balance of power would be maintained; competitors such as Iran and Russia would slowly be drained of resources; the very negative consequences of the first two scenarios would be avoided; and the local populations would largely find ways to muddle through, protected in part by their regional allies.

Under this scenario, Syrians and the region would await some structural change in the calculation of civil war in the years ahead to bring the war to an end, much as the denouement of the cold war led to the curtailing of the Lebanese civil war. If the spillover effects for other countries could be minimized, a continuing civil war scenario is troubling, to be sure, but less bad than the first two scenarios.

A political compromise forged among the various sides would easily be the best outcome for Syria and the region. The analogy for this scenario is Lebanon today: hardly an example of a fully functioning state, but certainly better than the plausible alternatives. Lebanon's national pact, struck at Taif, Saudi Arabia, in 1989, maintains the politics of corporate groups, albeit in somewhat altered form. Each ethno-confessional group gets its own recognition, protection, and slice of the patronage pie. The president must be a Maronite Christian, the prime minister a Sunni Muslim, the defense minister a Druze, the Speaker of Parliament a Shiite, and so on down the line. Parliament is split 50-50 between Muslims and Christians.

This is a recipe for gridlock, but each group at least has some guarantees and a stake in the system. It is ironic that Hezbollah is the major voice of reform inside Lebanon, wanting to do away with corporate representation in favor of a one-person, one-vote principle. Something like Lebanon's system would be the best transition out of civil war in Syria.

Currently, neither the regime nor the opposition has any interest in reaching a deal with the other side, and two requirements are mandatory for such a resolution. First, each side must fear that it could actually lose the civil war, with disastrous consequences for its own people. Second, the major powers must likewise agree to such an outcome and pressure their own sides; this is something that has not happened to date—and may never happen.

The Unconventional Option

The United States and its allies are wise to resist direct military involvement in Syria in the form of invasion, an air campaign, or a "no-fly zone" (which would quickly lead to direct military engagement). Likewise, Washington has been smart to resist providing advanced military hardware, such as anti-aircraft missiles, to an opposition with significant elements that would just as easily turn these weapons against American targets.

That said, the flow of funds and small arms to the opposition from various parties has been an important source of balancing in the civil war, preventing the regime thus far from winning outright. However, the turn toward a Chechnya strategy of using airpower to destroy urban pockets of rebellion does threaten the opposition with outright defeat and should be countered in smart ways.

The West was sometimes criticized for adopting a Machiavellian posture during the 1980s Iran-Iraq war by hoping that neither side won outright, but such criticism was not warranted. A clear victory for either side would have been a disaster for the region and the world. The same approach is warranted in Syria: working to prevent either side from winning a total victory so that both sides will be more inclined to reach a compromise.

To accomplish this, the United States and its allies must consider the use of unconventional warfare techniques undertaken directly by very small numbers of allied forces, not indigenous Syrian ones. For example, Turkish special forces, working secretly with their American ally, could surreptitiously shoot down a handful of Syrian jets that are attacking Syrian cities. That alone might ground the Syrian Air Force entirely; in addition it would settle the score with Syria for shooting down a Turkish military jet (and reportedly executing the pilots after the fact).

By carrying out small-scale unconventional warfare, the United States and its allies could work toward balancing the conflict and putting a fear into the Syrian regime that it might actually lose the civil war. At the same time, this strategy would not allow for independent military capacity-building by the opposition, which might give it the sense that it could win the war. Both sides must fear defeat.

This type of unconventional warfare campaign might also bring the Russians and Iranians to the table in an attempt to salvage something of their alliance with a new national-pact regime. They too must fear that the old regime could lose entirely.

Such a balancing approach might get us to the best plausible outcome—a negotiated solution—and it would likely prevent either of the two worst outcomes. It is not pretty, but it might actually work.

Critical Thinking

1. What should U.S. policy be toward the Syrian civil conflict?

2. Does the norm of the responsibility to protect apply to the Syrian civil conflict?

3. Why is Russia supporting the Assad regime?

Create Central

www.mhhe.com/createcentral

Internet References

Syrian Observatory for Human Rights
www.syriahr.com

Middle East Online
www.middle-east-online.com/english

The Correlates of War Project
www.correlates.correlatesofwar.com

Peace Research Institute at Oslo
www.prio.no

GLENN E. ROBINSON is an associate professor at the US Naval Postgraduate School.

Prepared by: Robert Weiner, *University of Massachusetts/Boston*

Article

State of Terror

What Happened When an Al Qaeda Affiliate Ruled in Mali

JON LEE ANDERSON

Learning Outcomes

After reading this article, you will be able to:

- Explain why it is difficult to make the transition to democracy in Mali.

- Explain how it was possible for the North African affiliate of Al Qaeda to gain control of northern Mali.

- Explain U.S. policy towards Mali.

On the spine of a hogback hill overlooking Bamako, the capital of the West African nation of Mali, is a green sliver of a park, decorated with effigies of Mali's historic explorers. On a recent visit, I stopped one piercingly hot morning to admire a bronze bust of a turbaned, bearded man set on a plinth. The nameplate was missing, but, judging from the man's wide brow and Arab features, it seemed likely that this was Ibn Battuta, the great Moroccan traveller, who journeyed through the Empire of Mali and visited its capital, near the River Niger, in 1352.

When Battuta arrived, the empire was at its zenith. Mali had converted to Islam, and Battuta, a Muslim, was impressed by the sight of boys bound in chains to force them to memorize the Koran. In his diaries, Battuta marvelled at the Niger's crocodiles, so ferocious that a local man stood guard when he went to urinate on the riverbank. A few decades earlier, Mali's emperor Mansa Müsä had crossed the Sahara on a pilgrimage to Mecca, with a vast entourage carrying thousands of gold bars; in his zeal, he gave away so many of them that the value of gold in Cairo's bazaar was depressed for years. The Emperor's largesse established Mali's reputation as a place of fabled bounty—a notion that never entirely vanished.

Yet Mali, the proud home of an ancient civilization, has never quite managed to maintain itself as a sovereign state. Slavery has dominated its history, with the lighter-skinned Arab-descended peoples of the north often in positions of control over the darker-skinned Africans of the south. (Battuta was offended by the sight of slave girls going naked in the capital.) Mali has also been attacked from outside: conquered, neglected, and then conquered again. A century after Battuta's visit, the empire was invaded by the Songhay people, and not long after that the Songhay collapsed, too—their warriors, armed with spears and arrows, overwhelmed by invading Moroccans with muskets. Mali lapsed into obscurity for two hundred years; the ancient northern city of Timbuktu, once a flourishing outpost for the exchange of salt, slaves, minerals, and cultural knowledge, was reduced to a mustering point for camel caravans crossing the Sahara.

In the park, another statue depicted a European man, holding a sword and wearing a fastidiously buttoned uniform and a peaked kepi hat. There was no nameplate there, either, so I asked a group of teen-agers sitting in the shade of a nearby tree if they knew whom the statue represented. They looked blank. One of them said, "A Frenchman?"

It was a fair guess. To judge from appearances, it may have been General Louis-Léon-César Faidherbe, who oversaw France's colonial expansion into Mali. The French regime lasted nearly eighty years, and ended only with the country's independence, in 1960. It had complicated effects. The French outlawed slavery—after decades of taking slaves themselves—but they tolerated it in the north, where it still persists. And they created the outline of a state in Mali, but also conscripted tens of thousands of its men to fight in the First and Second World Wars. It seemed odd that Malians continued to pay tribute to their former conquerors. Do statues of King Leopold still stand in Kinshasa? Maybe Malians were more sanguine than other Africans about their colonial legacy. Or perhaps, after centuries of subjugation, Mali has got used to being defined by whoever comes to dictate the terms of its existence—usually by force.

Early last year, the latest wave of conquerors came. First were the Tuaregs, traditional nomads, slavers, and warriors who had been fighting for decades to win the north for themselves. Days later, they were joined by an international group of Islamist radicals. In three months of combat, they seized the entire north and declared an independent state, which they called Azawad, from a Tuareg word meaning "the land of pasture." Before long, though, the occupiers began fighting among themselves for control. The Islamists won, and, as Azawad's new rulers, they espoused fealty to Al Qaeda and decreed strict adherence to Sharia law. It was a unique victory in the history of contemporary extremism. For decades, Al Qaeda had acted as a largely rootless and amorphous agent of terror. Now its brash new affiliate had secured itself a state.

The rules were harsh. Music and television were banned, graven images were destroyed, and men and women were obliged to wear conservative dress. Transgressors were publicly punished with amputation, whipping, and stoning. Three hundred thousand people fled. For ten months, as the Islamists secured their hold, the world powers dithered, and the United States pointedly stayed on the sidelines. Many Malians believed that their best hope was France. But its new President, François Hollande, had behaved diffidently in foreign affairs—even in Africa, where France has intervened militarily dozens of times since 1960. "I'm adamant," he said in October. "We will not put boots on the ground."

Just before Christmas, the United Nations passed a resolution authorizing an African-led military force to be trained and sent to Mali—but the mission would take nine months to prepare. The Islamists were buoyed. On January 10th, a force of about eight hundred barrelled south, and swiftly overran the town of Konna, close to a strategic airbase and two days' march from Bamako. Foreigners began to flee the capital, and embassy workers sent their children home. "There was real alarm," an American diplomat in Mali told me.

France, finally spurred into action, launched a series of blistering air strikes, and began airlifting in troops. Along with soldiers from several neighboring countries, the French retook Mali's northern towns in rapid succession. Except for one helicopter pilot, who was shot down on the first day of the campaign, there were no French casualties. In a few weeks the rebels had retreated.

In the end, Mali was rescued by the people who had occupied it for eight decades. But the insurgents were still in hiding, preparing to fight again. Worse, the essential cause of Mali's problems—the racial divide, which effectively split the country—persisted, and, with it, the unresolved question of Mali's identity. Was it one country or two? For rogue groups like Al Qaeda, broken states are appealing havens—and excellent bases from which to strike at the West. But what can hold a country together when centuries of occupation and bigotry have forced it apart?

A few weeks after the insurgents were chased out, Bamako, hundreds of miles to the south, showed few signs that the country had been at war. A state of emergency had been declared, and weddings and baptisms rescheduled for safe private places. But the schools had reopened, and the roads were crammed with cars and cheap Chinese motorbikes. In Bamako, the life styles of past centuries coexist with those of the present day. Amid the telecom ads and battered Peugeot taxis were horse carts carrying firewood. Outside the National Assembly, a line of redolent fetish stalls sold monkey skulls, desiccated parrots, and pots of lion fat. One featured skins of serval, a ferocious nocturnal wildcat, which had lent its name to the French military mission: Opération Serval.

Most people seemed consumed with everyday work. On the riverbank, women in bright print dresses tilled the soil, using every available square inch. Mali's population is more than eighty per cent Muslim, and the fighting had not interrupted devotions. At one workshop, where a team was painting prayer-time boards, the owner smiled and said, "Everybody wants to pray nowadays. This war is good for business."

Still, it was clear in Bamako that the war in the north was a symptom of Mali's larger problems. When the Islamists began their attacks, it was widely interpreted as a sign that the government was weak. Soon afterward, an Army captain named Amadou Sanogo staged a coup in Bamako, deposing the President. Under pressure, he quickly stepped aside to allow the speaker of parliament to become the interim leader, but he retained the real power, controlling the ministries of defense and of the interior.

In early February, two rival groups of soldiers, on opposite sides of the coup, got into a firefight outside a garrison in Bamako. Two bystanders were killed and several children injured, just months after a previous clash that had killed fourteen people. Civilians expressed their outrage to local reporters, but that was as far as things went. With the military in effective control of the country, its soldiers could do anything they wanted.

A couple of days after the clash, a soldier with a Kalashnikov stopped my car near the Presidential residence. The walls of the houses by the roadside had been sprayed with bullets. Ibrahim, a Malian who was with me, reminded me that the President's house had been attacked twice in 2012. It had happened first during the initial coup. Then, a few weeks after the interim President, Dioncounda Traoré, was sworn in, a mob of soldiers and civilians stormed the palace, stripped him naked, and beat him badly. "The soldiers did not behave properly," Ibrahim scowled. "They also took things—televisions, refrigerators, curtains. It was very shameful."

Now the Army was moving north to assist the French in holding the newly liberated towns. For many of these towns' residents, this was little comfort. The Army, made up mostly of black southerners, is deeply distrusted by the light-skinned inhabitants of the north. Some northerners are of Moroccan extraction, and consider themselves as much Arab as Malian; others are ethnically linked to the Tuaregs, who have been essentially at war with the south since independence. In the course of five decades, the Tuaregs have launched several uprisings, which have been brutally put down. These attempts at holding the country together had, in their merciless implementation, further estranged its two halves. Now, as the soldiers moved toward Timbuktu, they were again behaving badly; reports were coming back of drunkenness, beatings, and, in some towns, summary executions.

A large, poor state straddling the River Niger and bisected by the Sahel—the great drought belt that separates the forests of southern Africa from the vast Saharan wasteland—Mali has few conspicuous advantages. Like many former colonies, it has made fitful progress as an independent nation: its recent history contains a failed attempt at revolutionary socialism, an abortive merger with Senegal, a couple of border wars, three coups, and one long military dictatorship. Its first free elections came only in 1992.

But, until recently, Mali was regarded as a modest success among African states, having avoided the large-scale civil wars that ravaged many of its neighbors. Despite its large Muslim majority, it was renowned for a spirit of moderation and tolerance, with most people practicing the comparatively relaxed, cheerful Sufi brand of Islam. This has permitted a rich culture to flourish, and, for a number of Malians I talked with, the Islamists' cultural depredations seemed as offensive as their political dominance. The director of the National Museum, Samuel Sidibé, showed me a number of old wooden fetish gods that fleeing northerners had brought in, hoping that he would preserve them. He was still enraged. "The Islamists' ideology cannot be spread without destroying culture," he said. The museum, three large buildings situated in a splendid expanse of park, is a reminder that Mali is a place of long and diverse artistic traditions. Its rooms hold intricately woven and elaborately decorated textiles, gold jewelry and ornaments, and stylized sculptures of iron, wood, and terra cotta, representing a spectrum of human activity from farming to war and maternity.

The cultural scene is still thriving. The portrait photographer Malick Sidibé, who chronicled Bamako life in the nineteen-sixties, recently became the first photographer to win a lifetime-achievement award at the Venice Biennale. Malians are also fixtures on the world-music scene, with Ali Farka Touré, Salif Keita, and Amadou and Mariam especially renowned. Every year since 2001, Mali has hosted the Festival in the Desert, a three-day musical extravaganza at which foreign bands come to play alongside their Malian counterparts. Last year, the special guest was Bono.

In Mali, music is the maximum expression of public emotion, and musicians are essential for every kind of celebration. The jihadis' ban on music was especially devastating. Many musicians had deserted the north, including Khaira Arby, one of the country's best-known singers, whom I met one afternoon in the house she had rented in Bamako. A regal woman in her fifties, wearing a bold red-and-beige print dress, Arby was seated in a heavily upholstered living room, where she was rehearsing with several young guitarists. The aroma of burning frankincense filled the air. Children sat obliviously in front of a TV on the carpeted floor, watching a French version of "The X-Factor."

Arby told me that she was in Bamako when the jihadis took Timbuktu, and she had stayed there ever since. Musicians she knew in the north told her that the jihadis had destroyed their instruments and sent her a message. "They threatened to cut off my tongue," she said sourly. She continued to record new music, but she felt that she was in limbo, and wanted nothing more than to return home. So far, Timbuktu was too insecure for her to go back.

Not long after I arrived in Mali, I went to see a storyteller named Muhammad Djinni. In his seventies, with a worn, lined face, Djinni sat on a low stool in the alleyway of his house, wearing a dirty saffron smock. He was a butcher, like his father. But when he was young he had hung around old soldiers and his town's most renowned historian, and in this manner he had learned a great deal. People came to him to learn their tribal history, especially orphans, who were always eager to know their real origins.

"What history do you want to hear?" he asked me.

I asked him to tell me about the French.

"The dominations we have had, we didn't call for any of them," Djinni said. "They just came. When the French came, we were surprised. We didn't know why they had come. They would take your kids, and you wouldn't know where they were, sometimes for years. Later, you would see your son and know that he was a soldier."

Djinni was born in 1938, in the house he lived in now. "I grew up under the French—I saw many Frenchmen," he said. They had forced him to go to school when he was nine, he said, but he had run away after two months; his father wanted him to learn the family trade. Life under the French had been simple, Djinni recalled. "It was easy because we did everything they said. The French named chiefs for each area of Timbuktu, and when they wanted to do something they called them and told them what to do."

I asked what it was it like seeing the French return after fifty-three years.

"It was like a dream," Djinni said. "They came to free us, and removed the noose from our necks."

Many Malians seemed thrilled to have been rescued. On February 2nd, five days after French troops reached Timbuktu, Hollande was flown in for a brief, joyous reception in the dusty central square. In front of a cheering crowd, he cautioned that the Islamists still posed a threat, but declared, "We have liberated this town." For a moment, Hollande seemed to approach the grandeur of de Gaulle—a rarity for a man known during his electoral campaign as Mr. Normal. The intervention was a *coup de foudre* for Hollande, whose approval ratings had slid well below the halfway mark; sixty-three per cent of his citizens endorsed the invasion. In Mali, French troops were met by crowds waving the tricolor and chanting "*Merci, France.*" Malians began referring to the French President as Papa Hollande, and during his quick visit they gave him a camel as a token of gratitude. Hollande declared his appearance there to be "the most important day of my political life." The camel was left in Timbuktu, where its host family innocently made it into tagine.

I made my way to Timbuktu with a force of a hundred and fifty French marines, who drove from Bamako in a three-day convoy in order to relieve the troops there. The marines moved in a light column of armored personnel carriers and open jeeps, without the heavy anti-I.E.D. armor that U.S. Army vehicles have. They were watchful with Malian civilians—this was unfamiliar territory for most of them—but there was a great deal of joking around, too. The French in Mali, unlike the Americans in Iraq and Afghanistan, enjoyed the advantage of a common language, and some shared history, with the citizens. Along the way, soldiers broke open their ration boxes, some of them complete with canned Camembert, to distribute to children.

There was danger out there, even so. After the French strikes, Mali's Islamists fled to remote mountain hideouts to regroup. In the town of Gao, they had begun to strike back, sending suicide bombers on motorbikes. Outside Timbuktu, a sizable Islamist convoy had been spotted, and jihadis were suspected of hiding in the surrounding villages. A few more French soldiers had been killed as the campaign wore on, and the men in the convoy were taking no chances; at night our camps were like wagon trains, the vehicles pulled up in concentric circles, with sentinels on the perimeter.

However marginal and remote Mali may seem to Westerners, its travails exemplify the security problems posed by neglected places in the age of Islamist terror. With little effort, criminal and terrorist groups can seize countries, or parts of them—just as narcotraffickers have effectively taken over West African states like Guinea Bissau. Since the U.S. invasion of Iraq, the Islamist cause has grown more popular, and tensions have risen throughout the Maghreb, the northwestern region of North Africa. In 2006, a franchise of Al Qaeda formed in the area, calling itself Al Qaeda in the Islamic Maghreb (AQIM). It originated in the nineteen-nineties, in Algeria, where a decade-long Islamist insurgency was crushed at the cost of tens of thousands of lives. That confrontation left behind bitter memories and a core group of survivors, who gained further expertise in Iraq and Afghanistan and new adherents throughout the region.

The Arab Spring provided the jihadis with an unusual opportunity. Before 2011, the Mediterranean coast of Africa—the great borderland facing Europe from the south—was guarded by a series of secular states, where Islamists were ruthlessly pursued, jailed, and killed. Now only two of those states, Algeria and Morocco, remain, leaving a new, more chaotic political environment. In Libya last September, Islamist extremists mounted a murderous attack on the U.S. consulate in Benghazi. In April, Al Qaeda carried out a car-bombing attack against the French Embassy in Tripoli.

Mali, like the other weak, underpopulated states of the Sahel, is ideally situated for outlaws. Its northern territory is a hub for jihadis and smugglers for the same reason that Timbuktu was an ancient trading center: it offers easy access to Europe, and to the rest of West Africa and North Africa. From here, guns and insurgents and immigrants can be trafficked along the old caravan routes that once carried slaves, salt, and ivory. Even before the jihadis came, Azawad was a sanctuary for kidnappers, a place where government emissaries went to negotiate ransoms and get their people back.

But the West has largely ignored the region. On the French convoy, the officers grilled me politely about why the Americans had not got openly involved in Mali. Initially, their government had requested help in refuelling aircraft, and the U.S. had responded by asking France to repay its costs, of about seventeen million dollars. Although the request was later rescinded, it left the distinct impression that the Obama Administration was either extremely cautious or simply uninterested.

France's European allies had also done very little, and the young commander of the convoy, Captain Aurélien, said that the lack of solidarity had been disappointing. "This is a French area, because of history, but we don't have the resources to do it alone. We need a collective effort to defeat terrorism in this area." With Europe's economies as bad as they were, he doubted that help would come. "If the Islamists gain a foothold here, it will not be good," he said. "This is the doorstep to all of Europe, not only France."

These days, no bridge connects the lower part of Mali to Timbuktu, which sits on a desert plain about eight miles from the northern bank of the Niger. Travellers from the south must cross on a flat barge propelled by motor-powered canoes lashed alongside. Near the landing, on a bare riverbank, is a cluster of stalls made out of sticks, and when I arrived it was market day. Several Tuaregs were there, dressed in indigo robes and turbans, one wearing a sword in a leather scabbard and gold-rimmed sunglasses that concealed his eyes. Women carried dried hides on their heads. Small families, led by men with beards and wooden staffs, walked out of the desert to wait for the barge to come.

Timbuktu is a small, unlovely city in shades of brown and gray, a warren of low, flat-roofed homes made of mud or concrete. Interspersed are beehive-shaped tents covered with hides and scrap—the hovels of the nomadic Bella, former slaves who remain in serflike conditions, working as goatherds and as servants for their former owners. Other than one paved street, the roads are dirt. At the outer edges, the city peters out amid sand dunes and piles of uncollected refuse. In Timbuktu, as in many parts of Africa, plastic rubbish is so prevalent as to seem part of a new ecology.

Two weeks after its liberation, Timbuktu was subdued and wary. Many of the city's fifty-five thousand inhabitants had fled after the Islamist takeover, and more when the French invasion began. Perhaps a third of its prewar population remained. Most of the shop fronts in the downtown marketplace were shuttered, and homes stood empty. Like vandals, the Islamists had scrawled graffiti and painted over road signs that showed human and animal images, leaving messy swatches of brown paint. Hair salons, which in Mali advertise their services with portraits of men and women, had also been targeted. There was little electricity, and not much water; the Islamists had also destroyed the city's telecommunications system before they left.

There was no visible authority. The French military was based a couple of miles outside of town, at a small airport, and, except for several daily patrols in jeeps and armored cars, it was an unseen presence. Across the street from the grand mosque, a group of Malian soldiers lounged amid crumbling buildings in a colonial-era barracks. The town hall had been vandalized, its computers stolen or wrecked, its archives trashed. The only garbage truck had been shot up, and sat in front of the mayor's office like a broken yellow toy. The mayor, like many of the civic authorities, had run away.

The Malian Arabs made up much of Timbuktu's merchant class, and many of them had lived in Abaradjou, a neighborhood of well-built stone houses. Most of the homes had been ransacked; their doors and windows were gone, and their contents plundered. The shops had also been raided, the metal security doors yawning open, the interiors gutted. Virtually all the Malian Arabs, like the light-skinned Tuaregs, had fled when the French invasion began, making their way to refugee camps on the border with Mauritania or to villages in the desert. (The city's few Christians were already gone, having left to escape the Islamists.) Timbuktu was no longer a city of several races; except for a few stragglers, there were only black Malians left.

Hollande had declared that France's troops would begin withdrawing by April, leaving the area to an all-African force. But even though the city had been formally liberated, few people felt secure enough to return. Many feared that the Islamists were lying low until the French left. For the Arabs and Tuaregs, the threat was more immediate; they believed that their safety was at risk from the Malian Army.

There was only one hotel open, the Colombe, a dusty two-story structure, made of dun-colored brick, where the electricity and water ran for only two hours a day. In the evenings, the few paying guests shared the bar uneasily with Malian military men, armed and in uniform, who sat around drinking beer until they were in a sullen stupor.

My translator in Timbuktu was Idrissa, an unemployed tour guide in his twenties. He had learned English from a Peace Corps volunteer, before the organization pulled out of Mali, in April, 2012. Idrissa had earned a living taking intrepid Western travellers on expeditions into the desert, but he hadn't had any customers in a year. He wore a turban and an electric-green robe and carried a cell phone with a startlingly loud call-to-prayer ringtone.

In the ten months of Islamist rule, there had been an execution by firing squad, an amputation, and several whippings in Timbuktu. One impoverished couple had been whipped because they were not married, though they had a child and the woman was pregnant with another. A young woman was accused of fornication and given ninety-five lashes. In the central square, Idrissa had witnessed the beating of one of the jihadis' own men, who had been accused by his comrades of raping a young girl. The spectators loudly criticized the jihadis for a double standard. "Everyone was angry because they didn't kill him," Idrissa said. Afterward, the jihadis had gone on the local radio station and warned that anyone who spoke badly about their men would be killed.

Near the town center was an unfinished hotel complex, built in single-level faux-Islamic style, which Idrissa explained had been financed by Muammar Qaddafi. We climbed a sand dune behind the hotel and looked down into a natural depression that had been filled with water, forming a large pond. Idrissa said that until recently it had been dry, and that it was where the jihadis had executed a man who was accused of murdering someone in a nearby village. Idrissa, along with many other people, had been summoned by the Islamists to gather on the dunes above the killing ground; jihadis with guns kept them in place. As the victim's family stood by, they carried out the sentence. At the crucial moment, Idrissa had turned away, but he

had heard the gunshot, and then the guards shouting, "*Allahu Akbar!*"—"God is great."

On the day I visited, there were a dozen women and girls washing clothes at the edge of the pond. Some were standing waist deep in the water to clean themselves as they washed. As we walked around, little children danced and chanted, "Mali, França, Mali, França." It was a new kind of game for them. Their parents smiled and waved.

Timbuktu, at the edge of the vast Sahara, has served for centuries as a repository of culture. In 1510, the Moroccan geographer and scholar Leo Africanus visited, and described a city of veiled women and thatched mud dwellings, with a large mosque and an adjoining palace, and a thriving market in which European fabrics were sold and great amounts of gold traded. "The people of Timbuktu are of a peaceful nature," he wrote. "They have a custom of almost continuously walking about the city in the evening (except for those that sell gold), between 10 P.M. and 1 A.M., playing musical instruments and dancing." Timbuktu was ruled by an immensely rich warrior king, whom Africanus described as merciless yet cultivated: "There are in Timbuktu numerous judges, teachers, and priests, all properly appointed by the king. He greatly honors learning. Many handwritten books imported from Barbary are also sold. There is more profit made from this commerce than from all other merchandise."

That legacy persists. In the center of town is the Ahmed Baba Institute of Higher Learning and Islamic Research, a modern archive library—all clean lines and sand-colored planes, to suit the surroundings—where tens of thousands of the city's ancient documents were preserved. During the jihadis' invasion, all the researchers fled, and when Idrissa and I arrived the building was still abandoned. A custodian let us inside, in exchange for a tip, and allowed us to wander with him through floors of mostly empty rooms. He showed us in to a locked storeroom filled with ancient manuscripts: shelves of little yellowing bundles. The custodian explained that the room was intact because the jihadis had not known about it. Most of the other manuscripts had, like the city's several private collections, been hidden away by a network of librarians.

The jihadis had destroyed a few hundred documents that they found in the preservation room, where fragile pages were microfilmed and then preserved in specially made folders or boxes. The custodian led us out to a garden patio, where the charred remains of the documents lay. I was surprised to see that they had not been cleaned up; in the ash piles, there were scraps of ancient paper with identifiable calligraphy. The custodian said that no one from the center had returned, and so he was leaving everything as it was.

As well as a literary center, Timbuktu was known as the home of shrines to three hundred and thirty-three Sufi saints. The Islamists, who regarded the shrines as idolatrous, demolished as many as they could. In the Old City, a labyrinth of dirt alleys, was the grand mosque, built in 1327 by Mansa Müsä, the "gold" emperor. Idrissa pointed to piles of dirt outside the rear wall, and explained that they were the remains of the first two shrines that the jihadis attacked. He had been there that day. "They blocked the street both ways, held people back with Kalashnikovs, and destroyed them."

Before that, they had gone to the imam in the mosque, to inform him of their intentions. He had told them, "Go ahead. You are going to do what you are going to do." The imam had maintained this neutral policy throughout the jihadi occupation, and told his flock to go along with whatever they ordered, so as to avoid violence. Everything was God's will, he said.

At another mosque, the Sidi Yahya, a piece of sheet metal was propped up over an entrance to the back courtyard. There had been a door there, a massive wooden slab covered with geometric metal decorations, which had traditionally been kept sealed, in the belief that it led to a tomb of saints. Idrissa told me that the door was hundreds of years old, and that, according to an old superstition, if it was ever broken, the world would end. During the occupation, a squad of Islamists arrived with axes and destroyed the door. Afterward, they jeered at onlookers, "Has the world ended?"

In many respects, Azawad was regulated with almost parodic strictness; at one point, a spokesman decreed that tombs more than six inches tall would not be tolerated. But, compared with the harshest applications of Sharia law, the rule of the jihadis was not so stringent. At times, they seemed to be as concerned with establishing a government as with enforcing the holy writ.

On a nondescript street near the center of town is a small boutique hotel, La Maison, owned by a Frenchwoman. It is a simple two-story stone building, its rooms hung with Moroccan lamps and furnished with cushions and rugs. The jihadists evidently approved of La Maison's style, because they took it over during their stay and turned it into their Sharia court. I went there with Cicce, a local man who had been arrested by the jihadis and held for three days. When we arrived, a caretaker wordlessly brought out great old-fashioned keys and allowed us into a dusty downstairs room, with furniture stacked up, where some of the jihadis had lived. The upstairs had functioned as the court's chambers. Suspects were held under guard in the hotel's restaurant; the trials took place next door, in a spacious room with two rugs on the floor and a chest of drawers that was used to store evidence. A short coil of rope, used for tying suspects' hands, had been left behind.

Cicce had been arrested at a checkpoint, when jihadist fighters found a gris-gris—a traditional good-luck charm—in his glove compartment. "I thought my life was over from the moment they brought me in," he said. He showed me where he had knelt on one rug in front of three judges, who sat crosslegged on the other. He had explained that an elderly man to whom he had given a ride presented the gris-gris to him in gratitude. He had thrown it in his glove compartment and forgotten about it. The judges asked him again and again about the incident, but also discussed religion with him, as if to test his fealty. The trial had gone on for many hours, and he had said what he could and hoped they believed him. But they had treated him courteously, he said. When it was lunchtime, they invited him to share their food. Eventually, he had been released, and considered himself very lucky.

The Malians' reasons for complying with the Islamists were complex. After centuries of subservience, they understood when they had to accede in order to survive. The north of Mali is a neglected place, with no highway and few signs of development; what money flowed into the area in recent years came not from the government but from Qaddafi and from international charitable agencies. Kidnappers and other criminals operate with impunity in the surrounding desert. According to at least a few locals I talked to, the Islamists were tolerated because they brought order to Timbuktu. "The Islamists fought corruption, and the privilege of one citizen over another," one told me. "Whether the person is white or black, they are lashed with the same whip."

The jihadis had also made some effort to cultivate the city's residents. Although they silenced most journalists, they allowed a middle-aged radio reporter named Yehia Tandina to continue working. When I visited Tandina, at his house, he explained why. "The Islamists wanted to make propaganda through me," he said, matter-of-factly. "At the beginning, I had trouble with their fighters, but then they gave me an authorization." He showed me the document, on Al Qaeda letterhead: stamped by the "Sécurité Islamique," with a logo of a crossed Kalashnikov and sword, it was a tangible sign of the Islamists' attempt to acquire the trappings of official power.

We were sitting in his family room, which had cushions around the edges and a small desk with a Dell laptop. Outside, in an alley courtyard, women ate rice from a communal bowl. Tandina wore Western clothes: tan trousers and a blue longsleeved shirt. He explained that the first rebels who had arrived in Timbuktu were the Tuareg separatists, and there were no problems. "Then, on day two, the Islamists came," he recalled. He had asked the leader what he wanted. Naming the northern towns of Mali, he had said, "Timbuktu, Gao, and Kidal are Muslim towns, and we want to make Sharia in them. We are not asking. We are saying what we are doing, and we're here to make Sharia."

Tandina told me that afterward the people of Timbuktu, who were mostly moderate Muslims, had coped as well as they could. "We created a crisis committee, and if the Islamists wanted something to be done they went through the council," he said. "The council didn't want violence, and so it advised the people to do whatever the Islamists said. For example, the Islamists once wanted to collect money from each family for the electricity, and the crisis council went and politely collected the money from the people. There were also times when the council members disagreed. They wanted the boys and girls to have separate schools, and when we opposed them they said, 'O.K., then, no school.' So the schools closed after that."

One of the few schools to stay open was a religious academy, led by a marabout, or Koranic teacher, named Baba Moulay al Arby. A thin man of thirty-four, Arby received me graciously in his home, which doubled as his school, and showed me the downstairs room where he held classes. Dozens of wooden prayer tablets were stacked against the wall. In a marabout school, groups of children learn the

Koran by rote from the age of six to about fifteen. "Those who want to go on to learn other topics, like history, math, or geography, go on to a madrasah," Arby explained.

Arby, one of the few Arabs remaining in the city, had a mustache and wore a brown djellabah. His family had been in Timbuktu since the time of his great-great-great-grandfather, he emphasized; he was a descendant of the Prophet Muhammad. "I know Islam better than these Islamists, but I was afraid of them and disliked them, because they are violent," he said. "What they do in the name of Islam is just to trick the people." He said that he had stayed on "because I have the support of the people of Timbuktu. I teach their children."

His school had fared little better under the liberators. He normally had two hundred pupils, he said, but, since the French arrived and the town's Arabs had fled, his classes had been reduced to forty. "Most of the Arabs were afraid of the French and the Malian Army," he said. "They knew what they would do." The Arabs feared that their houses and businesses would be looted, he said; a new home he was building had been broken into and plundered.

Outside the city, an A.P. reporter named Rukmini Callimachi had found shallow graves containing the bodies of two Malian Arab men, their hands bound behind their backs. Both had been shot in the head. The day before I arrived, an Army unit had grabbed eight other Arab men off the streets, including an older man named Ali, who had given interviews to journalists after the French arrived. The men had not been heard of since, and most people believed that they had been murdered. When I mentioned the killings and disappearances to Arby, he responded carefully. "I don't think those killed were innocents," he said, and added, "Ali was not an Islamist, but his children were. The military have him, and I don't know what will happen to him. This kind of thing makes me afraid."

He was in touch with some of the Arabs who had fled into the surrounding desert, and many of them were hungry and increasingly desperate. "I have talked to the military and arranged to send them food," he said. "They were not with the Islamists, but they are afraid, and ashamed of what the Arabs did." He had sent his mother and sisters to Niger for safety, he said, but he believed that Timbuktu's "innocent" Arabs would eventually be able to return. The Arabs had owned most of the city's businesses and had provided most of the employment, and they were needed. "If the imams and the notables come back and give their excuses, the black people will excuse them," Arby said.

Two weeks after Hollande visited Timbuktu, Colonel Paul Gèze, the commander of the French invasion force, was getting ready to leave. In the dirt square where Hollande was greeted as a liberator—and where the Islamists carried out most of their whippings—the city's residents organized a celebration for him. Gèze, a burly man with a gray buzz cut, sat with a few fellow-officers and local dignitaries in plastic chairs that had been arranged for the V.I.P.s. A bunting in the red, yellow, and green of the Malian flag hung around the edges of the plaza, alongside a banner with the French tricolor that

read "Merci à La France et Les Pays Amis." Crowds of people had gathered noisily all around the square, where gendarmes and men with sticks sought to keep them in place. There were drummers, and women singing and laughing. A group of male dancers wearing cows' horns on their heads represented the butchers of Timbuktu, and one man with a black turban wore a hot-pink robe that was festooned, pirate style, with a great sword and crossed belts.

As the sun set, a kind of order prevailed, and the speeches began. A young Malian man performed a hip-hop verse he had composed in honor of the French. Commander Gèze smiled graciously throughout, and then he spoke, praising the people of Timbuktu and promising undying French support. As dancers came before the V.I.P.s to perform, my eyes were drawn to a man standing nearby. He was extremely tall and broad, with skin the color of old mahogany, and he wore a tunic made out of what looked like a grain sack. On the front were some words I couldn't make out, and on the back was an image of a man sexually penetrating a woman from behind.

While the dance troupe performed, the man began a performance of his own. He pulled out a large black dildo, strapped it around his waist, and began grinding lasciviously, wearing a lewd expression. Eventually, he stopped dancing and stood, slowly stroking his dildo. People in the crowd stared at him with fixed, unbelieving looks; others smiled, or put hands over their mouths to suppress laughs. It was plain that everyone was familiar with his performance, and that many people were delighted. A pair of officials flanked him and tried to get him to stop. He pushed them away and, with a mischievous look, continued dancing.

It seemed extraordinary that Malian society could accommodate such opposing impulses. For most of the past year, jihadis had meted out lashings in the town square, where a man now danced around pretending to masturbate—and no one seemed overly perturbed by it. Idrissa explained that the man was a griot, one of the hereditary West African figures who carry on the oral traditions of their communities, through storytelling, history, poetry, or music. This particular one, according to Idrissa, was a bit of a rascal. When I asked him to translate the words on the griot's tunic, he refused: "They are very bad words. You have seen the picture, so you can imagine."

A few days later, we found the griot in his home, an improvised mud dwelling in the dunes at the edge of town. His wife was peeling vegetables on a patio, accompanied by several young children. The griot, wearing trousers and a shirt, invited me to sit with him under a veranda of woven grass, where songbirds flitted in and out. I asked if they were his birds. He gave a surprised laugh and said no, they just lived there.

His name was Boubakar Traore, but he preferred to be called Chief Firga, after the name of his village. He had been born a *banya,* or slave, of the Songhay people, and had taken on the role of a highly sexed buffo as an escape. Around the time of his birth, forty-seven years ago, slavery had been formally outlawed in Mali, but in the north the tradition of hereditary servitude persisted. The slave trade in Mali had a tangled legacy. "At the beginning of all this history, there were the poor black people and then the rich, intelligent people," he said. "They would take you by force and make you work for them. If they lost their wealth, they would take you to the market and sell you. Later, some of the slaves got some money and learning, or became Muslims, and were able to buy their freedom."

When Boubakar was a young teen-ager, he came to understand that his parents were slaves, and that he was, too. He went to his owners, a local family called Heydara, and asked if he was their slave. Embarrassed, they told him that slavery was a thing of the past, and that they thought of him as a son. But he knew the truth. Later, he told his parents that by law they were free, but they were afraid and couldn't change. "They were slaves until the end of their lives," he said. "My parents' generation had no rights. My generation is the first to demand their rights."

Boubakar's father told him that he shouldn't live with bitterness, and encouraged him to become a griot, as his grandfather had been, and his grandfather's grandfather. Boubakar learned that if he donned his costume and went to the homes of his former masters they would pay him to go away. He had made a career of haunting them, appearing at their special occasions and brandishing the dildo. If necessary, he added, he blackmailed them: "I can tell stories about my former masters, and since I am their former slave, people will believe me. This gives me power over them." He had been successful, he felt. "They paid for this house," he said, smiling, and waved at his hovel. He said that he did it because he had resolved not to go through life concealing the fact that he had been a slave. Although he conceded that he was not a truly free man, because he lived off his former owners, he had found a way to make use of his heritage.

While the Islamists were in Timbuktu, Boubakar concealed his costume. The day I had seen him in the town square was his first day out in ten months. He had feared the Islamists, he said. "They cut off people's hands and told our women that their clothes were not suitable. I think they wanted slavery to come back."

Boubakar had ten children, and he was training several of them to be griots, including a couple of his daughters. When I expressed surprise at the idea of female griots, Boubakar explained how it worked. Former slave women wearing nothing underneath their dresses showed up at the social occasions of their former owners. "If they don't pay them, they throw up their dresses and try to sit on them." He laughed heartily.

Boubakar had no national history he could be proud of. Slavery was illegal now, but its social structures remained in place, and the state had provided no means of overcoming them. Being a griot, on the other hand, was something he could proudly pass on to his children; during his performances, he could make his former masters submit to him. But the larger question of belonging to Mali had no easy answer. Even Boubakar's former owners considered themselves Arabs, and did not feel truly Malian. Where did that leave him?

As I prepared to leave Timbuktu, President Obama announced that he had sent a hundred U.S. troops to neighboring Niger, to work at a new base for surveillance drones that would help the French track Al Qaeda. A week later, Chadian troops fighting with the French announced that two AQIM chiefs had been killed. But the Malian Army seemed far from ready to take over. In the main garrison, I found Colonel Kéba Sangaré, the newly arrived military commander for the city. A courteous man in his early forties, Sangaré insisted that people felt safer since his forces had arrived, and pointed out that the schools had reopened. "People now walk around freely, and Western journalists are able to go where they want. We're slowly moving into a new phase, in which we are depending on the support of the local population to defend against the insurgents."

In fact, he and his men relied far more on the French, which Sangaré acknowledged cheerfully. "We are not capable of taking care of all of Mali," he said. "We have a big land and a small population and Army, and we don't have the technology to see everywhere." His own area of command, he pointed out, was two hundred and twenty thousand square miles.

Sangaré's boss for Opération Serval was Captain Aurélien, my host on the convoy to Timbuktu. One morning, Aurélien came to the Hotel Colombe to speak with a group of reporters. Security had been reëstablished, but only wherever the French military was at any given time. He explained that the French would remain at the airport and had begun patrols around the city. He warned journalists against straying too far. "There may still be terrorists among the Arabs in the villages outside the city. I ask for your help in this matter—please don't go there."

When I mentioned that Hollande had said that his troops would hand off to the Malians by the end of April, Aurélien was noncommittal. "I am not sure about the dates," he said. "I think it's a little more open-ended." A few feet away, in the restaurant, a uniformed Malian soldier sat drinking beer in a determined way. It was eleven in the morning.

In Bamako, I went to discuss Mali's prospects with a well-regarded political activist named Tiébilé Dramé. A tall, affable man in his fifties, Dramé told me apologetically that he had not gone to the north since the French had ousted the Islamists. He had helped to organize fact-finding teams, but a recent trip he hoped to go on was cancelled for security reasons. When I told him about the disappearances and murders, he looked gloomy: "The country is in a real crisis, and too few people seem to be aware of the extent of it."

He thought that the Western powers had allowed the Islamist occupation to persist. "My conclusion until recently," he said, "was that the U.S. government did not believe that the situation in Mali represented a threat to international security—that's the bottom line, I believe." A few months earlier, he said, a senior U.S. diplomat in the region had bluntly asked whether AQIM represented a threat to the United States. "Can you imagine?" Dramé said, with a disbelieving look.

Across town, the U.S. Ambassador, Mary Beth Leonard, offered a robust defense of America's role in Mali. She explained that the government had given logistical support to the French mission, spent a hundred and twenty million dollars on humanitarian aid, and pledged an additional hundred million to a U.N.-backed stabilization program. (There were also reports of small numbers of Special Forces at work in Mali.) "But we cannot deal with the Malian military until after it holds elections and democracy is restored," Leonard added. "We are very firm on what we think Mali's priorities are. Governance is the main issue. Reconciliation is also near the top of the list." She believed that the Islamists could be defeated only if all the groups in the north, including the Tuaregs—who had broken their ties with the Islamists during the French invasion—were brought to the table. There was still a war to be fought, though, Leonard said: "AQIM did not prevail on the battlefield, so it will shift to asymmetrical tactics. We're all waiting to see what those tactics are going to be."

Soon after our meeting, AQIM said that it had executed one of the six French hostages its forces were holding in North Africa. A week later, in Timbuktu, a suicide bomber blew himself up outside the airport, killing a Malian soldier and wounding several others. Ten days after that, another bomber and two squads of gunmen attacked simultaneously at points around the city. One group sneaked into the back of the Hotel Colombe and opened fire on the rooms where several Westerners and Malian officials were lodged. It took forty-eight hours, and several more firefights, before the town was quiet again. On a visit to Mali in April, Laurent Fabius, the French Foreign Minister, announced that France intended to leave a permanent force of a thousand troops, in order to "fight terrorism."

In the end, the United States' hands-off strategy seemed to have had benefits. The U.S. had paid relatively little; the French did most of the work; and, for the moment, Al Qaeda was on the defensive. But the jihadis' ten months of statehood may have long consequences. The short existence of the Islamic state of Azawad showed extremists that the West is more fractious, and perhaps weaker, than before, and less eager to take on armed fighters in faraway lands. It seemed likely that those jihadis who survived might want to repeat their experiment, perhaps with greater sophistication. In the meantime, the extremists hiding out across the region have shown their ability to strike, as they did in Niger, where suicide bombers killed twenty-six people in May.

If Mali is to stop being an appealing haven for terrorists, it must somehow pull together as a cohesive nation. But a considerable challenge remains. Mali has few leaders of note, and no civic culture adequate to the task of demanding more of its politicians—not to mention better behavior from its Army. When I met Tiébilé Dramé, he told me that the country badly needed "a national discussion about how we can lay a new foundation for democracy. Equally important are the internal discussions we Malians need to have about how we are going to maintain peace and hold the country together." At the moment, he said, Mali was ruled by "a cohabitation between the coup and the constitution." There was a transitional President, but the rogue elements of the military still exercised enormous influence. "The truth is," he said, "no one really seems to be in charge of anything."

In the past few weeks, though, there have been some promising signs. New elections are scheduled for July 28th, and it is possible that they will help allay the problem of governance. On April 25th, the U.N. decided to deploy peacekeeping troops. In mid-June, Dramé announced that he had completed weeks of negotiations, in which the insurgent Tuaregs of the north had agreed to stand down and, for the moment, rejoin the Malian nation.

There are many countries with unresolved sectarian and ethnic differences, and some of them manage to move forward, however fitfully. With help, it should be possible for Mali, despite all its problems, to limp along. But a comprehensive political solution seems far off. Until one can be found, Mali's best hope may be its cultural verve and its spirit of tolerance, which, at least on the scale of a neighborhood or a town, can reconcile its people as politics cannot.

When I was in Bamako, public festivities were still on hold, but one Friday night the dimly lit Djandjo Club was open for business. The headliner was the Malian guitarist and singer Baba Salah, a thin man in his thirties, who wore a dark suit and a white shirt with a collar worthy of Billy Eckstine. As his band members set up on a nearby stage backed with mirrored glass, we sat and talked.

He was from Gao, he said, and though he and his wife and children were lucky enough to live in Bamako, many of his friends and relatives had endured life under the Islamists. "After a year of oppression, the people there had lost hope," he said. "It had become clear that no one had the ability to do anything about the situation. So it was an incredible relief when the French came." He was waiting a little while longer before he went back for a visit. "The cities are liberated, but the roads are not yet safe," he said. He was frightened when the jihadis charged south from Timbuktu. He was known to have spoken out against them, and he was concerned that they would make him stop playing. The melancholy title track of his latest album was "Dangay," which means "north." In the song, he asks listeners to pray for the people of the region.

The club began filling up with exuberantly dressed men and women, who waved and smiled at Baba Salah. Most of them drank Coca-Cola; a few drank beer. There seemed to be no stigma to adhering more or less zealously to the faith.

Baba Salah's guitar was a black-and-white Schechter, and before he got up to play he told me how he acquired it. In 2000, he was on his first trip outside Mali, touring with the great Oumou Sangaré, and in New York the songwriter Jackson Browne came to watch. "Afterward he called me over and said, 'You play like an angel. What can I do for you?'" Baba Salah told me. "I didn't know what to say. He said, 'It's O.K., I know what to do.' So later he flew from California to here. I was living like a bachelor in a really small room. We played together. He said to me, 'This is the guitar I play, and I want you to have it.'" Baba Salah had played Browne's guitar ever since.

Onstage, he picked up the guitar and said a few words about how happy he was that the north was free, which drew applause and joyful shouts. Then he and his band began playing. Their music, driven by his guitar, was a looping, riffing, electric sound that evoked the sixties underground yet was distinctly African. Baba Salah's singing was casual, comfortable, and as he launched into the first song people filled the floor to dance.

Critical Thinking

1. Why does Mali lack a national identity?
2. What are some of the difficulties associated with making the transition to democracy in Mali?
3. What is Al Qaeda in the Islamic Maghreb?

Create Central

www.mhhe.com/createcentral

Internet References

Office of the Coordinator for Counterterrorism
www.state.gov/s/ct

Africa Focus
www.AfricaFocus.org/

Mali Embassy, United States
www.maliembassy.us

Article

Prepared by: Robert Weiner, *University of Massachusetts/Boston*

No Chemical Weapons Use by Anyone: An Interview With OPCW Director-General Ahmet Üzümcü

DANIEL HORNER

Learning Outcomes

After reading this article, you will be able to:

- Explain the work of the Organization for the Prohibition of Chemical Weapons.

- Explain what the Chemical Weapons Convention is designed to do.

Ahmet Üzümcü took office as director-general of the Organisation for the Prohibition of Chemical Weapons (OPCW) on July 25, 2010. Immediately prior to that appointment, he served as the permanent representative of Turkey to the UN Office at Geneva. His previous career included two postings at NATO headquarters in Brussels.

Üzümcü spoke with *Arms Control Today* by telephone on December 19 from his office in The Hague. A large part of the interview dealt with concerns over Syria's reportedly large arsenal of chemical weapons, the prospect that those weapons would be used, and the OPCW's responsibilities, capabilities, and constraints with regard to that situation. The interview also covered issues that are likely to receive considerable attention at the upcoming review conference for the Chemical Weapons Convention (CWC), scheduled for April 8-19.

The interview was transcribed by Marcus Taylor. It has been edited for length and clarity. The text of the full interview is available at http://www.armscontrol.org/interviews.

ACT: *The CWC has now been in force for 15 years. In just a few words, could you summarize the ways in which you think the CWC regime has succeeded and the ways in which the potential of the treaty has not yet been realized?*

Üzümcü: The implementation of the Chemical Weapons Convention over the past 15 years has been successful, especially in the field of demilitarization. The level of destruction

of declared chemical weapons stockpiles has reached the level of 78 percent under the verification of the Technical Secretariat of the OPCW. I think this is a significant achievement, which needs to be acknowledged. It has required the allocation of a lot of resources by possessor states-parties, as well as by the organization itself.

Nevertheless, the deadline—the final extended deadline of April 29, 2012—was not met. But a decision by the conference of states-parties in November 2011 enabled the possessor states to continue the destruction activities with greater transparency and reporting.[1] So I think this decision was somehow a manifestation of the culture of cooperation and dialogue that has been developed over the past 15 years.

The decision was nearly by consensus, with one exception. I think the fact that the organization was able to take its decisions by consensus over the past 15 years with a few exceptions has been a clear demonstration of the evolving global cooperation on an important security issue, the destruction of chemical weapons, as well as the prevention of re-emergence [of chemical weapons]. This also shows to a great extent the strong political will that exists on the part of the states-parties to get rid of those chemical weapons for good and to collectively prevent their re-emergence through nonproliferation activities.

That in and of itself, I believe, is a big achievement. There are other areas in which we should do more, such as Article VI inspections, verification of the chemical industry, improvements in our on-site inspections and monitoring capabilities. I think the verification mechanism can be improved by selecting the most relevant sites to be inspected and making the inspections more consistent.

There are still discrepancies on import and export data provided by states-parties, which we try to reconcile. This requires a lot of effort. The states-parties as well as the Technical Secretariat should step up their efforts in this domain so that we can ensure a more effective nonproliferation or verification mechanism with a view to preventing the diversion—the possible diversion—of chemical activities.

On the assistance and protection activities under Article X, I think we have been focusing so far on building activities at the national level with individual states-parties. For the past one or two years, we have focused more on regional activities; from now on, we are encouraging states-parties to build regional training centers for that purpose. We are also cooperating and will cooperate with the European Union in the field of their regional centers of excellence. They will cover nuclear and biological [weapons], and we will support them in the chemical field.

I believe that it is in the interest of states-parties to develop regional capacities for emergency response because they are more effective. In case of emergencies, the time is extremely important. [Regional capacities] are actually more sustainable. Small states-parties will have no capability and no resources to support and sustain these kinds of capacities even if they are developed at a certain stage. Therefore, our aim is to build these capacities but make them sustainable in the future.

Finally, on the peaceful use of chemistry, I think the states-parties have agreed that more could be done, and this is a major incentive for a large number of states-parties that have no chemical weapons, no declarable chemical industries. They are more interested in capacity development activities or the peaceful use of chemistry, and we are offering a lot in this area. I think we also will be able to increase this type of activity in the future and to meet the expectations and needs of developing countries. This will enable us to keep them engaged in implementation of the convention.

On the national implementation part, half of states-parties still have no national legislation to enforce the convention. This is a major challenge for the future of the organization. Even if those countries have nothing to declare, I think it is in the interest of the international community and the overall membership of the OPCW to ensure global implementation of the convention because they may be used as transit countries. We have to be able to control these kinds of transfers of scheduled chemicals,[2] dubious materials, also in the context of counterterrorism efforts. Therefore, we have to actually encourage them to pass the necessary legislation, and we have been working on this. Now, we are going to follow a more tailored and specifically designed approach. But of course, any national legislation should cover key points of the convention.

So these are areas [in which] we should do more on. The review conference in April will provide an opportunity for states-parties, as well as the Technical Secretariat, to focus on achievements clearly, but also on the unfinished job for the organization, the way ahead.

ACT: *Thank you. You have laid out a lot of issues here, and I'm going to try to come back to many of them. But first I want to get to a very current topic, which is the situation in Syria. Syria is one of only eight countries not part of the CWC, and many governments are concerned that the Assad regime may use its sizable arsenal or that Syria's chemical stocks may be lost or stolen.*

In your December 7 statement, you said the OPCW's responsibilities "include the prevention of the use of chemical weapons by anyone." What responsibilities and what authorities does the OPCW have with regard to possible use of chemical weapons by states that are not parties to the convention? How is the OPCW working to prevent the use of chemical weapons by anyone in Syria?

Üzümcü: First of all, the situation in Syria, the reported existence of chemical weapons, is a stark reminder to the international community of the need of universality of the convention. There are eight countries that are not yet members or parties to the convention. I think this case clearly shows that the lack of full universality would prevent full and effective implementation of the convention and the overall objective of eliminating those weapons for good and preventing their re-emergence.

In our statement dated 7 December, we wanted to point out actually that the Chemical Weapons Convention has the overall mandate. When one looks at its preamble, it says the elimination of chemical weapons universally from the world and prevention of their use. So, it doesn't say from states-parties or excluding states [that are] not parties. We have the overall mandate to oversee or watch the global situation in this respect.

Although we may not have the mechanisms to enforce it with regard to states not party, I think this should not prevent us from commenting on the potential security risks deriving from the existence of such weapons in one part of the world or another. When we say "by anyone," we wanted to make clear that either opposition or government forces should not use such weapons under any circumstances. That's the purpose of it, and it is the same for nonstate actors. So this was a very general statement, in my view, expressing our principled position on this matter.

ACT: *You have noted the possibility that the UN secretary-general could request the assistance of the OPCW in investigating the alleged use of chemical weapons. What capabilities and what expertise can the OPCW bring to the table when it comes to securing and destroying chemical weapons in Syria? Do you currently have the personnel, equipment, and financial resources to respond promptly to a request?*

Üzümcü: In the relationship agreement between the OPCW and the United Nations, which goes back to the year 2000, and in the Verification Annex to the CWC, there are provisions that require the OPCW to put its resources at the disposal of the UN secretary-general for the conduct of an investigation of alleged use involving a state not party. The recently concluded supplementary arrangement between the two organizations provides the modalities for the implementation of these provisions. If it happens in the case of Syria, clearly the secretary-general could ask us to do it; and if the security situation permits, we would be able [to carry this out]. We have the technical expertise to do it, to send some experts to verify whether such an allegation was valid or not.

On the destruction of chemical weapons, it depends on the different scenarios, of course. But let's say that if we were actually asked by the Syrian government, if the Syrian government decided to join the convention, we would be able to provide some expertise. This doesn't mean that we would actually be able to go and destroy those weapons; we don't have the technical means in place. The destruction of chemical weapons is quite a complex operation. Billions of dollars have been spent in the past by possessor states. This would require some equipment to be put in place. But primarily, the situation has to improve, and I don't think we can operate in a conflict zone. We depend on the UN safety and security regulations, and we should have a green light from them. The priority at the present should be to secure those weapons in order to prevent any access or use.

ACT: *So if there was an allegation of use by Syria, you would be able to investigate in the countries allegedly attacked, but you would not be able to go into Syria regardless of whatever authority you have through the convention or the UN secretary-general? Or would you be able to somehow get additional authority to go into Syria? Is there a way to do that?*

Üzümcü: Actually, the wording of the convention is that the UN secretary-general could request the investigation of alleged use involving a state not party, and whether we are able to go into Syria or not would totally depend on the political situation as well as the actual situation on the ground. Therefore, I cannot predict how it would unfold and whether we would be able to practically operate on the territory of Syria. So it is actually unpredictable, I would say.

ACT: *What preparations are you making for eventual Syrian accession to the CWC and for OPCW on-site inspections in Syria, perhaps even before formal accession? For example, have you had any contact with the Syrian opposition, the Assad government, or other Syrian organizations?*

Üzümcü: We haven't had any contact with the Syrian opposition and, other than the letters that we sent to and received from the Syrian foreign minister, we haven't had any contact [with the Syrian government]. Having said that, I think the Technical Secretariat has the capacity to conduct technical assistance visits or inspections once there would be legal grounds for that, either through accession or by decisions to be made by the policymaking organs of the Technical Secretariat and if the situation on the ground also permits.

Our experts are fully capable of identifying the chemical weapons and providing advice on the security, how they should be secured and so on, also what kind of methods should be applied for their eventual destruction. So in terms of technical capabilities, I think we can provide some advice and expertise, but clearly the protection and security of chemical weapons is a national responsibility for states-parties that are possessor states. The method for destruction is a national sovereign decision as well. There are some methods that are prohibited, such as dumping in the sea or burying them and so on. But apart from that, it's a national decision to choose the destruction methods.

ACT: *You mentioned your correspondence with the Syrian foreign minister. I've seen your letter to him. Is his letter to you public? Is it available on your website?*

Üzümcü: Actually, we didn't make public the response letter; but basically it says that Syria will not use chemical weapons,

if it has any, under any circumstances. It also says one should focus on the potential use of such weapons by the opposition groups and has allegations about other states in the region and elsewhere.

ACT: *[With regard to the upcoming review conference], in what areas do you think it will actually lay out some new policy or give some specific charge to the states-parties and the Technical Secretariat?*

Üzümcü: I mentioned earlier that 78 percent of the weapons stocks were destroyed, and until the destruction is complete, it will remain a priority to the OPCW. We expect that close to 99 percent of those weapons will be destroyed by the time of the following review conference in 2018. Therefore, I believe that, for the next five years following the April conference, there will be a transitional period during which we should use the opportunity to adapt the organization.

This means the adaptation of the Technical Secretariat, but also the other organs. For instance, there is the discussion about the improvement of the Executive Council proceedings, methodologies, and so on, which I hope will be done in the coming months. But the adaptation pertains to the Technical Secretariat structure, too.

In terms of deliverables, I think we should go beyond the verification of destruction. I mentioned earlier the improvement of the verification mechanism under Article VI of the convention. There are other areas, for instance, chemical safety and security. On chemical safety and security, this organization has been conducting some new activities over the past three years; and this is an area where we can deepen our activities in collaboration with fellow institutions, such as the chemical industry, chemical industry associations, as well as others.

Another area is to improve our capacity-building activities. In the advisory panel report prepared by the panel chair, Rolf Ekéus, one and a half years ago, there is a mention of the future of the organization [and its potential role as] a repository of knowledge and expertise in the field of chemical weapons. Now I think that is a quite good determination because I don't think any other organization—many states-parties will not be able to maintain such expertise because it will not be a priority anymore.

Nevertheless, there still will be risks of the use of toxic chemicals by nonstate actors or the discovery of old and abandoned chemical weapons. [Also there are] some countries that are not members at the moment but may become members and posses chemical weapons. Such expertise will be required in the future, and this is the only organization that can do it. Therefore, I think the Technical Secretariat should be able to maintain such expertise in the future.

There are challenges. We have a tenure policy that limits the term of the staff members, and I don't know whether the states-parties will consider to remove it, but this is one of the challenges. Then we will have to develop, I believe, some kind of training capacity by the organization. There are some projects that we have in mind and we want to submit to the consideration of states-parties during this transitional phase, showing that we are prepared to meet the challenges in the future once the destruction will be complete, hopefully by 2017, 2018—or nearly complete, I should say.

ACT: *Now let's go to some specific issues. You alluded earlier to the question of the April 29, 2012, deadline and the fact that Russia and the United States did not meet that deadline. As you said, the 2011 decision did not declare those countries to be in violation of the treaty, but requires them to regularly submit detailed plans for their ongoing destruction activities and imposes reporting, transparency, and monitoring requirements for the ongoing destruction work. So, are those requirements being fully implemented?*

Üzümcü: Yes, they are. The decision taken at the [2011 conference of states-parties] is being implemented. By this decision, the possessor states-parties are expected to complete the destruction in the shortest time possible without setting a new deadline. So they submitted their destruction plans, which were approved by the policymaking organs of the organization, and they in fact are complying with the reporting requirements and other obligations. I, as the director-general, have been tasked to report regularly to the Executive Council meetings, as well as the conference of states-parties, and to provide my own evaluation of the progress made and whether the states-parties concerned have made necessary efforts to accelerate the destruction process or not. So this process is very much under way.

ACT: *In your view, has the issue of the 2012 deadline been settled, or will it be a contentious issue at the upcoming review conference and beyond?*

Üzümcü: Actually, I think, from a legal perspective, it is settled. But I believe that the states-parties will observe the situation, the progress, and will continue to urge the states-parties concerned, the possessor states, to try to accelerate the destruction process, because from their point of view, some genuine efforts should be seen by them and demonstrated by the possessor states. The decision itself clearly states what efforts should be made.

I know that the United States is making some efforts to accelerate the process, as well as the Russian Federation. There are technical challenges and other difficulties, but their responsibility, in my view, is to demonstrate they are making those efforts.

ACT: *The third state-party that did not meet the April 2012 deadline was Libya. What is happening now to get Libya's stockpile destruction program restarted, and when might OPCW inspectors return to Libya?*

Üzümcü: Our inspectors have been to Libya three times since the crisis there was over, and they were able to inspect the storage site, and they were able to inspect the newly found, the previously undeclared weapons. And I've been to Tripoli myself. We expect the Libyans to resume the destruction of the bulk sulfur mustard—half of which was destroyed earlier, before the crisis erupted [in early 2011]—some time early [in 2013] and

under the verification of the OPCW inspectors. The equipment is functioning, and the issue is how to ensure the security of our inspectors and that the accommodation premises will be ready. They are being prepared for that. I think this is feasible and [the destruction of the bulk sulfur mustard] could be completed in a space of two months maximum.

I was talking about the bulk of sulfur mustard in large containers. As to the newly found weapons that consist of artillery shells and a few aerial bombs filled with sulfur mustard, it will take a little longer because they will need some new equipment to destroy them and explode them in a detonation chamber, which they need to procure with the support of some states-parties as well as the Technical Secretariat. It may take a little more than a year to deploy it and to start destruction, but what I should stress is that the Libyan authorities are very cooperative, very transparent, and willing to go ahead with the elimination of those weapons.

ACT: *So the inspectors have been just to check the declarations, but they will actually be on the ground on a permanent basis early [in 2013] so that this can resume. Is that correct?*

Üzümcü: Yes, the destruction of bulk sulfur mustard could resume because, as I said earlier, the equipment that was broken in February 2011 is now repaired and functional. Provided that we have the necessary security measures in place and the UN has given a green light for our inspectors to travel to this part of Libya, I think that the destruction could resume anytime soon. The inspectors will have to verify the destruction during the whole process.

ACT: *Okay, let's go back to an issue that you had mentioned before about industry verification and related issues and the question of the future of the OPCW. You said that the focus of the OPCW will have to shift from destruction to preventing the re-emergence of chemical weapons. How can the current industry verification regime be adapted so that it can meet this challenge? In particular, what has been achieved so far in increasing the OPCW's ability to monitor the so-called other chemical production facilities, and what more needs to be done?*

Üzümcü: I think the overall balance, which was struck during the negotiations of the convention, will be upset not due to the failure of the implementation, but rather due to its success, in the coming years. The initial balance was between the elimination of chemical weapons and the industrial verification on the one hand and the rest of the activities emanating from the convention on the other. Now, since the destruction activities will be completed, let's say in a few years' time, then there will be a need to strike a new balance.

And the new balance, I believe, can be actually achieved between the verification, under Article VI mainly, which would aim at preventing the re-emergence of chemical weapons on the one hand, and the rest of activities on the other. Therefore, this will give us an opportunity to reinforce this verification mechanism. I don't think the numbers and caps for each state-party could be changed. There is a cap of 20 inspections per year for other chemical production facilities, or Schedule 3 facilities, and this cap will be in place. On the other hand, we could improve the efficiency or effectiveness of those inspections, the selection methodology, and their conduct and also improve the declaration system. We have made some progress in this respect on declarations—more accurate, more timely declarations, as well as on the evaluations so far.

But I think we should do more [not only] by educating and training the states-parties about doing this declaration in a more proper, more accurate way, but also in our own capacity to evaluate them. So I think there is still work to be done in order to improve this verification mechanism in collaboration with the chemical industry. The chemical industry is our main partner in this domain, and they are willing to cooperate further with us. The [Ekéus] advisory panel report recommended that we should establish a joint working group with the chemical industry. We are working on that and, following the review conference, we want to somehow informally, but still by establishing a mechanism, have a permanent, regular dialogue and cooperation with the chemical industry on this and other relevant issues.

Another related issue is scientific and technological development. There are several new discoveries and inventions, which may have some implications for the verification mechanism of the convention. And we have a Scientific Advisory Board, and they have been working on the convergence between biology and chemistry for some time, on sampling analysis and other issues. So that is an area that needs to be taken into account by states-parties. And the Scientific Advisory Board provided its input to the review conference very recently, as well as [the International Union of Pure and Applied Chemistry], which jointly organized with the OPCW a workshop providing its own inputs into the process. I think it is on their website as well as ours. So the technological scientific development system is something that we should clearly bear in mind.

Another area in which we were not that active is education and outreach. We have realized that we cannot achieve the goals of the convention only through verification mechanisms and prevention, or nonproliferation, activities. We need to raise awareness among the relevant communities, the scientific communities as well as the relevant educational institutions. So we are in the process of collaborating with some partners to produce some educational materials, e-learning modules, so that we can reach out to universities, even high schools. Soon we will invite some chemistry teachers from high schools so that we can inform them about the goals of the convention and disseminate the necessary information to raise awareness and also to raise awareness among the chemical industry as well as the scientific community about the risks, which might be associated with handling the dual-use chemical material.

ACT: *The primary goal of the CWC has been a "world free of chemical weapons." What is the OPCW doing to bring in the remaining eight countries [Angola, Egypt, Israel, Myanmar, North Korea, Somalia, South Sudan, and Syria] into the treaty regime to reach this goal of universality that you mentioned earlier?*

Üzümcü: Universality, I think, is one of the key objectives of the OPCW. It has been so for many years. I think having a membership of 188 countries is a big achievement, but it is not enough; and as I said, Syria is a reminder of that. Recently, the UN secretary-general and I have written letters to the heads of state and government of those eight countries that are outside of the realm of the convention.

We have been approaching those countries for several years. It is likely that three countries—Angola, South Sudan, and Myanmar—may join the convention some time during 2013, hopefully. We have been sending some delegations to Myanmar; the second one will go in early February. We have proposed similar assistance to South Sudan, which is a new independent state, and to Angola. We see that there shouldn't be any problem for them to join the convention. So we understand that it has not been a matter of priority so far, but they have shown some interest, and we encourage them to do it as early as possible.

As for the remaining three countries in the Middle East, including Syria, we were hopeful that this WMD [weapons of mass destruction]-free-zone conference would be held before the end of [2012]. Now it is postponed.[3] We know that it is going to be the beginning of a process, and we hope that this process will pave the way for universality of the Chemical Weapons Convention.

Our position has been that CWC membership should not be linked to any other processes and it should be addressed on its own merits. We think that the possession of chemical weapons should be repudiated by any country, irrespective of any other process. We know that the countries in the Middle East relate this issue to regional security concerns, as well as the nuclear issue, and we hope that the hurdles will be removed during the process of [establishing] a WMD-free zone in the Middle East.

ACT: *Thank you very much.*

Endnotes

1. See Daniel Horner, "Accord Reached on CWC's 2012 Deadline," *Arms Control Today,* January/February 2012.

2. The CWC requires states-parties to declare chemical industry facilities that produce or use chemicals of concern to the convention. These chemicals are grouped into "schedules" based on the risk they pose of violating the convention's conditions. Schedule 1 chemicals and precursors pose a "high risk" and are rarely used for peaceful purposes. States-parties may not retain these chemicals except in small quantities for research, medical, pharmaceutical, or defensive use. Many Schedule 1 chemicals have been stockpiled as chemical weapons. Schedule 2 chemicals are toxic chemicals that pose a "significant risk" and are precursors to the production of Schedule 1 or Schedule 2 chemicals. These chemicals are not produced in large quantities for commercial or other peaceful purposes. Schedule 3 chemicals are usually produced in large quantities for purposes not prohibited by the CWC, but still pose

a risk to the convention. Some of these chemicals have been stockpiled as chemical weapons.

3. At the 2010 Nuclear Nonproliferation Treaty Review Conference, the parties to that treaty agreed to hold a conference in 2012 on establishing a Middle Eastern zone free of weapons of mass destruction. The conference had been tentatively scheduled to take place in December in Helsinki, but the key countries involved in organizing the meeting announced in November that the meeting was being postponed. They did not set a new date. See Kelsey Davenport and Daniel Horner, "Meeting on Middle East WMD Postponed," *Arms Control Today,* December 2012.

Critical Thinking

1. What are some of the problems that the OPCW will encounter in Syria in verifying the destruction of the chemical weapons arsenal there?

2. What is the relationship between the OPCW and the Chemical Weapons Convention?

Create Central

www.mhhe.com/createcentral

Internet References

Chemical Weapons Convention
 www.opc.org/chemicalweapons-convention
Organization for the Prevention of Chemical Weapons
 www.opcw.org
1925 Geneva Protocol Prohibiting the Use of Asphyxiating Gases
 www.un.org/Disarmamaent/WMD/Bio/pdf/Status_Protocol.pdf.

Prepared by: Robert Weiner, *University of Massachusetts/Boston*

Article

Reducing the Global Nuclear Risk

SIDNEY D. DRELL, GEORGE P. SHULTZ, AND STEVEN P. ANDREASEN

Learning Outcomes

After reading this article, you will be able to:

- Explain the risks associated with the safety and the security of the nuclear enterprise.

- Explain what is meant by nuclear enterprise.

The times we live in are dangerous for many reasons. Prominent among them is the existence of a global nuclear enterprise made up of weapons that can cause damage of unimaginable proportions and power plants at which accidents can have severe, essentially unpredictable consequences for human life. For all of its utility and promise, the nuclear enterprise is unique in the enormity of the vast quantities of destructive energy that can be released through blast, heat, and radioactivity.

To get a better grip on the state of the nuclear enterprise, we convened a group of prominent experts at Stanford University's Hoover Institution. The group included experts on nuclear weapons, power plants, regulatory experience, public perceptions, and policy. This essay summarizes their views and conclusions.

We begin with the most reassuring outcome of our deliberations: It's the sense generally held that the U.S. nuclear enterprise currently meets very high standards in its commitment to safety and security. That has not always been the case in all aspects of the U.S. nuclear enterprise. But safety begins at home, and while the U.S. will need to remain focused to guard against nuclear risks, the picture here looks relatively good.

Our greatest concern is that the same cannot be said of the nuclear enterprise globally. Governments, international organizations, industry, and media must recognize and address the nuclear challenges and mounting risks posed by a rapidly changing world.

The biggest concerns with nuclear safety and security are in countries relatively new to the nuclear enterprise, and the potential loss of control to terrorist or criminal gangs of the fissile material that exists in such abundance around the world. In a number of countries, confidence in civil nuclear energy production was severely shaken in the spring of 2011 by the Fukushima nuclear reactor plant disaster. And in the military sphere, the doctrine of deterrence that remains primarily dependent on nuclear weapons is seen in decline due to the importance of nonstate actors such as al-Qaeda and terrorist affiliates that seek destruction for destruction's sake. We have two nuclear tigers by the tail.

When risks and consequences are unknown, undervalued, or ignored, our nation and the world are dangerously vulnerable. Nowhere is this risk/consequence calculation more relevant than with respect to the nucleus of the atom.

From Hiroshima to Fukushima

The nuclear enterprise was introduced to the world by the shock of the devastation produced by two atomic bombs hitting Hiroshima and Nagasaki. Modern nuclear weapons are far more powerful than those early bombs, which presented their own hazards. Early research depended on a program of atmospheric testing of nuclear weapons. In the early years following World War II, the impact and the amount of radioactive fallout in the atmosphere generated by above-ground nuclear explosions was not fully appreciated. During those years, the United States and the Soviet Union conducted several hundred tests in the atmosphere that created fallout.

A serious regulatory weak point from that time still exists in many places today, as the Fukushima disaster clearly indicates. The U.S. Atomic Energy Commission (AEC) was initially assigned conflicting responsibilities: to create an arsenal of nuclear weapons for the United States to confront a growing nuclear-armed Soviet threat; and, at the same time, to ensure public safety from the effects of radioactive fallout. The AEC was faced with the same conundrum with regard to civilian nuclear power generation. It was charged with promoting civilian nuclear power and simultaneously protecting the public.

Progress came in 1963 with the negotiation and signing of the Limited Test Ban Treaty (LTBT) banning all nuclear explosive testing in the atmosphere (initially by the United States, the Soviet Union, and the United Kingdom). With the successful safety record of the U.S. nuclear weapons program, domestic anxiety about nuclear weapons receded somewhat. Meanwhile, public attitudes toward nuclear weapons reflected recognition of their key role in establishing a more stable nuclear deterrent posture in the confrontation with the Soviet Union.

The positive record on safety of the nuclear weapons enterprise in the United States—there have been accidents involving nuclear weapons, but none that led to the release of nuclear energy—was the result of a strong effort and continuing commitment to include safety as a primary criterion in new weapons designs, as well as careful production, handling, and deployment procedures. The key to the health of today's nuclear weapons enterprise is confidence in the safety of its operations and in the protection of special nuclear materials against theft. One can imagine how different the situation would be today if there had been a recognized theft of material sufficient for a bomb, or if one of the two four-megaton bombs that fell from a disabled B-52 Strategic Air Command bomber overflying Goldsboro, North Carolina, in 1961 had detonated. In that event, a single switch in the arming sequence of one of the bombs, by remaining in its "off" position while the aircraft was disintegrating, was all that prevented a full-yield nuclear explosion. A close call indeed.

In the 26 years since the meltdown of the nuclear reactor at Chernobyl in Soviet-era Ukraine, the nuclear power industry has strengthened its safety practices. Over the past decade, growing concerns about global warming and energy independence have actually strengthened support for nuclear energy in the United States and many nations around the world. Yet despite these trends, the civil nuclear enterprise remains fragile. Following Fukushima, opinion polls gave stark evidence of the public's deep fears of the invisible force of nuclear radiation, shown by public opposition to the construction of new nuclear power plants in close proximity. It is not simply a matter of getting better information to the public but of actually educating the public about the true nature of nuclear radiation and its risks. Of course, the immediate task of the nuclear power component of the enterprise is to strive for the best possible safety record. The overriding objective could not be more clear: no more Fukushimas.

Another issue that must be resolved involves the continued effectiveness of a policy of deterrence that remains primarily dependent on nuclear weapons, and the hazards these weapons pose due to the spread of nuclear technology and material. There is growing apprehension about the determination of terrorists to get their hands on weapons or, for that matter, on the special nuclear material—plutonium and highly enriched uranium that fuels them in the most challenging step toward developing a weapon.

The global effects of a regional war between nuclear-armed adversaries such as India and Pakistan would also wield an enormous impact, potentially involving radioactive fallout at large distances caused by a limited number of nuclear explosions.

This is true as well for nuclear radiation from a reactor explosion—fall-out at large distances would have a serious societal impact on the nuclear enterprise. There is little understanding of the reality and potential danger of consequences if such an event were to occur halfway around the world. An effort should be made to prepare the public by providing information on how to respond to such an event.

An active nuclear diplomacy has grown out of the Cold War efforts to regulate testing and reduce superpower nuclear arsenals. There is now a welcome focus on rolling back nuclear weapons proliferation. Additional important measures include the Nunn-Lugar program, started in 1991 to reduce the nuclear arsenal of the former Soviet Union. Such initiatives have led to greater investment by the United States and other governments in better security for nuclear weapons and material globally, including billions of dollars through the G8 Global Partnership Against the Spread of Weapons and Materials of Mass Destruction. The commitment to improving security of all dangerous nuclear material on the globe within four years was made by 47 world leaders who met with President Obama in Washington, D.C., in April 2010; this commitment was reconfirmed in March 2012 at the Nuclear Security Summit in Seoul, South Korea. Many specific commitments made in 2010 relating to the removal of nuclear materials and conversion of nuclear research reactors from highly enriched uranium to low-enriched uranium fuel have already been accomplished, along with increasing levels of voluntary commitments from a diverse set of states, improving prospects for achieving the four-year goal.

Three Principles

It is evident that globally, the nuclear enterprise faces new and increasingly difficult challenges. Successful leadership in national security policy will require a continuous, diligent, and multinational assessment of these newly emerging risks and consequences. In view of the seriousness of the potentially deadly consequences associated with nuclear weapons and nuclear power, we emphasize the importance of three guiding principles for efforts to reduce those risks globally:

First, the calculations used to assess nuclear risks in both the military and the civil sectors are fallible. Accurately analyzing events where we have little data, identifying every variable associated with risk, and the possibility of a single variable that goes dangerously wrong are all factors that complicate risk calculations. Governments, industry, and concerned citizens must constantly reexamine the assumptions on which safety and security measures, emergency preparations, and nuclear energy production are based. When dealing with very low-probability and high-consequence operations, we typically have little data as a basis for making quantitative analyses. It is therefore difficult to assess the risk of a nuclear accident and what would contribute to it, and to identify effective steps to reduce that risk.

In this context, it is possible that a single variable could exceed expectations, go dangerously wrong, and simply overwhelm safety systems and the risk assessments on which those systems were built. This is what happened in 2011 when an earthquake, followed by a tsunami—both of which exceeded expectations based on history—overwhelmed the Fukushima complex, breaching a number of safeguards that had been built into the plant and triggering reactor core meltdowns and radiation leaks. This in turn exposed the human factor, which is hard to assess and can dramatically change the risk equation. Cultural habits and regulatory inadequacy inhibited rapid decision-making and crisis management in the Fukushima disaster.

A more nefarious example of the human factor would be a determined nuclear terrorist attack specifically targeting either the military or civilian component of the nuclear enterprise.

Second, risks associated with nuclear weapons and nuclear power will likely grow substantially as nuclear weapons and civilian nuclear energy production technology spread in unstable regions of the world where the potential for conflict is high. States that are new to the nuclear enterprise may not have effective nuclear safeguards to secure nuclear weapons and materials—including a developed fabric of early warning systems and nuclear confidence-building measures that could increase warning and decision time for leaders in a crisis—or the capability to safely manage and regulate the construction and operation of new civilian reactors. Hence there is a growing risk of accidents, mistakes, or miscalculations involving nuclear weapons, and of regional wars or nuclear terrorism. The consequences would be horrific: A Hiroshima-size nuclear bomb detonated in a major city could kill a half-million people and result in $1 trillion in direct economic damage.

On the civil side of the nuclear ledger, the sobering paradox is this: While an accident would be considerably less devastating than the detonation of a nuclear weapon, the risk of an accident occurring is probably higher. Currently, 1.4 billion people live without electricity, and by 2030 the global demand for energy is projected to rise by about 25 percent. With the added need to minimize carbon emissions, nuclear power reactors will become increasingly attractive alternative sources for electric power, especially for developing nations. These countries, in turn, will need to meet the challenge of developing appropriate governmental institutions and the infrastructure, expertise, and experience to support nuclear power efforts with a suitably high standard of safety. As the world witnessed in Fukushima, a nuclear power plant accident can lead to the spread of dangerous radiation, massive civil dislocations, and billions of dollars in cleanup costs. Such an event can also fuel widespread public skepticism about nuclear institutions and technology.

Some developed nations—notably Germany—have interpreted the Fukushima accident as proof that they should abandon nuclear power altogether, primarily by prolonging the life of existing nuclear reactors while phasing out nuclear-produced electricity and developing alternative energy sources.

Third, we need to understand that no nation is immune from risks involving nuclear weapons and nuclear power within their borders. There were 32 so-called "Broken Arrow" accidents—nuclear accidents that do not pose a danger of an outbreak of nuclear war—involving U.S. weapons between 1950 and 1980, mostly involving U.S. Strategic Air Command bombers and earlier bomb designs not yet incorporating modern nuclear detonation safety designs. The U.S. no longer maintains a nuclear-armed, in-air strategic bomber force, and the record of incidents is greatly reduced. In several cases, accidents such as the North Carolina bomber incident came dangerously close to triggering catastrophes, with disaster averted simply by luck.

The United States has had an admirable safety record in the area of civil nuclear power since the 1979 Three Mile Island accident in Pennsylvania, yet safety concerns persist. One of the critical assumptions in the design of the Fukushima reactor complex was that, if electrical power was lost at the plant and back-up generators failed, power could be restored within a few hours. The combined one-two punch of the earthquake and tsunami, however, made the necessary repairs impossible. In the United States today, some nuclear power reactors are designed with a comparably short window for restoring power. After Fukushima, this is an issue that deserves action—especially in light of our own Hurricane Katrina experience, which rendered many affected areas inaccessible for days in 2005, and the August 2011 East Coast earthquake that shook the North Anna nuclear power plant in Mineral, Virginia, beyond expectations based on previous geological activity.

Reducing Risks

To reduce these nuclear risks, we offer four related recommendations that should be adopted by the nuclear enterprise, both military and civilian, in the United States and abroad.

First, the reduction of nuclear risks requires every level of the nuclear enterprise and related military and civilian organizations to embrace the importance of safety and security as an overarching operating rule. This is not as easy as it sounds. To a war fighter, more safety and control can mean less reliability and availability and greater costs. For a company or utility involved in the construction or operation of a nuclear power plant, more safety and security can mean greater regulation and higher costs.

But the absence of a culture of safety and security, in which priorities and meaningful standards are set and rigorous discipline and accountability are enforced, is perhaps the most reliable indicator of an impending disaster. In August 2007, after a B-52 bomber loaded with six nuclear-tipped cruise missiles flew from North Dakota to Louisiana without anyone realizing there were live weapons on board, then Secretary of Defense Robert Gates fired both the military and civilian heads of the U.S. Air Force. His action was an example of setting the right priorities and enforcing accountability, but the reality of the incident shows that greater incorporation of a safety and security culture is needed.

Second, independent regulation of the nuclear enterprise is crucial to setting and enforcing the safety and security rule. In the United States today, the nuclear regulatory system—in particular, the Nuclear Regulatory Commission (NRC)—is credited with setting a uniquely high standard for independent regulation of the civil nuclear power sector. This is one of the keys to a successful and safe nuclear program. Effective regulation is even more crucial when there are strong incentives to keep operating costs down and keep an aging nuclear reactor fleet in operation, a combination that could create conditions for a catastrophic nuclear power plant failure. Careful attention is required to protect the NRC from regulatory capture by vested interests in government and industry, the latter of which funds a high percentage of the NRC's budget.

In too many countries, strong, independent regulatory agencies are not the norm. The independent watchdog organization advising the Japanese government was working with Japanese utilities to influence public opinion in favor of nuclear power.

Strengthening the International Atomic Energy Agency (IAEA) so that it can play a greater role in civil nuclear safety and security would also help reduce risks, and will require substantially greater authorities to address both safety and security, and most importantly, more resources for an agency whose budget is only €333 million, with only one-tenth of that total devoted to nuclear safety and security. In addition, exporting "best practices" of the U.S. Nuclear Regulatory Commission—that is, lessons of nuclear regulation, oversight, and safety learned over many decades to other countries would pay a huge safety dividend.

Third, independent peer review should be incorporated into all aspects of the nuclear enterprise. On the weapons side, independent experts in the United States—from both within and outside the various concerned organizations—are relied on to review or "red team" each other, rigorously challenging and debating weapons and systems safety, and communicating these points up and down the line. The Institute of Nuclear Power Operations (INPO) provides strong peer review and oversight of the civil nuclear sector in the United States. Its global counterpart, the World Association of Nuclear Operators (WANO), should give a higher priority to further strengthening its safety operations, in particular its peer review process, learning from the experiences of the United States and other nations. Strong outside peer review—combined with an enhanced capacity to arrange fines based on incidents occurring in far distant countries—would help states entering the world of high-consequence operations to develop a culture and standard needed to achieve an exemplary safety record.

Beyond these recommendations, the military and civilian nuclear communities can and should learn from each other. A periodic dialogue structured around assessing and reducing the risks surrounding the nuclear enterprise would be valuable, both in the United States and abroad, and could be organized by governments or academia (as was done in the conference at Stanford). An analysis of the probabilities of undesired events and ways to minimize them, including lessons learned from accidents such as Fukushima as well as "close call" incidents, should be put on the front burner, as should consequence management—that is, what to do if a nuclear incident were to occur.

An informed public is also an essential element in responding to a nuclear crisis. Greater public awareness and understanding of nuclear risks and consequences can lead to greater public preparation to handle post-disaster challenges.

Fourth, progress on all aspects of nuclear threat reduction should be organized around a clear goal: a global effort to reduce reliance on nuclear weapons, prevent their spread into potentially dangerous hands, and ultimately end them as a threat to the world. A step-by-step process—along the lines proposed by George Shultz, William Perry, Henry Kissinger, and Sam Nunn in a series of *Wall Street Journal* essays—and demonstrated progress toward realizing the vision of a world free of nuclear weapons will build the kind of international trust and broad cooperation required to effectively address today's threats—and prevent tomorrow's catastrophe.

Our bottom line: Since the risks posed by the nuclear enterprise are so high, no reasonable effort should be spared to ensure safety and security. That must be the rule in dealing with events of very low probability but potentially catastrophic consequences.

Critical Thinking

1. What solutions do the authors offer to reduce the risks associated with nuclear safety and security?

2. How can states be persuaded to reduce their reliance on nuclear weapons?

3. What can be done to prevent nuclear weapons from falling into the hands of terrorists?

Create Central

www.mhhe.com/createcentral

Internet References

Nuclear Threat Initiative
www.nti.org/about/
International Atomic Energy Agency (IAEA)
www.iaea.com
Bulletin of Atomic Scientists
www.the bulletin.org

SIDNEY D. DRELL is a senior fellow at the Hoover Institution, where George P. Shultz is Thomas W. and Susan B. Ford distinguished fellow. Steven P. Andreasen is a lecturer at the Humphrey School of Public Affairs at the University of Minnesota.

Drell et al., Sidney D. From *Policy Review*, October/November 2012, pp. 15–22. Copyright © 2012 by Sidney D. Drell, George P. Schultz, and Steven P. Andreasen. Reprinted by permission of Policy Review, a publication of Hoover Institution, Stanford University, and the authors.

Unit 5

UNIT

Prepared by: Robert Weiner, *University of Massachusetts/Boston*

International Organization, International Law, and Human Security

The United Nations (UN) was created as the successor to the failed League of Nations. The main purpose of the UN was to prevent another world war, through the application of the principles of collective security. Collective security was based on the idea that the organized power of the international community was supposed to be sufficient to deter and punish aggression. The Charter of the United Nations, which envisaged the construction of a just world order, was based on the principle of liberal internationalism. In the second decade of the twenty-first century, the UN continues to support the promotion of important world order values, such as peace, economic security, the protection of human rights, and the protection of the environment, most of which fall under the rubric of human security.

The UN itself was embedded in a realist conception of international order, based on the Westphalian system, which rested on the fundamental principle of the primacy of state sovereignty. However, while the UN represents a political approach to the problem of world order, there exists a network of specialized agencies affiliated with the UN, which represent a more functional approach to world order. Classic functionalism was based on the hope that technical and economic cooperation between states in international agencies would result in a pooling of sovereignty which would eliminate nationalism as the root of war in the international system. Specialized agencies, ranging from the International Telecommunications Union to UNESCO (the United Nations Educational, Scientific, and Cultural Organization) constitute part of the UN family. Although some of the specialized I agencies have been able to focus on technical issues, it has not been possible to prevent the politicization of agencies like UNESCO.

The primacy of state sovereignty has now been challenged by the doctrine of humanitarian intervention, which states that the sovereignty of the individual should take precedence over the sovereignty of the state when gross and mass violations of human rights take place. The central question is at what point should the "international community" engage in military intervention, for example, in the case of Syria in 2013, according to the norm of the "Responsibility to Protect." According to the London-based Syrian Human Rights Observatory, it is estimated that more than 100,000 people have been killed in the civil conflict that has taken place in Syria since 2011. Other questions associated with the issue are who should authorize humanitarian military intervention? How can one make sure that such military intervention is not designed to serve the national interest of a single state or a group of states? Who determines when a threshold has been crossed by the government of a state that has committed gross and mass violations of human rights against its own people that should trigger military intervention? How many people have to be killed before such an intervention occurs? In an effort to deal with this problem, the international community has moved in the direction of holding political leaders who commit war crimes, crimes against humanity, and genocide accountable for their actions. For example, in 1993, the UN Security Council created the ad hoc Tribunal for the Former Yugoslavia. In 2011, after eluding capture for 16 years, the Bosnian Serbian general Ratko Mladic, who had been indicted by the tribunal for the commission of genocide in the Bosnian town of Srebrenica, was arrested by the Serbian government and turned over to the Tribunal for trial, a trial which was still underway in 2013.

More problems in international humanitarian and international human rights law have been raised by the revolution in military technology which has resulted in the creation of drones and autonomous weapons systems which have the capacity to become killer robots. Questions have been raised about the collateral damage caused by drones when innocent civilians suffer casualties in the course of a military strike or are erroneously targeted. The main question that has arisen in connection with the use of killer robots in future wars whether such machines can make the decision to kill human beings autonomously without any human input. As robotic military technology spreads to states throughout the world, the necessity for the development of international standards and ethics to ensure the use of robot soldiers in a just fashion becomes more urgent.

Article

Prepared by: Robert Weiner, *University of Massachusetts/Boston*

Law and Ethics for Robot Soldiers

KENNETH ANDERSON AND MATTHEW WAXMAN

Learning Outcomes

After reading this article, you will be able to:

- Explain the relationship between military technology and the ethical behavior of autonomous weapons systems.

- Discuss the objections to the use of killer robots.

A LETHAL SENTRY ROBOT designed for perimeter protection, able to detect shapes and motions, and combined with computational technologies to analyze and differentiate enemy threats from friendly or innocuous objects—and shoot at the hostiles. A drone aircraft, not only unmanned but programmed to independently rove and hunt prey, perhaps even tracking enemy fighters who have been previously "painted and marked" by military forces on the ground. Robots individually too small and mobile to be easily stopped, but capable of swarming and assembling themselves at the final moment of attack into a much larger weapon. These (and many more) are among the ripening fruits of automation in weapons design. Some are here or close at hand, such as the lethal sentry robot designed in South Korea. Others lie ahead in a future less and less distant.

Lethal autonomous machines will inevitably enter the future battlefield—but they will do so incrementally, one small step at a time. The combination of "inevitable" and "incremental" development raises not only complex strategic and operational questions but also profound legal and ethical ones. Inevitability comes from both supply-side and demand-side factors. Advances in sensor and computational technologies will supply "smarter" machines that can be programmed to kill or destroy, while the increasing tempo of military operations and political pressures to protect one's own personnel and civilian persons and property will demand continuing research, development, and deployment. The process will be incremental because nonlethal robotic systems (already proliferating on the battlefield, after all) can be fitted in their successive generations with both self-defensive and offensive technologies. As lethal systems are initially deployed, they may include humans in the decision-making loop, at least as a fail-safe—but as both the decision-making power of machines and the tempo of operations potentially increase, that human role will likely slowly diminish.

Recognizing the inevitable but incremental evolution of these technologies is key to addressing the legal and ethical dilemmas associated with them; U.S. policy for resolving such dilemmas should be built upon these assumptions. The certain yet gradual development and deployment of these systems, as well as the humanitarian advantages created by the precision of some systems, make some proposed responses—such as prohibitory treaties—unworkable as well as ethically questionable.

Those same features also make it imperative, though, that the United States resist its own impulses toward secrecy and reticence with respect to military technologies and recognize that the interests those tendencies serve are counterbalanced here by interests in shaping the normative terrain—i.e., the contours of international law as well as international expectations about appropriate conduct on which the United States government and others will operate militarily as technology evolves. Just as development of autonomous weapon systems will be incremental, so too will development of norms about acceptable systems and uses be incremental. The United States must act, however, before international expectations about these technologies harden around the views of those who would impose unrealistic, ineffective, or dangerous prohibitions—or those who would prefer few or no constraints at all.

Incremental Automation of Drones

THE INCREMENTAL MARCH toward automated lethal technologies of the future, and the legal and ethical challenges that accompany it, can be illustrated by looking at today's drone aircraft. Unmanned drones piloted from afar are already a significant component of the United States' arsenal. At this writing, close to one in three U.S. Air Force aircraft is remotely piloted (though this number also includes many tiny tactical surveillance drones). The drone proportion will only grow. Yet current drone military aircraft are not autonomous in the firing of weapons—the weapon must be fired in real time by a human controller. So far there are no known plans or, apparently in the view of military, reasons to take the human out of the weapon firing loop.

Nor are today's drones truly autonomous as aircraft. They require human pilots and flight support personnel in real time, even when they are located far away. They *are*, however, increasingly automated in their flight functions: self-landing capabilities, for example, and particularly automation to the point that a single controller can run many drone aircraft at once, increasing efficiency considerably. The automation of flight is gradually increasing as sensors and aircraft control through computer programming improves.

Some believe that the next generations of jet fighters won't be manned. Or that manned fighters will join with unmanned.

Looking to the future, some observers believe that one of the next generations of jet fighter aircraft will no longer be manned, or at least that manned fighter aircraft will be joined by unmanned aircraft. Drone aircraft might gradually become capable of higher speeds, torques, g-forces, and other stresses than those a human pilot can endure (and perhaps at a cheaper cost as well). Given that speed in every sense—including turning and twisting in flight, reaction and decision times—is an advantage, design will emphasize automating as many of these functions as possible, in competition with the enemy's systems.

Just as the aircraft might have to be maneuvered far too quickly for detailed human control of its movements, so too the weapons—against other aircraft, drones, or anti-aircraft systems—might have to be utilized at the same speeds in order to match the beyond-human speed of the aircraft's own systems (as well as the enemy aircraft's similarly automated counter-systems). In similar ways, defense systems on modern U.S. naval vessels have long been able to target incoming missiles automatically, with humans monitoring the system's operation, because human decision-making processes are too slow to deal with multiple, inbound, high-speed missiles. Some military operators regard many emerging automated weapons systems as merely a more sophisticated form of "fire and forget" self-guided missiles. And because contemporary fighter aircraft are designed not only for air-to-air combat, but for ground attack missions as well, design changes that reduce the role of the human controller of the aircraft platform may shade into automation of the weapons directed at ground targets, too.

Although current remotely piloted drones, on the one hand, and future autonomous weapons, on the other, are based on different technologies and operational imperatives, they generate some overlapping concerns about their ethical legitimacy and lawfulness. Today's arguments over the legality of remotely piloted, unmanned aircraft in their various missions (especially targeted killing operations, and concerns that the United States is using technology to shift risk from its own personnel onto remote-area civilian populations) presage the arguments that already loom over weapons systems that exhibit emerging features of autonomy. Those arguments also offer lessons to guide short- and long-term U.S. policy toward autonomous weapons generally, including systems that are otherwise quite different.

Automated-Arms Racing?

THESE ISSUES ARE easiest to imagine in the airpower context. But in other battlefield contexts, too, the United States and other sophisticated military powers (and eventually unsophisticated powers and nonstate actors, as such technologies become commodified and offered for licit or illicit sale) will find increasingly automated lethal systems more and more attractive. Moreover, as artificial intelligence improves, weapons systems will evolve from robotic "automation"—the execution of precisely pre-programmed actions or sequences in a well-defined and controlled environment—toward genuine "autonomy," meaning the robot is capable of generating actions to adapt to changing and unpredictable environments.

Take efforts to protect peacekeepers facing the threat of snipers or ambush in an urban environment: Small mobile robots with weapons could act as roving scouts for the human soldiers, with "intermediate" automation—the robot might be pre-programmed to look for certain enemy weapon signatures and to alert a human operator, who then decides whether or not to pull the trigger. In the next iteration, the system might be set with the human being not required to give an affirmative command, but instead merely deciding whether to override and veto a machine-initiated attack. That human decision-maker also might not be a soldier on site, but an off-battlefield, remote robot-controller.

It will soon become clear that the communications link between human and weapon system could be jammed or hacked (and in addition, speed and the complications of the pursuit algorithms may seem better left to the machine itself, especially once the technology moves to many small, swarming, lightly armed robots). One technological response will be to reduce the vulnerability of the communications link by severing it, thus making the robot dependent upon executing its own programming, or even genuinely autonomous.

Aside from conventional war on conventional battlefields, covert or special operations will involve their own evolution toward incrementally autonomous systems. Consider intelligence gathering in the months preceding the raid on Osama bin Laden's compound. Tiny surveillance robots equipped with facial recognition technology might have helped affirmatively identify bin Laden much earlier. It is not a large step to weaponize such systems and then perhaps go the next step to allow them to act autonomously, perhaps initially with a human remote-observer as a fail-safe, but with very little time to override programmed commands.

These examples have all been stylized to sound precise and carefully controlled. At some point in the near future, however, someone—China, Russia, or someone else—will likely design, build, and deploy (or sell) an autonomous weapon system for battlefield use that is programmed to target something—say a person or position—that is firing a weapon and is positively identified as hostile rather than friendly. A weapon system programmed, that is, to do one thing: identify the locus of enemy fire and fire back. It thus would lack the ability altogether to take account of civilian presence and any likely collateral damage.

Quite apart from the security and war-fighting implications, the U.S. government would have grave legal and humanitarian concerns about such a foreign system offered for sale on the international arms markets, let alone deployed and used. Yet the United States would then find itself in a peculiar situation—potentially facing a weapon system on the battlefield that conveys significant advantages to its user, but which the United States would not deploy itself because (for reasons described below) it does not believe it is a legal weapon. The United States will have to come up with technological counters and defenses such as development of smaller, more mobile, armed robots able to "hide" as well as "hunt" on their own.

The implication is that the arms race in battlefield robots will be more than simply a race for ever more autonomous weapons systems. More likely, it will mostly be a race for ways to counter and defend against them—partly through technical means, but also partly through the tools of international norms and diplomacy, provided, however, that those norms are not over-invested with hopes that cannot realistically be met.

Legal and Ethical Requirements

THE LEGAL AND ethical evaluation of a new weapons system is nothing new. It is a long-standing requirement of the laws of war, one taken seriously by U.S. military lawyers. In recent years, U.S. military judge advocates have rejected proposed new weapons as incompatible with the laws of war, including blinding laser weapons and, reportedly, various cutting edge cyber-technologies that might constitute weapons for purposes of the laws of war. But arguments over the legitimacy of particular weapons (or their legitimate use) go back to the beginnings of debate over the laws and ethics of war: the legitimacy, for example, of poison, the crossbow, submarines, aerial bombardment, antipersonnel landmines, chemical and biological weapons, and nuclear weapons. In that historical context, debate over autonomous robotic weapons—the conditions of their lawfulness as weapons and the conditions of their lawful use—is nothing novel.

Likewise, there is nothing novel in the sorts of responses autonomous weapons systems will generate. On the one hand, emergence of a new weapon often sparks an insistence in some quarters that the weapon is ethically and legally abhorrent and should be prohibited by law. On the other hand, the historical reality is that if a new weapon system greatly advantages a side, the tendency is for it gradually to be adopted by others perceiving they can benefit from it, too. In some cases, legal prohibitions on the weapon system as such erode, as happened with submarines and airplanes; what survives is typically legal rules for the *use* of the new weapon, with greater or lesser specificity. In a few cases (including some very important ones), legal prohibitions on the weapon as such gain hold. The ban on poison gas, for example, has survived in one form or another with very considerable effectiveness throughout the 20th century.

Where in the long history of new weapons and their ethical and legal regulation will autonomous robotic weapons fit?

Where in this long history of new weapons and their ethical and legal regulation will autonomous robotic weapons fit? What are the features of autonomous robotic weapons that raise ethical and legal concerns? How should they be addressed, as a matter of law and process? By treaty, for example, or by some other means?

One answer to these questions is: wait and see. It is too early to know where the technology will go, so the debate over ethical and legal principles for robotic autonomous weapons should be deferred until a system is at hand. Otherwise it is just an exercise in science fiction and fantasy.

But that wait-and-see view is shortsighted and mistaken. Not all the important innovations in autonomous weapons are so far off. Some are possible now or will be in the near term, and some of them raise serious questions of law and ethics even at their current research and development stage.

Moreover, looking to the long term, technology and weapons innovation does not take place in a vacuum. The time to take into account law and ethics to inform and govern autonomous weapons systems is now, before technologies and weapons development have become "hardened" in a particular path and their design architecture becomes difficult or even impossible to change. Otherwise, the risk is that technology and innovation alone, unleavened by ethics and law at the front end of the innovation process, let slip the robots of war.

This is also the time—before ethical and legal understandings of autonomous weapon systems likewise become hardened in the eyes of key constituents of the international system—to propose and defend a framework for evaluating them that advances simultaneously strategic and moral interests. What might such a framework look like? Consider the traditional legal and ethical paradigm to which autonomous weapons systems must conform, and then the major objections and responses being advanced today by critics of autonomous weapons.

A Legal and Ethical Framework

THE BASELINE LEGAL and ethical principles governing the introduction of any new weapon are distinction (or discrimination) and proportionality. Distinction says that for a weapon to be lawful, it must be capable of being aimed at lawful targets, in a way that discriminates between military targets and civilians and their objects. Although most law-of-war concerns about discrimination run to the *use* of a weapon—Is it being used with no serious care in aiming it?—in extreme cases, a weapon itself might be regarded as inherently indiscriminate. Any autonomous robot weapon system will have to possess the ability to be aimed, or aim itself, at an acceptable legal level of discrimination.

Proportionality adds that even if a weapon meets the test of distinction, any actual use of a weapon must also involve an evaluation that sets the anticipated military advantage to be gained against the anticipated civilian harm (to civilian persons or objects). The harm to civilians must not be excessive relative to the expected military gain. While easy to state in the abstract, this evaluation for taking into account civilian collateral damage is difficult for many reasons. While everyone agrees that civilian harm should not be excessive in relation to military advantages gained, the comparison is apples and oranges. Although there is a general sense that excess can be determined in truly gross cases, there is no accepted formula that gives determinate outcomes in specific cases; it is at bottom a judgment rather than a calculus. Nonetheless, it is a fundamental requirement of the law and ethics of war that any military operation undertake this judgment, and that must be true of any autonomous weapon system's programming as well.

These are daunting legal and ethical hurdles if the aim is to create a true "robot soldier." One way to think about the requirements of the "ethical robot soldier," however, is to ask what we would require of an ethical human soldier performing the same function.

Some leading roboticists have been studying ways in which machine programming might eventually capture the two fundamental principles of distinction and proportionality. As for programming distinction, one could theoretically start with fixed lists of lawful targets—for example, programmed targets could include persons or weapons that are firing at the robot—and gradually build upwards toward inductive reasoning about characteristics of lawful targets not already on the list. Proportionality, for programming purposes, is a relative judgment: Measure anticipated civilian harm and measure military advantage; subtract and measure the balance against some determined standard of "excessive"; if excessive, do not attack an otherwise lawful target. Difficult as these calculations seem to any experienced law-of-war lawyer, they are nevertheless the fundamental conditions that the ethically designed and programmed robot soldier would have to satisfy and therefore what a programming development effort must take into account. The ethical and legal engineering matter every bit as much as the mechanical or software engineering.

Four Objections

IF THIS IS the optimistic vision of the robot soldier of, say, decades from now, it is subject already to four main grounds of objection. The first is a general empirical skepticism that machine programming could ever reach the point of satisfying the fundamental ethical and legal principles of distinction and proportionality. Artificial intelligence has overpromised before. Once into the weeds of the judgments that these broad principles imply, the requisite intuition, cognition, and judgment look ever more marvelous—if not downright chimerical when attributed to a future machine.

This skepticism is essentially factual, a question of how technology evolves over decades. Noted, it is quite possible that fully autonomous weapons will never achieve the ability to meet these standards, even far into the future. Yet we do not want to rule out such possibilities—including the development of technologies of war that, by turning decision chains over to machines, might indeed reduce risks to civilians by making targeting more precise and firing decisions more controlled, especially compared to human soldiers whose failings might be exacerbated by fear, vengeance, or other emotions.

It is true that relying on the promise of computer analytics and artificial intelligence risks pushing us down a slippery slope, propelled by the promise of future technology to overcome human failings rather than addressing them directly. If

forever unmet, it becomes magical thinking, not technological promise. Even so, articulation of the tests of lawfulness that autonomous systems must ultimately meet helps channel technological development toward the law of war's protective ends.

A second objection is a categorical moral one which says that it is simply wrong per se to take the human moral agent entirely out of the firing loop. A machine, no matter how good, cannot completely replace the presence of a true moral agent in the form of a human being possessed of a conscience and the faculty of moral judgment (even if flawed in human ways). In that regard, the title of this essay is deliberately provocative in pairing "robot" and "soldier," because, on this objection, such a pairing is precisely what should never be attempted.

This is a difficult argument to engage, since it stops with a moral principle that one either accepts or not. Moreover, it raises a further question as to what constitutes the tipping point into impermissible autonomy, given that the automation of weapons functions is likely to occur in incremental steps.

The third objection holds that autonomous weapons systems that remove the human being from the firing loop are unacceptable because they undermine the possibility of holding anyone accountable for what, if done by a human soldier, might be a war crime. If the decision to fire is made by a machine, who should be held responsible for mistakes? The soldier who allowed the weapon system to be used and make a bad decision? The commander who chose to employ it on the battlefield? The engineer or designer who programmed it in the first place?

One objection holds that autonomous weapons systems undermine the possibility of holding anyone accountable.

This is an objection particularly salient to those who put significant faith in laws-of-war accountability by mechanisms of individual criminal liability, whether through international tribunals or other judicial mechanisms. But post-hoc judicial accountability in war is just one of many mechanisms for promoting and enforcing compliance with the laws of war, and its global effectiveness is far from clear. Devotion to individual criminal liability as the presumptive mechanism of accountability risks blocking development of machine systems that would, if successful, reduce actual harm to civilians on or near the battlefield.

Finally, the long-run development of autonomous weapon systems faces the objection that, by removing one's human soldiers from risk and reducing harm to civilians through greater precision, the disincentive to resort to armed force is diminished. The result might be a greater propensity to use military force and wage war.

As a moral matter, this objection is subject to a moral counter-objection. Why not just forgo all easily obtained protections for civilians or soldiers in war for fear that without holding these humans "hostage," so to speak, political leaders would be tempted to resort to war more than they ought? Moreover, as an empirical matter, this objection is not so special to autonomous weapons. Precisely the same objection can be raised with respect to remotely piloted drones—and, generally, with respect to any technological development that either reduces risk to one's own forces or, especially perversely, reduces risk to civilians, because it invites more frequent recourse to force.

These four objections run to the whole enterprise of building the autonomous robot soldier, and important debates could be held around each of them. Whatever their merits in theory, however, they all face a practical difficulty: the incremental way autonomous weapon systems will develop. After all, these objections are often voiced as though there was likely to be some determinate, ascertainable point when the human-controlled system becomes the machine-controlled one. It seems far more likely, however, that the evolution of weapons technology will be gradual, slowly and indistinctly eroding the role of the human in the firing loop. And crucially, the role of real-time human decision-making will be phased out in some military contexts in order to address some technological or strategic issue unrelated to autonomy, such as the speed of the system's response. "Incrementality" does not by itself render any of these universal objections wrong per se—but it does suggest that there is another kind of discussion to be had about regulation of weapons systems undergoing gradual, step-by-step change.

International Treaties and Incremental Evolution

CRITICS SOMETIMES PORTRAY the United States as engaged in relentless, heedless pursuit of technological advantage—whether in drones or other robotic weapons systems—that will inevitably be fleeting as other countries mimic, steal, or reverse engineer its technologies. According to this view, if the United States would quit pursuing these technologies, the genie might remain in the bottle or at least emerge much more slowly and in any case under greater restraint.

This is almost certainly wrong, in part because the technologies at issue—drone aircraft or driverless cars, for example—are going to spread with respect to general use far outside of military applications. They are already doing so faster than many observers of technology would have guessed. And the decision architectures that would govern firing a weapon are not so completely removed from those of, say, an elder-care robot engaged in home-assisted living programmed to decide when to take emergency action.

Moreover, even with respect to militarily specific applications of autonomous robotics advances, critics worrying that the United States is spurring a new arms race overlook just how many military-technological advances result from U.S. efforts to find technological "fixes" to successive forms of violation of the basic laws of war committed by its adversaries. A challenge for the United States and its allies is that it is typically easier and faster for nonstate adversaries to come up with new behaviors that violate the laws of war to gain advantage than it is to come up with new technological counters.

In part because it is also easier and faster for states that are competitively engaged with the United States to deploy systems that are, in the U.S. view, ethically and legally deficient, the United States *does* have a strong interest in seeing that development and deployment of autonomous battlefield robots be regulated, legally and ethically. Moreover, critics are right to argue that even if U.S. abstention from this new arms race alone would not prevent the proliferation of new destructive technologies, it would nonetheless be reckless for the United States to pursue them without a strategy for responding to other states' or actors' use for military ends. That strategy necessarily includes a role for normative constraints.

These observations—and alarm at the apparent development of an arms race around these emerging and future weapons—lead many today to believe that an important part of the solution lies in some form of multilateral treaty. A proposed treaty might be "regulatory," restricting acceptable weapons systems or regulating their acceptable use (in the manner, for example, that certain sections of the Chemical Weapons Convention or Biological Weapons Convention regulate the monitoring and reporting of dual use chemical or biological precursors). Alternatively, a treaty might be flatly "prohibitory"; some advocacy groups have already moved to the point of calling for international conventions that would essentially ban autonomous weapons systems altogether, along the lines of the Ottawa Convention banning antipersonnel landmines.

Ambitions for multilateral treaty regulation (of either kind) in this context are misguided for several reasons. To start with, limitations on autonomous military technologies, although quite likely to find wide superficial acceptance among nonfighting states and some nongovernmental groups and actors, will have little traction with states whose practice matters most, whether they admit to this or not. Israel might well be the first state to deploy a genuinely autonomous weapon system, but for strategic reasons not reveal it until actually used in battle. Some states, particularly Asian allies worried about a rising and militarily assertive China, may want the United States to be more aggressive, not less, in adopting the latest technologies, given that their future adversary is likely to have fewer scruples about the legality or ethics of its own autonomous weapon systems. America's key Asian allies might well favor nearly any technological development that extends the reach and impact of U.S. forces or enhances their own ability to counter adversary capabilities.

Even states and groups inclined to support treaty prohibitions or limitations will find it difficult to reach agreement on scope or workable definitions because lethal autonomy will be introduced incrementally. As battlefield machines become smarter and faster, and the real-time human role in controlling them gradually recedes, agreeing on what constitutes a prohibited autonomous weapon will likely be unattainable. Moreover, no one should forget that there are serious humanitarian risks to prohibition, given the possibility that autonomous weapons systems could in the long run be more discriminating and ethically preferable to alternatives. Blanket prohibition precludes the possibility of such benefits. And, of course, there are the endemic challenges of compliance—the collective action problems of failure and defection that afflict all such treaty regimes.

Principles, Policies, and Processes

NEVERTHELESS, THE DANGERS associated with evolving autonomous robotic weapons are very real, and the United States has a serious interest in guiding development in this context of international norms. By international norms we do not mean new binding legal rules only—whether treaty rules or customary international law—but instead widely held expectations about legally or ethically appropriate conduct, whether formally binding or not. Among the reasons the United States should care is that such norms are important for guiding and constraining its internal practices, such as R&D and eventual deployment of autonomous lethal systems it regards as legal. They help earn and sustain necessary buy-in from the officers and lawyers who would actually use or authorize such systems in the field. They assist in establishing common standards among the United States and its partners and allies to promote cooperation and permit joint operations. And they raise the political and diplomatic costs to adversaries of developing, selling, or using autonomous lethal systems that run afoul of these standards.

A better approach than treaties is the gradual development of internal state norms and best practices.

A better approach than treaties for addressing these systems is the gradual development of internal state norms and best practices. Worked out incrementally, debated, and applied to the weapons development processes of the United States, they can be carried outwards to discussions with others around the world. This requires long-term, sustained effort combining internal ethical and legal scrutiny—including specific principles, policies, and processes—and external diplomacy.

To be successful, the United States government would have to resist two extreme instincts. It would have to resist its own instincts to hunker down behind secrecy and avoid discussing and defending even guiding principles. It would also have to refuse to cede the moral high ground to critics of autonomous lethal systems, opponents demanding some grand international treaty or multilateral regime to regulate or even prohibit them.

The United States government should instead carefully and continuously develop internal norms, principles, and practices that it believes are correct for the design and implementation of such systems. It should also prepare to articulate clearly to the world the fundamental legal and moral principles by which all parties ought to judge autonomous weapons, whether those of the United States or those of others.

The core, baseline principles can and should be drawn and adapted from the customary law-of-war framework: distinction and proportionality. A system must be capable of being aimed at lawful targets—distinction—but how good must that capability be in any particular circumstance? The legal threshold has historically depended in part upon the general state of aiming

technology, as well as the intended use. Proportionality, for its part, requires that any use of a weapon must take into account collateral harm to civilians. This rules out systems that simply identify and aim at other weapons without taking civilians into account—but once again, what is the standard of care for an autonomous lethal system in any particular "proportionality" circumstance? This is partly a technical issue of designing systems capable of discerning and estimating civilian harm, but also partly an ethical issue of attaching weights to the variables at stake.

The U.S. must develop a set of principles to regulate and govern advanced autonomous weapons.

These questions move from overarching ethical and legal principles to processes that make sure these principles are concretely taken into account—not just down the road at the deployment stage but much earlier, during the R&D stage. It will not work to go forward with design and only afterwards, seeing the technology, to decide what changes need to be made in order to make the system's decision-making conform to legal requirements. By then it may be too late. Engineering designs will have been set for both hardware and software; significant national investment into R&D already undertaken that will be hard to write off on ethical or legal grounds; and national prestige might be in play. This would be true of the United States but also other states developing such systems. Legal review by that stage would tend to be one of justification at the back end, rather than seeking best practices at the front end.

The United States must develop a set of principles to regulate and govern advanced autonomous weapons not just to guide its own systems, but also to effectively assess the systems of other states. This requires that the United States work to bring along its partners and allies—including NATO members and technologically advanced Asian allies—by developing common understandings of norms and best practices as the technology evolves in often small steps. Just as development of autonomous weapon systems will be incremental, so too will development of norms about acceptable systems and uses.

Internal processes should therefore be combined with public articulation of overarching policies. Various vehicles for declaring policy might be utilized over time—perhaps directives by the secretary of defense—followed by periodic statements explaining the legal rationale behind decisions about R&D and deployment of weapon technologies. The United States has taken a similar approach in the recent past to other controversial technologies, most notably cluster munitions and landmines, by declaring commitment to specific standards that balance operational necessities with humanitarian imperatives.

To be sure, this proposal risks papering over enormous practical and policy difficulties. The natural tendency of the U.S. national security community—likewise that of other major state powers—will be to discuss little or nothing, for fear of revealing capabilities or programming to adversaries, as well as inviting industrial espionage and reverse engineering of systems. Policy statements will necessarily be more general and less factually specific than critics would like. Furthermore, one might reasonably question not only whether broad principles such as distinction and proportionality can be machine-coded at all but also whether they can be meaningfully discussed publicly if the relevant facts might well be distinguishable only in terms of digital ones and zeroes buried deep in computer code.

These concerns are real, but there are at least two mitigating solutions. First, as noted, the United States will need to resist its own impulses toward secrecy and reticence with respect to military technologies, recognizing that the interests those tendencies serve are counterbalanced here by interests in shaping the normative terrain on which it and others will operate militarily as technology quickly evolves. The legitimacy of such inevitably controversial systems in the public and international view matters too. It is better that the United States work to set global standards than let other states or groups set them.

Of course, there are limits to transparency here, on account of both secrecy concerns and the practical limits of persuading skeptical audiences about the internal and undisclosed decision-making capacities of rapidly evolving robotic systems. A second part of the solution is therefore to emphasize the internal processes by which the United States considers, develops, and tests its weapon systems. Legal review of any new weapon system is required as a matter of international law; the U.S. military would conduct it in any event. Even when the United States cannot disclose publicly the details of its automated systems and their internal programming, however, it should be quite open about its vetting procedures, both at the R&D stage and at the deployment stage, including the standards and metrics it uses.

Although the United States cannot be too public about the results of such tests, it should be prepared to share them with its close military allies as part of an effort to establish common standards. Looking more speculatively ahead, the standards the United States applies internally in developing its systems might eventually form the basis of export control standards. As other countries develop their own autonomous lethal systems, the United States can lead in forging a common export control regime and standards of acceptable autonomous weapons available on international markets.

A Traditional Approach To a New Challenge

IN THE END, one might still raise an entirely different objection altogether to these proposals: That the United States should not unnecessarily constrain itself in advance through a set of normative commitments, given vast uncertainties about the technology and future security environment. Better cautiously to wait, the argument might go, and avoid binding itself to one or another legal or ethical interpretation until it needs to. This fails to appreciate, however, that while significant deployment of highly autonomous systems may be far off, R&D decisions are already upon us. Moreover, shaping international norms is a long-term process, and unless the United States and its allies accept some risk in starting it now, they may lose the opportunity to do so later.

In the end, all of this is a rather traditional approach—relying on the gradual evolution and adaptation of long-standing law-of-war principles. The challenges are scarcely novel.

Some view these automated technology developments as a crisis for the laws of war. But provided we start now to incorporate ethical and legal norms into weapons design, the incremental movement from automation to genuine machine autonomy already underway might well be made to serve the ends of law on the battlefield.

Critical Thinking

1. Should the use unilaterally set standards for robotic warfare or should standards be determined by an unternational conference?

2. Has the use of drones in Pakistan and Yemen been a success?why or why not?

3. Should killer robots have the power of life and death over human beings?

Create Central

www.mhhe.com/createcentral

Internet References

Report of the Special Rapporteur on extrajudicial, summary or arbitrary executions, Christ of Heynes
www.ohchr.org/Documents/HRBodies/HRCouncil/Regular/Session23/A-HRC

Between a Drone and Al-Qaeda
www.hrw.org/sites/default/files/reports/yemen1013_ForUpload-0.pdf

Report of the Special Rapporteur on extrajudicial, summary, or arbitrary executions
http://justsecurity.org/wp-content/uploads/2013/10/UN-Special-Rapporteur-Extrajudicial

KENNETH ANDERSON is a law professor at American University. Matthew Waxman is a professor at Columbia Law School and adjunct senior fellow at the Council on Foreign Relations. Both are members of the Hoover Institution Task Force on National Security and Law.

Prepared by: Robert Weiner, *University of Massachusetts/Boston*

Article

General Mladic in the Hague: A Report on Evil in Europe—and Justice Delayed

Michael Dobbs

Learning Outcomes

After reading this article, you will be able to:

• Explain why it is so difficult to prosecute war criminals for genocide.

• Explain why there is the feeling that the war criminals and ethnic cleansers won in Srebrenica.

Shortly after he was first charged with crimes against humanity in July 1995, Ratko Mladic was asked what it felt like to be branded "a war criminal" by an international court. The Bosnian Serb military commander seethed with a mixture of barely controlled anger and contempt as he rejected the "idiotic accusations."

"My people or I were not the first to start that war," he insisted, veins popping from his bloated red face. "I don't recognize any trials except the trial of my own people."

Seventeen years later, the once all-powerful general finally appeared this spring before the International Criminal Tribunal for the former Yugoslavia in The Hague to answer charges of genocide, persecution, extermination, unlawful attacks on civilians, and hostage-taking. Partially paralyzed on his right side and looking older than his 70 years, he is physically much diminished, a shadow of the man who became known as the "butcher of the Balkans" for the campaign of terror he waged against Bosnia's non-Serb population. But he is recognizably the same person—proud, willful, and completely unrepentant.

Mladic flashed the thumbs-up sign as he entered the courtroom in May, nodded approvingly as he listened to some of the charges against him, and even clapped his hands when the prosecutor played audio clips of him bullying United Nations peacekeepers and ordering the shelling of civilian areas of Sarajevo. "It was as if he was saying that everything that he did was completely justified," Jasmina Mujkanovic, whose father was killed in the infamous Omarska concentration camp, told me.

Together with victim representatives like her, I was seated in the public gallery of the tribunal's high-tech courtroom. We could see everything that was going on, but we were separated from the accused by a thick pane of glass. It was probably just as well, as the mother of one of his victims found it impossible to restrain herself in the presence of their tormentor and made insulting gestures. Mladic replied with a threat, slowly drawing a finger across his throat.

Mladic's alleged crimes represent the greatest evil that has been perpetrated in Europe since World War II: the ethnic cleansing of hundreds of thousands of Bosnian Muslims, culminating in the coldblooded execution of more than 7,000 prisoners in Srebrenica. The West may have closed its eyes to worse atrocities in the past 70 years, but none in its civilizational backyard, a mere stone's throw from where the Holocaust laid bare Europe's pretensions to enlightenment. Which is exactly why I have spent the past 10 months investigating the case and traveling through the former Yugoslavia—interviewing victims, witnesses, and perpetrators—identifying what we now know about these atrocities and trying to uncover what we still don't two decades later.

Watching Mladic finally appear in court, I couldn't help thinking about another much-anticipated war crimes case, 50 years ago. Adolf Eichmann went on trial in Jerusalem in 1961 accused of crimes against humanity for his involvement in the Nazis' murder of 6 million Jews. The most celebrated chronicler of the Eichmann trial was, of course, Hannah Arendt, who wrote a series of articles for the *New Yorker* that were eventually turned into a book, *Eichmann in Jerusalem*. The book was subtitled "A Report on the Banality of Evil," a phrase that sought to explain how the ordinary, harmless-looking bureaucrat in the dock had committed such monstrous, out-of-the-ordinary crimes.

Mladic was never a harmless-looking bureaucrat. He was a general born for command who got his hands dirty—and bloody—on the battlefield. He was not simply a cog in the machinery of genocide: He set the machinery in motion and supervised every aspect of its operation. In the words of the late Richard Holbrooke, the United States diplomat who helped bring the three-and-a-half-year Bosnian conflict to an end, Mladic was "one of those lethal combinations that history thrusts up occasionally—a charismatic murderer."

But even that does not fully explain Mladic and his motivations. When I lived in Belgrade during the final years of Tito's dictatorship in the late 1970s, I did not consciously divide my friends into Serb or Croat, Muslim or Christian. No one did. Tito's insistence on "brotherhood and unity," enforced when necessary by the army and secret police, along with collective pride in his refusal to kowtow to foreign powers, resulted in a sort of ethnic harmony. Even if it was imposed from above, that system had its true believers—Mladic among them. So as I finally had a chance to look into Mladic's piercing blue eyes, I tried to understand how a man who ritualistically swore to defend Tito's achievements could have ordered the coldblooded execution of thousands.

Indeed, however unambiguous the basic facts of Srebrenica—mass graves leave little room for moral interpretation—the hearings before the Yugoslav war crimes tribunal have already revealed significant gaps in our understanding of what happened. For example, according to Bosnian Serb military documents and testimony from key participants gathered in preparation for his trial, Mladic did not at first even intend to capture the town of Srebrenica. His initial goal was to create "an unbearable situation" for its inhabitants, forcing them to leave of their own accord. Only when he met no effective resistance from U.N. troops defending the internationally recognized "safe area" did he request approval from Bosnian Serb President Radovan Karadzic to order the final "takeover of Srebrenica." In other words, this was a mass murder born of opportunism.

Since his capture after years on the run, Mladic has appeared nearly a dozen times in public, at various pretrial hearings and then, in mid-May, at the long-awaited start of his trial. Never has he shown any remorse. Yet, revered by his supporters as a mythical, godlike figure whose image remains plastered on walls all over Serbia, he has by turns also appeared rambling, defiant, domineering, melodramatic, conciliatory, argumentative, and seemingly on the verge of tears. At one point during an appearance last October, he pleaded for an additional five minutes with his wife. Mladic seems banal only in that some of his reactions have been so predictably human.

None of this makes his case unique. It is simply the latest step in the world's attempt to bring closure to its most awful crimes. Half a century after the Eichmann trial, if the arc of history is bending at all toward justice, it is doing so not because we have finally recognized the true nature of evil and thereby exorcised it, but because we have continued the painstaking work of uncovering who did what when—and finding a way, however laborious, frustrating, or belated, of punishing them for it.

Seeking insights into Mladic's life, last November I tracked down the man who sheltered him for more than five years in an obscure village on the flat Danubian plain north of Belgrade. Known to his friends as Brane, **Branislav Mladic** is Ratko's second cousin. Their grandfathers were brothers, Serbs from the mountainous region of Bosnia known as Herzegovina.

With a thin, angular face and stubble of gray beard, Brane bears little outward resemblance to his famous relative, except for the same darting eyes and abrupt, no-nonsense manner. A bachelor, he lives by himself in a ramshackle farmhouse, with a few chickens, sheep, and pigs roaming about the courtyard. He made clear he disliked the United States ("The Americans attacked the Serbs for no reason"), but he agreed to talk to me—his first extended interview with an American reporter—because I had been introduced by a friend of a friend. Such connections count for everything in Serbia.

As Brane told the story, through a mist of tobacco smoke and repeated shots of slivovitz, the potent plum brandy that is the Serbian national drink, in early 2006 Ratko showed up on his doorstep in the village of Lazarevo in the middle of the night. By this time, he had become a vagrant, living in a series of borrowed apartments, a wanted man with a $5 million reward on his head from the United States government. First indicted for crimes against humanity in 1995, shortly after Srebrenica, Mladic lived more or less openly in Belgrade until 2002, when the Serbian parliament adopted a law belatedly promising to cooperate with the Yugoslav war crimes tribunal.

"Do you know who I am?" he whispered to Brane, before ordering him to turn off the porch light. Brane was shocked by his cousin's appearance, but recognized his voice, which still had the timbre of a man accustomed to being instantly obeyed. Although their paths had separated, Brane had followed Ratko's exploits as the legendary general who stood up for Serbian minorities, first in Croatia in 1991 and then in Bosnia, during the brutal war that ended with the 1995 Dayton peace agreement and the de facto partition of the country into mini-states controlled by Serbs, Muslims, and Croats.

Over the next few years, Ratko and Brane settled into a fixed routine, living in separate rooms across the small farmyard. In the early morning, before Brane headed off to the fields, they would drink coffee together. Ratko spoke about his father, Nedo, a member of Tito's communist partisans killed during World War II by Croatian nationalists allied with Hitler. Ratko described how he went looking for his father's grave in the mountains of Bosnia-Herzegovina. He eventually found an old Muslim who showed him the place where his father was buried. The grave had been washed away by a mountain stream, but Ratko told Brane that he was so grateful for this information that he "spared" the Muslim village of Bradina from Serbian assault during the war.

Several of Mladic's relatives, including Brane's father, Dusan, ended up in the rich agricultural region of Vojvodina after World War II. As former partisans, they were encouraged to occupy land that had been cleansed of Swabian Germans. At school, they were taught that ethnic differences no longer mattered in the brave new Yugoslavia being forged by Tito. At home, they clung to the traditions they had brought with them from the inhospitable Herzegovinian mountains, as well as the memory of defending themselves from their enemies, whether Germans, Muslims, or Croats.

Unlike Brane, a former factory worker who turned to farming when the Yugoslav economy fell apart following the collapse of communism, Ratko rose through the ranks of the Yugoslav army, serving in Kosovo and Macedonia. As a professional military officer, he had a strong incentive to embrace the Titoist idea of loyalty to an overarching nation. He described himself as a "Yugoslav," or "South Slav," rather than as a "Serb," in official censuses. At the same time, he was always very aware of his ethnic identity. With its Serb-dominated officer corps, the Yugoslav People's Army was one of the most efficient channels of upward mobility for peasant families like the Mladices who bore the brunt of the fighting in World War II.

Born in 1942 according to official Yugoslav records—1943 according to family lore—Mladic was a child of a conflict that was a struggle for national liberation, political revolution, and civil war all rolled into one. His very name, Ratko, derives from *rat,* Serbian for "war," and he spent his entire professional life preparing for war against the enemies of Tito's Yugoslavia. The war, when it came, was against internal enemies rather than external ones, but the ideological mindset was much the same. As Mladic saw it, Croats and Bosnian Muslims became proxies for Germans and Ottoman Turks, the peoples who had inflicted so much suffering on his Serbian ancestors.

Although Brane refuses to speak ill of his celebrated relative, living alone with Ratko for five years cannot have been easy. As Mladic has shown in The Hague, he is a controlling person, given to angry outbursts when he fails to get his way. Whether he is commanding an 80,000-man army, dealing with a roomful of judges and lawyers, or living at home with his hermit cousin, he must always be the focus of attention.

According to Brane, Ratko whiled away the time watching television and reading newspapers. For exercise, he would occasionally walk around the farmhouse late at night, once he was sure that all the neighbors had gone to bed. Brane let him have the keys to his ramshackle Volkswagen Polo, but he does not think his cousin ever used it. The former general liked to reminisce about his exploits during the war in Bosnia, but steered clear of controversial topics, such as the killings in Srebrenica. Brane accepted his cousin's explanation that "everything that I did was for one purpose only—to defend the rights of the Serbian people."

One evening in January 2011, Brane returned home from the fields to find Ratko slumped over in the bath, paralyzed on the right side of his body. "For four days, he could not get up. He could not go to the toilet. He could not move," Brane told me. Afraid to summon a doctor who might report Mladic to the authorities, Brane treated his cousin with heart medicine he was able to scrounge from a pharmacy.

Recovering from a stroke without proper medical attention may have weakened Mladic's resolve to evade capture and transfer to The Hague. He yearned for contact with family, particularly his son, Darko. A few weeks after what his lawyer said was a third stroke, in May 2011, Mladic demanded to see his grandchildren, who were visiting another relative in Lazarevo. Darko brought his 10-year-old daughter and 5-year-old son to Brane's place, on the pretext of looking at the animals. Mladic stared at the children through the curtains of his room as they petted the pigs, but did not actually greet them.

"I told him it was a big mistake, but he wouldn't listen," Brane recalled.

The Serbian police were monitoring Darko's movements, hoping that he would eventually lead them to his father. Less than a week later, on May 26, they broke through the gates of the farmyard. Mladic had a loaded pistol nearby, but made little attempt to reach it. Sick and feeble, he was psychologically ready to embark on a new stage of his never-ending war.

There could scarcely be a greater contrast between the fugitive who meekly surrendered to police a year ago and the warlord who determined with a wave of his finger whether a prisoner would live or die. "I am giving your life to you as a gift," he told a frightened young man captured by Serbian troops a week after the fall of Srebrenica. "Don't go back to the front. Next time, there won't be any forgiveness."

A former silver-mining town with a population of close to 40,000 before the war, Srebrenica was one of several Muslim-controlled enclaves in eastern Bosnia that survived the initial Serbian onslaught in 1992. Declared a "safe area" by the United Nations in 1993, it was an obstacle to Serbian control of the strategically important Drina River valley separating Bosnia from Serbia. After Muslim fighters based in Srebrenica mounted raids against nearby Serbian villages, destroying dozens of homes and killing hundreds of Serbs, Mladic swore to take revenge.

Video of Mladic's triumphant entry into Srebrenica on July 11, 1995, captures a man intent on controlling every detail of the operation. He is commanding general, platoon leader, traffic cop, political commentator, and movie producer all rolled into one. "Film that," he shouts to the cameraman. "Take down that Muslim street sign," he tells someone else, addressing his subordinates as "dumb fucks." When he comes across a U.N. vehicle stuck in a ditch, he personally supervises its recovery.

"The boss can't stop commanding for five minutes," jokes a member of his staff on another occasion, when Mladic's back is briefly turned.

"You know how he is," laments another.

Mladic's penchant for micromanagement is one reason it is impossible to imagine that the brutal executions of more than 7,000 Srebrenica men and boys between July 12 and 15 could possibly have happened without his knowledge and express instructions. Evidence presented at The Hague strongly suggests that Mladic ordered the executions, which were then supervised by a trusted aide, Col. Ljubisa Beara.

Mladic personally oversaw the separation of Muslim male refugees from women and children outside the gates of the U.N. military compound in Srebrenica. He was also present when thousands of Muslim men attempting to flee across the mountains to government-held territory were captured by Bosnian Serb forces. Mladic promised the refugees they would be exchanged for Serbian prisoners. Instead, they were loaded into buses and taken to execution sites, where they were mowed down by firing squad.

In the fall of 1995, when the outside world began to learn the horror of what had happened at Srebrenica, Mladic mobilized the resources of the Bosnian Serb army to cover up the crime. His subordinates used bulldozers and dump trucks to dig up at least four mass graves containing the bones of Srebrenica victims and scatter the remains in dozens of secondary graves in remote valleys of eastern Bosnia. Unfortunately for Mladic, United States spy satellites recorded the attempted deception in detail, enabling investigators to locate the secondary graves and use DNA samples to identify victims.

The biggest remaining mystery is not whether Mladic ordered the massacre or how it was carried out, but why.

The biggest remaining mystery is not whether Mladic ordered the massacre or how it was carried out, but why. Plenty of evidence shows that he always had a ruthless, hands-on streak. Intercepted phone calls show that he stood on the hills above Sarajevo in May 1992 personally directing Serbian artillery fire. "Don't let them sleep. . . . Drive them crazy," he ordered at the beginning of the siege. "Shoot at Pofalici [a predominantly Muslim neighborhood]. There is not much Serb population there. . . . Fire one more salvo at the Presidency [headquarters of the Muslim-led Bosnian government]." His willfulness and determination to win at all costs caused him to commit acts that most of us would consider war crimes.

There is, however, an important distinction between shelling a city, even indiscriminately, and murdering 7,000 prisoners in the space of three days. Unlike Karadzic, his nominal superior, Mladic was not a Serbian nationalist, at least not initially. He disapproved of the Chetnik paramilitaries who ran riot in Bosnia at the beginning of the war, and he attempted to build a professional army. Bosnian Serb records show that Mladic urged his comrades to restrain their territorial ambitions and avoid a strategy of ethnic cleansing, which would be impossible to justify to international public opinion. Speaking to a session of the Bosnian Serb assembly on May 12, 1992, Mladic chillingly warned that such a policy "would be genocide."

So what happened? A study of the trial record of top Mladic associates shows that Mladic's thinking changed in several important ways between 1992 and 1995. First, he blamed the Muslims and the Croats for breaking up his beloved Yugoslavia in the dramatic years of 1991 and 1992, with the assistance of Western countries, notably Germany, which had been quick to recognize the new republics of Slovenia and Croatia. "We were a happy country with happy peoples, and we had a good life," he told Dutch peacekeepers in Srebrenica, "until Muslims began listening to what [European leaders] and the Western mafia were telling them."

As Mladic sees it, Yugoslavia was destroyed by the same forces that tore the country apart during World War II, when many Croat and Muslim politicians allied themselves with Nazi Germany. Yugoslavia's breakup left nearly 2 million Serbs stranded in the newly independent states of Croatia and Bosnia, easy prey for politicians intent on stirring up memories of World War II atrocities.

Then there was the logic of the war itself. Serbian atrocities against Muslims led to Muslim atrocities against Serbs (though on nowhere near the same scale). Before U.N. peacekeepers arrived in 1993, the Muslim defenders of Srebrenica had raided nearby Serbian villages in search of food, destroying property and killing civilians. And Mladic lived by a very simple code, the same code that had guided so many of his ancestors: kill or be killed. He justified the mass killing of Srebrenica Muslims by pointing to the crimes allegedly committed against Serbs.

Finally, as the international community failed to intervene, Mladic became ever more contemptuous of the West and ever more convinced of his own invincibility. In 1992, he was still concerned about how the world would react to mass killings and expulsions of non-Serbs. By 1995, he had lost all sense of restraint. A pile of U.N. resolutions that were never implemented, along with the fecklessness of Western leaders, convinced him that he could get away with anything. He was fully in control of the situation in his own country, and nobody could challenge him. NATO had become a joke.

"Are they going to bomb us?" he asked rhetorically shortly after Srebrenica fell. "No way!" (In fact, a massive NATO bombing campaign began a few days later, laying the groundwork for the Dayton peace negotiations.)

I've come to conclude that Mladic is a prime example of Lord Acton's dictum that "Power tends to corrupt, and absolute power corrupts absolutely." By the summer of 1995, he had become the master of his little universe, cut off from political reality. Surrounded by sycophants who dared not contradict him, he became a victim of his own propaganda, comparing himself to heroes in Serbian history who had gained immortality by "fighting the Turks"—a term he used to disparage Bosnian Muslims, who, to Mladic, had committed the unforgivable historical sin of aiding the Ottomans who ruled over Bosnia for more than four centuries and crushed a series of rebellions by Orthodox Serbs. "We present this city to the Serbian people as a gift," he announced grandly the day Srebrenica fell. "Finally, the time has come to take revenge on the Turks."

Like Bosnia itself, Srebrenica today is a town divided. Several thousand Muslim refugees have returned to their homes, but they have little contact with their Serbian neighbors. "We nod at each other, but we don't drink coffee together," said Samedin Malkic. Out of the 27 boys in his high school class, only three survived, and only Malkic came back to Srebrenica. "It is a ghost town," he told me sadly. "You don't see a single person you know."

Two decades after the start of the Bosnian war, it is hard to escape the feeling that the war criminals and ethnic cleansers won. There is a painful sense on both sides of the ethnic divide that Srebrenica's former comity will never be restored. "It was like a little America here before the war," said Zejneba Ustic, another Muslim returnee. "We had everything we needed. Today, there is no work. The factories are nearly all closed. The economy has collapsed."

Two decades after the start of the Bosnian war, it is hard to escape the feeling that the war criminals and ethnic cleansers won.

It is sobering to think that a communist dictator did a better job—at least in the short term—of reconciling ethnic groups and building a functioning economy than the Western democracies that took responsibility for Bosnia after the Dayton peace agreement. Tito promoted his "brotherhood and unity" ideology by throwing dissenters into prison and forcibly suppressing any real debate about the ethnic bloodletting triggered by World War II. He forced Bosnians to forget their hatreds—or at least pretend to forget. The West, by contrast, is encouraging them to remember, even if this complicates the process of reconciliation.

The Yugoslav war crimes tribunal set itself the goal of creating an objective historical record on which all reasonable people should be able to agree, based on impartial experts' meticulous documentation of the Srebrenica massacre and other atrocities. Unfortunately, this has not prevented nationalists on all sides from challenging the evidence the court has assembled and promoting alternative, ethnic-centered versions of history. For a taste of these often-outlandish conspiracy theories, you need look no further than the comments section of my blog about the Mladic trial on FOREIGN POLICY'S website, where, for example, "experts" funded by the Bosnian Serb statelet REPUB-LIKA Srpska explain away the mass graves of Srebrenica victims by insisting that they contain the remains of Muslims "killed in combat" rather than executed prisoners.

The start of Mladic's trial was supposed to represent both a crowning moment and a decisive test for the system of international justice that, we should remember, grew out of a humiliating failure to act. Formed in May 1993 as a half-measure by the United States and other Western governments that were unwilling to intervene militarily to stop the bloodshed in the former Yugoslavia, the court was widely viewed as an empty gesture toward the victims of a terrible war, the product of one more meaningless U.N. resolution. Indeed, in its early phase, the tribunal was noteworthy primarily for its powerlessness. Indicted war criminals continued to lead almost normal lives, seemingly immune from justice. Mladic attended weddings and soccer games, and he even went skiing at an Olympic resort near Sarajevo frequented by NATO peacekeepers. Even after he was stripped of official protection in 2002, he was still able to benefit from a support network of retired army officers as he moved from one hiding place to another.

Meanwhile, the Yugoslav war crimes tribunal spawned a network of special courts for Rwanda, Sierra Leone, Cambodia, and East Timor, in addition to the International Criminal Court, which has been hearing Darfur-related cases. The tribunal's first big breakthrough came in 2001 with the transfer to The Hague of former Serbian leader Slobodan Milosevic to face charges of crimes against humanity in Croatia, Bosnia, and Kosovo. Milosevic's trial ended inconclusively in March 2006 when the defendant was found dead in his cell following a massive heart attack.

In an attempt to avoid a repetition of the unsatisfactory ending of the Milosevic case, prosecutors have eliminated 90 incidents from the list of accusations against Mladic. The slimmed-down indictment still includes 106 separate charges, however, including two counts of genocide, revolving around the 1995 Srebrenica massacre and a massive campaign of ethnic cleansing elsewhere in Bosnia. The trial is likely to take at least two years—once, that is, it actually gets going. On only its second day, the presiding judge announced an indefinite suspension, possibly for months, because of "significant disclosure errors" by prosecutors, who had failed to share tens of thousands of documents with Mladic's defense team. It was not a reassuring sign. It took the Yugoslav war crimes tribunal the better part of two decades to bring its most high-profile target to justice, only to bungle the grand opening.

To this point, the tribunal's greatest service has been the promotion of the notion of individual responsibility over the pernicious doctrine of collective guilt. The Bosnian atrocities were made possible in the first place because men like Mladic sought revenge against entire communities for crimes committed "against the Serbian people" by Muslims and Croats. Similarly, Mladic has sought to depict the criminal case against him as a conspiracy by the United States and other NATO countries to discredit the entire "Serbian nation."

"I am not defending myself," he told the court in one of his pretrial hearings. "I am still defending both the Republika Srpska and Serbia and the whole people there."

The presiding judge was quick to set the record straight. "You are charged before this tribunal . . . no one else, not a republic, not a people," he told the old man in the dock. "I would urge you to defend yourself as an accused, rather than to defend persons, entities, organizations which are not accused before this tribunal."

Mladic is right that the trial is about more than just him. But for Bosnia to escape the vicious cycle of hatred begetting more hatred, the judge's approach to history must triumph over Mladic's. As much as we may want to divine the nature of evil, it is more important that we first resolve this one case.

Critical Thinking

1. Do you agree that the international system of justice has failed in dealing with genocide in Srebrenica in the former Yugoslavia?

2. What factors limit the ability of international criminal tribunals to apprehend war criminals?

3. Is the norm of the Responsibility to Protect a viable option to the failure of states to punish genocide?

Create Central

www.mhhe.com/createcentral

Internet References

The International Criminal Tribunal for the Former Yugoslavia
www.icty.org

Genocide Watch Home Page
www.genocidewatch.org

The IInternational Criminal Tribunal for Rwanda
www.unictr.org

U.S. Holocaust Museum
www.ushmm.org

The International Criminal Court
www.icc-cpi.int/Menus/icc

MICHAEL DOBBS, who covered the former Yugoslavia for the *Washington Post*, is a research fellow at the United States Holocaust Memorial Museum and author of the blog *Mladic in The Hague* at ForeignPolicy.com.

Prepared by: Robert Weiner, *University of Massachusetts/Boston*

Article

Why UNESCO Is a Critical Tool for Twenty-First Century Diplomacy

AMBASSADOR DAVID T. KILLION

Learning Outcomes

After reading this article, you will be able to:

- Discuss the goals of UNESCO.
- Explain why the U.S. disengaged from UNESCO.

Fletcher Forum

Many organizations within the United Nations system address education as a key component of international development. What specific role does UNESCO play?

Ambassador Killion

It is true that there are several players in the field of global education, but it's not quite as chaotic as it might appear. The different players agree on the overall goals—for example, completion of primary school—and each organization within the UN system brings a slightly different focus and expertise to the table.

UNESCO is a key piece of the UN puzzle. By concentrating on capacity building, policy implementation, and promotion of best practices at the country level, it advances the UN's goal of strengthening country capacity to deliver effective, equitable and inclusive education. UNESCO also places a particular emphasis on helping countries protect those at risk of social exclusion. Issues of inclusion have similar themes across the globe, but play out differently in each country. UNESCO plays a critical role in building relationships with education ministries and other stakeholders to identify the barriers and opportunities for inclusion, which is essential to ensuring the right to education.

UNESCO also brings a unique perspective to education because of its joint focus on peace and sustainable development. The organization is preparing students to take on twenty-first century problems—from climate change to extremism—by building their critical thinking skills and teaching them the value of freedom and tolerance.

Fletcher Forum

UNESCO's Education Sector aims to tackle a very broad range of education issues, from literacy to post-conflict education to incorporating information communication technologies (ICT) in the classroom. What specific education issues does the U.S. Mission prioritize and why?

We try to take a holistic view and support initiatives that can impact a whole range of education issues. Take Open Educational Resources (OERs), for instance. The term "Open Educational Resources," coined at UNESCO in 2002, refers to teaching, learning, or research materials that have been released under an intellectual property license that allows for their free use and repurposing by others. The U.S. Mission has made it a priority to support UNESCO's work in OERs because they have the potential to transform global education by dramatically increasing access to quality educational materials not only for students in the U.S., but also for girls in Afghanistan and Pakistan, and for children in South Sudan.

The U.S. Mission also sees girls' and women's education as a critical priority because it is connected to so many other education and social issues. You cannot address literacy, for example, without putting a special focus on girls' and women's education because two-thirds of the 775 million people that lack basic literacy skills are women. Literate women are more likely to send their children to school, especially their daughters. They are also more likely to attain a higher socioeconomic status and to raise healthier children. Women and girls are central to the future of global education, which is why we have worked to support UNESCO's work in this area, including through the Global Partnership for Girls' and Women's Education and the U.S.-funded report "From Access to Equality: Empowering Girls and Women through Literacy and Secondary Education."

Finally, we recognize UNESCO's value in cultivating peace and countering extremism through education, so we have initiated or supported programs that aim to ensure that the lessons of the Holocaust are not forgotten or repeated, to combat racism and discrimination among youths, and to prevent homophobic bullying in schools.

Fletcher Forum

Former Secretary of State Hillary Clinton visited UNESCO in 2011 to help kick off "The Global Partnership for Girls' and Women's Education," a new public-private initiative designed to strengthen educational opportunities for girls and women around the globe. Has this public-private partnership been successful thus far?

Killion

The Global Partnership for Girls' and Women's Education has definitely been a highlight of my tenure as Ambassador. The Partnership is a perfect example of how public-private partnerships can be used to address our most pressing educational challenges. Currently, three companies are working within the framework of the Partnership to empower girls and women in five African countries: the GEMS Foundation in Kenya and Lesotho, the Packard Foundation in Ethiopia and Tanzania, and Procter & Gamble (P&G) in Senegal.

The Procter & Gamble-funded example is particularly compelling. P&G partnered with UNESCO to develop a special cause-related marketing campaign for girls' education. The "Always" marketing campaign features the UNESCO logo on certain packs of Always feminine products sold in France and a few other pilot countries. Each time a pack is purchased, Always donates a portion of the proceeds to help fund UNESCO education campaigns in Senegal. The campaign has been wildly successful, boosting sales for P&G and generating significant funds for UNESCO. These funds have allowed the organization to create close to 200 new classrooms, train 100 literacy teachers, and reach 10,000 Senegalese girls and women. But it's really just the beginning because the Always-UNESCO campaign is being extended to Eastern Europe and potentially Brazil, China and the United States. The additional revenue generated will be used to fund new girls' education projects in Africa, starting in Kenya. The Procter & Gamble initiative has prompted other American companies to consider joining the Partnership, so I hope we will see it expand this year.

The Global Partnership for Girls' and Women's Education reflects the changing landscape of development assistance, which is shifting away from a Westphalian state-centric model where governments and intergovernmental organizations are the only actors that count. The reality is that official development aid from developed countries is falling, and is likely to stagnate in Least Developed Countries. However, non-state actors, including private companies, are ready and able to play a more important role in international development. Regarding girls' and women's education, the private sector understands that educating girls generates economic growth and development, which in turn creates new markets. Private companies are ready to invest not just resources, but also ideas, energy, and technology to educate women and girls. But they cannot go it alone if they want to have large-scale impact, which is why they want to work with UNESCO. UNESCO connects them to education ministries and experts, allowing them to fine tune their projects and then scale them up for maximum impact.

Fletcher Forum

In September 2012, UN Secretary-General Ban Ki-moon announced the "Global Education First" initiative in order to invigorate the global movement for education. Does "Global Education First" challenge UNESCO's education mandate? How does the new initiative reflect on the progress or inertia of the education-related Millennium Development Goals (MDGs) intended to be completed by 2015?

Killion

The Global Education First Initiative (GEFI) does not challenge UNESCO's mandate in any way. In fact, it strengthens UNESCO's agenda by giving it visibility at the highest levels of the UN system. In recognition of its lead role in education, UNESCO's Director-General Irina Bokova was selected to serve as the Executive Secretary of the GEFI Steering Committee.

GEFI provides a platform to strengthen education advocacy efforts that are already underway and to ensure that education is a top priority on national agendas. It is important to see that the Global Education First Initiative was designed to highlight effective approaches and encourage action. It was created to help us make that last big push towards achieving the Millennium Development Goals but also look toward the future, post-2015.

Fletcher Forum

"Education for All" (EFA), a global initiative established in 2000 at the World Economic Forum, is also set to conclude in 2015. EFA seeks to provide quality education for children, youth, and adults around the world. Yet, UNESCO reports suggest that at the current rate of progress, most African countries will not reach the EFA goals by 2015. Are these broad goals regarding education ever achievable, given the range of contexts in which the programs are implemented? If so, what should be done now in order to get closer to achieving the EFA goals by 2015?

Killion

There are two ways to look at the objectives of Education for All and the Millennium Development Goals. If we look at them in terms of measurable outcomes, there is no doubt that we are at risk of not achieving the targets that were set. But we can also view them as aspirational, as initiatives that inspired dialogue, kick-started action, and resulted in real progress. It is easy to see 2015 as an end, but in doing so we remove the aspiration and risk giving in to disillusionment and complacency. Instead, we need to think of 2015 in terms of progress along a trajectory and rededicate ourselves to another set of goals post-2015.

Of course, EFA has taught us some valuable lessons about global education. Perhaps the most valuable lesson is that achieving equitable education is an ongoing process, and one that requires constant focus and persistence by every country in the world, including developed countries. EFA has also shown us the value of exchanging information and ideas. This exchange will become all the more important as the global economy evolves and the previous delineations between high- and low-income countries shift along with it.

The United States Mission to UNESCO has been a strong supporter of EFA at UNESCO on both a policy level and a pro- grammatic level. We have advocated for a multi-stakeholder approach and encouraged UNESCO to partner with non-gov- ernmental organizations and the private sector to address edu- cation challenges. We have also supported literacy and human rights education programs, including Holocaust education and Teaching Respect For All.

Fletcher Forum

The United States is working with Brazil and UNESCO to com- bat racism in schools worldwide through the "Teaching Respect for All" program. What sparked the creation of this partner- ship, and why Brazil? Why is it worthwhile to invest in anti- racism programming when more concrete educational needs also exist, such as poor infrastructure and a lack of textbooks?

Killion

The Teaching Respect for All program is so important not only because of its goal of promoting tolerance through education, but also because it shows that the United States is ready to play a leading role internationally in combating racism. After the World Conference Against Racism in 2001 was hijacked and transformed into an "anti-American, anti-Israel circus,"[1] the United States remained disengaged from the UN's work against racism. Teaching Respect for All, spearheaded by Assistant Secretary for International Organization Affairs Esther Brim- mer, effectively ended this period of disengagement and sig- naled the United States' commitment to working with the UN and other key countries to address the scourge of racial and ethnic intolerance. UNESCO and Brazil were natural partners: Brazil because of its shared struggle to overcome the legacy of slavery and UNESCO because of its leadership in international education and past work on tolerance education.

Concretely, Teaching Respect for All is mapping existing anti-racism programs worldwide, collecting best practices, and developing a curriculum framework for anti-racism and toler- ance that countries can adapt to their respective contexts and needs. UNESCO is in the process of selecting the first pilot countries for the program and its goal is to start testing the curriculum framework this year. While I can understand the temptation to view tolerance education as less urgent than infra- structure, they are equally critical. Ethnic tensions continue to destabilize societies and lead to conflict, from the Sahel to Burma. And racism and discrimination continue to take a toll on education objectives and economic prosperity, even in the developed world. Education is essential to addressing these issues; it allows us to reach children in their formative years and to inoculate them against intolerance and extremism.

In addition to its anti-racism programming, UNESCO is working to combat anti-Semitism and ensure that the tragic lessons of the Holocaust are understood, not just in Europe or the United States, but all over the world. UNESCO's Holo- caust Remembrance Education program is relatively new, but, thanks in part to U.S. support, it has already made important strides. Last year UNESCO held a high-level conference in South Africa to discuss how studying the Holocaust can pre- vent genocide in Africa. Thanks to this conference, countries like Rwanda and the Republic of Congo are working with UNESCO to integrate Holocaust education into their national curricula. UNESCO will be holding a similar conference in Latin America in May. UNESCO is also partnering with the Georg Eckert Institute for International Textbook Research on a mapping project that will make it possible to compare repre- sentations of the Holocaust in school textbooks and national curricula. This will help UNESCO and its partners understand where they need to focus their efforts. I've made it a priority to support UNESCO's Holocaust Education program because I believe it is essential to stemming the rise of anti-Semitism and helping people understand the roots of genocide.

Fletcher Forum

In 2011, the United States ceased all extra budgetary contri- butions to UNESCO in accordance with American legislation dating back to the 1990s after Palestine became the organiza- tion's 195th full member. How has the United States continued to engage UNESCO in the wake of this decision?

Killion

Despite the cessation of funding, the United States has remained very engaged in support of U.S. goals at UNESCO through the force of good ideas and creative partnerships. For example, we proposed a resolution to establish an International Day of Jazz, which quickly drew thirty-one co-sponsors and was adopted enthusiastically by UNESCO's 195-member General Conference. Jazz Day events draw musicians from around the world together through the freedom of music with the support of private sec- tor sponsors. We have also encouraged UNESCO to establish more public-private partnerships and have helped them to reach out to U.S. companies and foundations interested in supporting UNESCO's work. For instance, we helped broker the partnership between UNESCO and Procter & Gamble and are encouraging expanded partnerships with Microsoft, Intel, and Google. We also work with American UNESCO Goodwill Ambassadors—jazz legend Herbie Hancock, Oscar-winning actor Forest Whitaker, and Holocaust survivor Sam Pisar—to promote UNESCO values in support of freedom, human rights, and peace-building.

We succeeded in this work, and UNESCO and its member delegations appreciate our positive engagement. UNESCO is an organization that values collaboration and cooperation in support of common goals. However, I will not deny that it has become more difficult for the U.S. Mission as the funding crisis has continued and forced UNESCO to both slash spending and cut positions, which are never popular measures. We under- stand this frustration at the U.S Mission and are working very hard to support the Obama administration's efforts to resolve this issue as soon as possible.

Fletcher Forum

What are the impacts of U.S. disengagement with UNESCO, from both a programmatic and a political perspective?

Killion

U.S. disengagement poses a number of risks. It undermines the President's commitment to a strong multilateral approach in foreign policy. It also paves the way for other powers to play a more powerful role at UNESCO. In the wake of the U.S. funding cut-off, a number of countries including Qatar, Saudi Arabia, Oman, Turkey and Indonesia, made large donations to help cover the shortfall, boosting their stature in the organization. China made a major contribution as well. We work with these countries on many issues; however, we do have significant differences of opinion on some very important policy areas. If the United States disengages from UNESCO, it is possible that other countries might in the future seek to dial back UNESCO's unique work in freedom of expression or Internet freedom, for example.

From a programmatic standpoint, U.S. disengagement would mean that we lose the opportunity to help instill democratic values in UNESCO's educational, cultural, communications, and cultural programs.

Fletcher Forum

Finally, from a legislative standpoint, what would it take for the United States to resume budgetary contributions to UNESCO?

Killion

In order for the United States to resume funding, Congress would have to amend the current law barring U.S. funding to UNESCO. The President included funding for UNESCO in his FY 2014 budget, which was delivered to the Congress on April 10. The President's FY 2014 budget also contained a request for an amendment to the current legislative prohibitions against UNESCO funding.

The United States will lose its vote in UNESCO's General Conference meeting in November of this year if U.S. funding is not restored.

With a lot at stake for our foreign policy, the clock is literally running out for the U.S. at UNESCO.

Endnote

1. See my mentor Tom Lantos' account of the disastrous conference at: http://www.humanrightsvoices.org/assets/attachments/articles/568_durban_debacle.pdf.

Critical Thinking

1. Why did the U.S. suspend its funding to UNESCO?
2. Why is the politicization of UNESCO harmful to its mission?
3. Should the U.S. withdraw from UNESCO? why or why not?

Create Central

www.mhhe.com/createcentral

Internet References

UNESCO
   www.unesco.org
U.S. National Commission for UNESCO
   http://www.state.gov/p/10/unesco/
United Nations
   www.un.org

AMBASSADOR DAVID T. KILLION was nominated as U.S. Permanent Representative to the United Nations Educational, Scientific, and Cultural Organization (UNESCO) with the rank of Ambassador on June 25, 2009 by President Obama and was sworn into office on August 12, 2009. Since his appointment, Ambassador Killion has worked tirelessly to advance U.S. national interests at UNESCO and to help the organization to fulfill its mandate of building peace, eradicating poverty, and achieving sustainable development solutions for today's global challenges. Prior to his appointment, Ambassador Killion served as a Senior Professional Staff Member of the House Committee on Foreign Affairs and was the Committee's top expert on International Organizations and State Department Operations.

Unit 6

UNIT

Prepared by: Robert Weiner, *University of Massachusetts/Boston*

International Political Economy

International political economy is an important component of study in the field of international relations, especially in the highly globalized and interdependent world of the 21st century. Consequently, the collapse of the U.S. credit and housing market in 2009 upended a number of major U.S. banks and financial houses, with far-reaching global implications for advanced economies as well as emerging markets in the multipolar international system. The central question revolves around the extent and nature of the global recovery from the recession of 2009 and whether developments such as the crisis in the Eurozone can threaten what some view as a rather fragile recovery. The U.S. national debt crisis in 2013 also underscored the fragility of the global economy and efforts at recovery. Some bright spots in the global economy were evidenced by recent African economic growth, and the progress that the UN had experienced in achieving the goal of cutting global poverty in half by 2015. The UN had made some gains in promoting eight millennial development goals by 2015, making inroads in cutting poverty, reducing hunger, promoting gender equality in primary education, and increasing access to safe drinking water. However, as 2015 approached, more work needed to be done in setting up a strategy to work on the implementation of post-2015 millennial development goals.

In 2013, the Eurozone crisis that has now roiled the area since 2010, jeopardizing not only the future of the Eurozone, but perhaps even the future of the European Union Itself, seemed to have abated somewhat. Critical to the recovery of the global economy is a successful resolution of the crisis in the Eurozone, which began in 2010 and continued to threaten the future not just of the Eurozone, but the European Union itself. By the fall of 2013, the Eurozone crisis still posed a threat to the viability of the European organization that had kept the peace in Western Europe since the end of the Second World War. The crisis of the euro could also be seen as a political crisis rather than just a financial crisis. The European Union may have expanded too quickly after the end of the Cold War by admitting the former communist states of Eastern Europe. The European Union might have focused more on deeper integration, rather than a broadening of its membership. The euro was designed as a common currency to serve as the basis of the further monetary integration of the organization, but without a sufficient institutional structure to follow up on the next steps toward economic integration. Moreover, nationalism still was an important factor preventing all of the members of the European from adopting the common currency. For example, the United Kingdom and several other members of the European Union opted not to join the Eurozone, and the United Kingdom was scheduled to hold a referendum on its relationship to the European Union. In the final analysis, the Eurozone crisis boiled down to a question of the extent to which the members of the European Union had developed a sense of European identity.

Article

Prepared by: Robert Weiner, *University of Massachusetts/Boston*

Own the Goals: What the Millennium Development Goals Have Accomplished

JOHN W. MCARTHUR

Learning Outcomes

After reading this article, you will be able to:

- Explain the purpose of the Millennium Development Goals.
- Discuss the progress that has been made in achieving the eight Millennium Development Goals.

For more than a decade, the Millennium Development Goals—a set of time-bound targets agreed on by heads of state in 2000—have unified, galvanized, and expanded efforts to help the world's poorest people. The overarching vision of cutting the amount of extreme poverty worldwide in half by 2015, anchored in a series of specific goals, has drawn attention and resources to otherwise forgotten issues. The MDGs have mobilized government and business leaders to donate tens of billions of dollars to life-saving tools, such as antiretroviral drugs and modern mosquito nets. The goals have promoted cooperation among public, private, and nongovernmental organizations (NGOs), providing a common language and bringing together disparate actors. In his 2008 address to the UN General Assembly, the philanthropist Bill Gates called the goals "the best idea for focusing the world on fighting global poverty that I have ever seen."

The goals will expire on December 31, 2015, and the debate over what should come next is now in full swing. This year, a high-level UN panel, co-chaired by British Prime Minister David Cameron, Liberian President Ellen Johnson Sirleaf, and Indonesian President Susilo Bambang Yudhoyono, will put forward its recommendations for a new agenda. The United States and other members of the UN General Assembly will then consider these recommendations, with growing powers, such as Brazil, China, India, and Nigeria, undoubtedly playing a major role in forging any new agreement. But prior to deciding on a new framework, the world community must evaluate exactly what the MDG effort has achieved so far.

Working on a Dream

The MDGs are not a monolithic policy following a single trajectory. Ultimately, they are nothing more than goals, established by world leaders and subsequently reaffirmed on multiple occasions. The MDGs were not born with a plan, a budget, or a specific mapping out of responsibilities. Many think of the MDGs as the UN's goals, since the agreements were established at UN summits and UN officials have generally led the follow-up efforts for coordination and reporting. But the reality is much more complicated. No single individual or organization is responsible for achieving the MDGs. Instead, countless public, private, and nonprofit actors—working together and independently, in developed and developing countries—have furthered the goals. Amid this complexity, the achievements toward reaching the MDGs are all the more impressive. The goals have brought the diffuse international development community closer together.

Before the MDGs were crafted, there was no common framework for promoting global development. After the Cold War ended, many rich countries cut their foreign aid budgets and turned their focus inward, on domestic priorities. In the United States, for example, the foreign aid budget hit an all-time low in 1997, at 0.09 percent of gross national income. Meanwhile, throughout the 1990s, institutions such as the World Bank and the International Monetary Fund (IMF) encouraged developed and developing countries to scale back spending on public programs—in the name of government efficiency—as a condition for receiving support.

The results were troubling. Africa suffered a generation of stagnation, with rising poverty and child deaths and drops in life expectancy. Economic crises and the threat of growing inequality plagued Asia and Latin America. The antiglobalization movement gained such force that in November and December 1999, at what has come to be called "the Battle in Seattle," street protesters forced the World Trade Organization to cancel major meetings midstream.

The suspicions on the part of civil society carried over into policy debates. In the late 1990s, the Organization for Economic Cooperation and Development proposed "international development goal" benchmarks for donor efforts. The oecd's proposal was later co-signed by leaders of the IMF, the World Bank, and the UN. In response, Konrad Raiser, then head of the World Council of Churches, hardly a firebreathing radical, wrote UN Secretary-General KofiAnnan to convey astonishment and disappointment that Annan had endorsed a "propaganda exercise for international finance institutions whose policies are widely held to be at the root of many of the most grave social problems facing the poor all over the world."

That proposal never got off the ground, but the international community made other progress in the lead-up to 2000 that helped set the groundwork for the MDGs. Most notably, G-8 leaders took a major step forward when they crafted a debt-cancellation policy at their 1999 summit in Cologne, Germany. Under this new policy, countries could receive debt relief on the condition that they allocated savings to education or health. This helped reorient governments toward spending in social sectors after many years of cutbacks.

At the 2000 UN Millennium Summit, which was the largest gathering of world leaders to date, heads of state accepted that they needed to work together to assist the world's poorest people. Looking at the challenges of the new century, all the UN member states agreed on a set of measurable, time-bound targets in the Millennium Declaration. In 2001, these targets were organized into eight MDGs: eradicate extreme poverty and hunger; achieve universal primary education; promote gender equality and empower women; reduce child mortality; improve maternal health; combat hiv/aids, malaria, and other diseases; ensure environmental sustainability; and forge global partnerships among different countries and actors to achieve development goals. Each goal was further broken down into more specific targets. For example, the first goal involves cutting in half "between 1990 and 2015, the proportion of people whose income is less than $1 a day."

In practical terms, the MDGs were actually launched in March 2002, at the UN International Conference on Financing for Development, in Monterrey, Mexico. The attendees, including heads of state, finance ministers, and foreign ministers, agreed that developed countries should step in with support mechanisms and adequate financial aid to help poor countries committed to good governance meet the MDG targets. Crucially, leaders set a benchmark for burden sharing when they urged "developed countries that have not done so to make concrete efforts towards the target of 0.7 percent of gross national income (gni) as official development assistance to developing countries." At the time of the conference, the 22 official oecd donor countries allocated an average of 0.22 percent of gni to aid. Thus, working toward a 0.7 target implied more than tripling total global support. The Monterrey conference established the MDGs as the first global framework anchored in an explicit, mutually-agreed-on partnership between developed and developing countries.

The Global Conversation

These historic intergovernmental agreements have inspired much debate. Some NGO leaders, including participants in the annual World Social Forum, distrusted any agreement that involved international financial institutions and was negotiated behind closed doors. Human rights activists were dismayed that the MDGs excluded targets for good governance, which they considered a contributor to development and a key outcome unto itself. Some environmental activists were bothered by the narrow formulation of the targets, which ignored major issues, such as climate change, land degradation, ocean management, and air pollution.

To be sure, the MDG framework is imperfect. Several issues, such as gender equality and environmental sustainability, are defined too narrowly. The education goal is limited to the completion of primary school, overlooking concerns about the quality of learning and secondary school enrollment levels. In addition, some academics, such as the economist William Easterly, argue that the remarkable ambition of the goals is unfair to the poorest countries, which have the furthest to go to meet the targets, and minimizes what progress those countries do achieve. Sure enough, if the child survival goal were to cut mortality by half, instead of by two-thirds, 72 developing countries would already have met the target by 2011. Instead, the two-thirds goal has been achieved by only 20 developing countries so far. In addition, the MDGs' emphasis on human development issues, such as education and health, sometimes downplays the importance of investments in energy and infrastructure that support economic growth and job creation.

Nonetheless, the framework has provided a global rallying point. In 2002, with a mandate from Annan and Mark Malloch Brown, then the administrator of the UN Development Program, the economist Jeffrey Sachs launched the UN Millennium Project, which brought together hundreds of experts from around the world from academia, business, government, and civil-society organizations to construct policy plans for achieving the goals. Sachs also tirelessly lobbied government leaders in both developed and developing countries to expand key programs, especially in health and agriculture, in order to meet the MDG targets.

In the lead-up to the 2005 G-8 summit, in Gleneagles, Scotland, advocacy organizations worldwide championed the MDGs. In developing countries, NGO leaders, such as Amina Mohammed, Kumi Naidoo, and Salil Shetty, encouraged civil-society leaders to hold their governments accountable for meeting the goals. In developed countries, organizations such as one, co-founded by the activist Jamie Drummond, the rock star Bono, and others, petitioned politicians and conducted public awareness campaigns to demand that world leaders step up their efforts to meet the targets. At the summit, British Prime Minister Tony Blair and Gordon Brown, then British chancellor of the exchequer, put the MDGs and foreign aid commitments at the top of the agenda. Leaders at Gleneagles committed to increasing global aid by $50 billion by 2010 and set the groundwork for larger commitments to be made by 2015. However, one powerful player on the world stage, the United States, remained hesitant to embrace the MDG agenda.

Players on the Bench

U.S. President George W. Bush launched the Millennium Challenge initiative in 2002, promising a 50 percent increase in U.S. foreign aid within three years, with money going to countries committed to good governance. The initiative drew inspiration from the MDGs, as the name suggests, but confusingly, it did not directly link to the targets. Ten months later, in his 2003 State of the Union address, Bush launched the President's Emergency Plan for AIDS Relief, which has dramatically improved access to AIDS treatment in the developing world. This program was in many ways in line with the MDG effort but did not explicitly link to the goals. Bush even endorsed the UN Millennium Declaration and the Monterrey agreements, but he refused to support the MDGs, largely because his administration viewed them as UN-dictated aid quotas.

Holding a similar view, State Department officials regularly claimed that they supported the targets of the Millennium Declaration but not the MDGs, despite the fact that the MDG targets were drawn directly from the Millennium Declaration. U.S.-UN tensions over the Iraq war were a critical backdrop, with the Bush administration reticent to support a major UN initiative. Washington's aversion was so strong that many U.S. advocacy groups avoided using the term "Millennium Development Goals" for fear of losing influence. When John Bolton became the U.S. ambassador to the UN in August 2005, one of his first actions was to suggest deleting all references to the MDGs in the drafted agreement of the upcoming UN World Summit. The subsequent uproar from other countries and U.S. media outlets forced Washington to modify its position. In his summit speech, Bush finally endorsed the MDGs, using the phrase "Millennium Development Goals" publicly for the first time.

By refusing to directly engage with the MDGs in their early years, the United States missed an opportunity to highlight its contributions to development efforts and foster international goodwill. In the early years of this century, the United States helped revolutionize global health, a central pillar of the MDGs, first through Bush's AIDS initiative and later through efforts on malaria and other deadly diseases. Furthermore, by resisting a project on which most of the world was actively collaborating, Washington missed easy opportunities to build political capital for solving much thornier and divisive international issues.

Diplomatic tensions have subsided under the Obama administration, which has given much stronger rhetorical support to the MDGs and has continued the previous administration's basic development policies, in addition to launching a major initiative to reduce poverty by supporting small farms around the world. Nevertheless, many officials in Washington remain either skeptical or disengaged when it comes to the MDGs, most likely because of a long-standing aversion to fixed foreign aid spending, especially when defined by an international agreement. This fear, however, is baseless. The MDGs do not dictate any aid commitments, and the only related figure, the

0.7 aid target, which countries agreed to work toward in Monterrey in 2002, was endorsed by Bush. It was only later that some countries, such as the United Kingdom, made timetables to meet this aid target.

The World Bank has similarly missed out. Although the bank has championed the framework at senior political levels, it has not adequately facilitated MDG efforts on the ground. Early resistance was in part due to bureaucratic resentment of the UN for its having been given such a prominent role on development issues. In addition, as an institution dominated by economists, the bank is prone to prioritize economic reforms over investment in social sectors. Even more, there is widespread distrust among the bank's staff that donor countries will provide adequate financing for the MDGs. Such concerns are not without merit, as the G-8 ended up falling more than $10 billion short on its Africa pledges for 2010 alone.

Nevertheless, the bank, as a main interlocutor with the developing world, should have helped poor countries assess how they could achieve the MDGs and sounded the alarm about donor financing gaps. Furthermore, the bank has a self-serving reason to get onboard: the MDGs spurred a major budgetary expansion for the International Development Association, the branch of the bank devoted to supporting the poorest countries. Fortunately, the United States and the World Bank are coming around on the MDGs, attracted by the proven success of the framework.

It's a Small World After All

As of late 2010, five years before the deadline, the world had already met the overarching MDG of cutting extreme poverty by half. The estimated share of the developing-world population living on less than $1.25 per day (the technical MDG measurement of extreme poverty) had dropped from 43 percent in 1990 to roughly 21 percent in 2010. This statistic is somewhat skewed by progress that was under way in China and other Asian countries long before the MDGs were adopted. The framework is not solely responsible for all of the advancements of the past 12 years. Many other forces, such as the expansion of global markets and the creation of groundbreaking health and communications technologies, have helped the developing world. Moreover, the goals relating to hunger, sanitation, and the environment have not been met. Poverty reduction, however, has progressed in every region since 2000. Even excluding China from the global calculation, the world's share of impoverished people fell from 37 percent in 1990 to 25 percent in 2008, and forthcoming data should show an even greater drop.

Most important, the MDGs have kick-started progress where it was lacking, especially in Africa, where unprecedented economic growth and poverty reduction are now taking place. From 1981 to 1999, extreme poverty in sub-Saharan Africa rose from 52 percent of the population to 58 percent. But since the launch of the MDGs, it has declined sharply, to 48 percent in 2008. Much of this was likely driven by MDG-backed investments in healthier and better-educated work forces in the region. The global MDG campaign has also prompted support for small subsistence and cash-crop farms, which has boosted growth in many low-income countries, such as Malawi.

Primary education rates have increased around the world, too, with South Asia and sub-Saharan Africa experiencing particularly big jumps in enrollment. Much of this has been the result of funding from MDG-linked initiatives, such as the Global Partnership for Education, launched in 2002 by the World Bank and other development organizations to help poor countries "address the large gaps they face in meeting education MDG 2 and 3, in areas of policy, capacity, data, finance." These same efforts have helped nearly every world region achieve gender parity in classrooms.

The greatest MDG successes undoubtedly concern health. The MDGs have invigorated multilateral institutions, such as the GAVI Alliance (formerly called the Global Alliance for Vaccines and Immunization), which seeks to achieve MDGs "by focusing on performance, outcomes and results." The goals have also inspired a huge increase in private-sector aid. Ray Chambers, a respected philanthropist and co-founder of a New York private equity firm, first learned of the goals in 2005. Since then, working with Sachs and others, Chambers has coordinated a worldwide coalition of policy, business, and NGO leaders in an effort to help the developing world meet the goal for malarial treatment and prevention. Thanks in part to this global effort, malaria-related mortality has dropped by approximately 25 percent since 2000, with most of those gains probably occurring since 2005. Many pharmaceutical companies have also put forth major efforts to make their medicines more widely available in poor countries, and new initiatives are continuing to take shape. The MDG Health Alliance, founded in 2011, is comprised of business and NGO leaders around the world working toward the MDG health targets, including the elimination of mother-to-child HIV transmission.

The combined results of these campaigns are remarkable. For example, in Senegal, child mortality has plummeted by half since 2000. In Cambodia, it has dropped by 60 percent. Rwanda has recorded a 10 percent average annual reduction since 2000, one of the fastest declines in history. Even China has seen a significant decrease in child deaths, possibly because the expanded global emphasis on health has encouraged the country's policymakers to pay more attention to relevant issues. Overall, despite rapid global population growth, there has been a decrease in children dying worldwide before their fifth birthdays, from 11.7 million in 1990 to 9.4 million in 2000 and 6.8 million in 2011.

No issue has been more closely interconnected with the MDGs than the HIV/AIDS treatment campaign. In 2000, nearly 30 million people were infected, the vast majority in Africa, where only approximately 10,000 people were in treatment and over 1 million people were dying every year from the disease. The next year, the head of the U.S. Agency for International Development publicly deemed large-scale AIDS treatment in Africa impossible. Undeterred, Annan launched the Global Fund to Fight AIDS, Tuberculosis and Malaria, which aims to achieve "long-term outcome and impact results related to the Millennium Development Goals."

Spurred by the launch of the MDGs, Jim Yong Kim, then head of the World Health Organization's HIV/AIDS department,

introduced the "3 by 5" initiative in 2003, which aimed to have three million people living with AIDS in the developing world receiving treatment by 2005. By the end of 2005, only 1.3 million people were receiving treatment—fewer than half of the target. But thanks to the interwoven AIDS-MDG campaign, the notion of service delivery targets has sunk in globally, helping expand AIDS treatment by orders of magnitude: also in 2005, the G-8 and the UN General Assembly endorsed a target of universal access to treatment by 2010, backed by major financial commitments. The MDG movement has expanded the world's ambitions in tackling health crises and made extraordinary progress. In 2011, more than 8 million people worldwide were receiving AIDS treatment.

Next-Generation Goals

The MDGs have proved that with concentration and effort, even the most persistent global problems can be tackled. The post-2015 goals should remain focused on eliminating the multiple dimensions of extreme poverty, but they also need to address emerging global realities. These new challenges include the worsening environmental pressures affecting the livelihoods of hundreds of millions of people, the growing number of middle-income countries with tremendous internal poverty challenges, and rapidly spreading noncommunicable diseases.

The new goals also need to be matched with resources. Without the Monterrey agreements of 2002 and the financial commitments made at the Gleneagles summit in 2005, the MDGs might well have faded from the international agenda. It is crucial that the post-2015 negotiations not be left solely to foreign and development ministries. Finance ministries will need an equal say on many of the most central issues and therefore need to be included from the beginning. Other relevant ministries, such as those that deal with health and environmental issues, should be consulted regularly. Additionally, in preparation for 2015, multilateral organizations, such as the World Bank and UN agencies, should conduct independent external reviews of their contributions to the MDGs and identify benchmarks for post-2015 success based on the results. And the United States needs to join the international community in making a solid commitment to long-term, goal-oriented foreign aid.

The MDGs have helped mobilize and guide development efforts by emphasizing outcomes. They have encouraged world leaders to tackle multiple dimensions of poverty at the same time and have provided a standard that advocates on the ground can hold their governments to. Even in countries where politicians might not directly credit the MDGs, the global effort has informed local perspectives and priorities. The goals have improved the lives of hundreds of millions of people. They have shown how much can be achieved when ambitious and specific targets are matched with rigorous thinking, serious resources, and a collaborative global spirit.

Looking forward, the next generation of goals should maintain the accessible simplicity that has allowed the MDGs to succeed and also facilitate the creation of better accountability mechanisms both within and across governments. In addition, the new goals need to give low-and middle-income countries a greater voice in shaping the agenda. Most important, momentum matters. Just as progress in individual MDG areas has inspired other campaigns, so work done now, in the final stretch, will affect what happens in the future. The results achieved by 2015 will mark an endpoint, but even more, they will provide a springboard for the next generation of goals. There is no time to lose.

Critical Thinking

1. Why was the Bush administration hesitant to embrace the Millennium Development Goals?

2. Why was the World Bank averse to supporting the Millennium Development Goals?

3. What are the obstacles faced in realizing all eight of the Millennium Development Goals?

Create Central

www.mhhe.com/createcentral

Internet References

Millennium Development Goals
www.un.org/millenniumgoals

MDG Task Force Report 2013
www.un.org/milleniumgoals/2013_Gap-Report/MDG%20GAP%20 Task%20Force%2013

World Bank
www.worldbank.org

JOHN W. MCARTHUR is a Senior Fellow at the Fung Global Institute and the UN Foundation and a Nonresident Senior Fellow at the Brookings Institution. From 2002 to 2006, he was Manager and Deputy Director of the UN Millennium Project. Follow him on Twitter @mcarthur.

McArthur, John W. From *Foreign Affairs*, March/April 2013, pp. 152–162. Copyright © 2013 by Council on Foreign Relations, Inc. Reprinted by permission of Foreign Affairs. www.ForeignAffairs.com

Prepared by: Robert Weiner, *University of Massachusetts/Boston*

Article

Africa's Economic Boom: Why the Pessimists and the Optimists Are Both Right

Devarajan Shantayanan and Wolfgang Fengler

Learning Outcomes

After reading this article, you will be able to:

- Explain why optimists and pessimists are both right about Africa's economic development.

- Explain the importance of human capital in Africa's economic growth.

- Discuss the role of external intervention in Africa's economic development.

Talk to experts, academics, or businesspeople about the economies of sub-Saharan Africa and you are likely to hear one of two narratives. The first is optimistic: Africa's moment is just around the corner, or has already arrived. Reasons for hope abound. Despite the global economic crisis, the region's GDP has grown rapidly, averaging almost five percent a year since 2000, and is expected to rise even faster in the years ahead. Many countries, not just the resource-rich ones, have participated in the boom: indeed, 20 states in sub-Saharan Africa that do not produce oil managed average GDP growth rates of four percent or higher between 1998 and 2008. Meanwhile, the region has begun attracting serious amounts of private capital; at $50 billion a year, such flows now exceed foreign aid.

At the same time, poverty is declining. Since 1996, the average poverty rate in sub-Saharan African countries has fallen by about one percentage point a year, and between 2005 and 2008, the portion of Africans in the region living on less than $1.25 a day fell for the first time, from 52 percent to 48 percent. If the region's stable countries continue growing at the average rates they have enjoyed for the last decade, most of them will reach a per capita gross national income of $1,000 by 2025, which the World Bank classifies as "middle income." The region has also made great strides in education and health care. Between 2000

and 2008, secondary school enrollment increased by nearly 50 percent, and over the past decade, life expectancy has increased by about ten percent.

The second narrative is more pessimistic. It casts doubt on the durability of Africa's growth and notes the depressing persistence of its economic troubles. Like the first view, this one is also justified by compelling evidence. For one thing, Africa's recent growth has largely followed rising commodity prices, and commodities make up the overwhelming share of its exports—never a stable prospect. Indeed, the pessimists argue that Africa is simply riding a commodities wave that is bound to crest and fall and that the region has not yet made the kind of fundamental economic changes that would protect it when the downturn arrives. The manufacturing sector in sub-Saharan Africa, for example, currently accounts for the same small share of overall GDP that it did in the 1970s. What's more, despite the overall decline in poverty, some rapidly growing countries, such as Burkina Faso, Mozambique, and Tanzania, have barely managed to reduce their poverty rates. And although most of Africa's civil wars have ended, political instability remains widespread: in the past year alone, Guinea-Bissau and Mali suffered coups d'état, renewed violence rocked the eastern Democratic Republic of the Congo, and fighting flared on the border between South Sudan and Sudan. At present, about a third of sub-Saharan African countries are in the throes of violent conflict.

More mundane problems also take a heavy toll. Much of Africa suffers from rampant corruption, and most of its infrastructure is in poor condition. Many governments struggle to provide basic services: teachers in Tanzania's public primary schools are absent 23 percent of the time, and government-employed doctors in Senegal spend an average of only 39 minutes a day seeing patients. Such deficiencies will become only more pronounced as Africa's population booms.

And then there's the fact that African countries, especially those that are rich in resources, often fall prey to what the

Africa's Economic Boom: Why the Pessimists and the Optimists Are Both Right by Devarajan Shantayanan and Wolfgang Fengler

155

economist Daron Acemoglu and the political scientist James Robinson have termed "extractive institutions": policies and practices that are designed to capture the wealth and resources of a society for the benefit of a small but politically powerful elite. One result is staggering inequality, the effects of which are often masked by positive growth statistics.

What should one make of all the contradictory evidence? At first glance, these two narratives seem irreconcilable. It turns out, however, that both are right, or at least reflect aspects of a more complex reality, which neither fully captures. The skeptics focus so much on the region's commodity exports that they fail to grasp the extent to which its recent growth is a result of economic reforms (many of which were necessitated by the misguided policies of the past). The optimists, meanwhile, underestimate the degree to which the region's remaining problems—such as sclerotic institutions, low levels of education, and substandard health care—reflect government failures that will be very difficult to overcome because they are deeply rooted in political conflict.

However, even if both narratives are reductive, the optimists' view of Africa's future is ultimately closer to the mark and more likely to be borne out by developments in the coming decades. Africa will continue to face daunting obstacles on its ongoing path to prosperity, especially when it comes to improving its human capital: the education, skills, and health of its population. But the success of recent reforms and the increased openness of its societies, fueled in part by new information and communications technologies, give Africa a good chance of enjoying sustained growth and poverty reduction in the decades to come.

Bouncing Back

After several lost decades, during which debt, disease, famine, and war held back Africa's development, things began to improve in the late 1990s. So far, the gains have proved durable. Despite the global financial crisis of 2008 and its lingering effects, the economies of sub-Saharan Africa grew at an average of 4.7 percent a year between 2000 and 2011. This robust performance has resulted in the first overall decline in the region's poverty rate since the 1970s, from 58 percent in 1999 to 47.5 percent in 2008. These positive trends have been widespread, with every part of the region benefiting. And the change in fortunes has not been limited to certain kinds of economies: oil exporters such as Angola and Nigeria have boomed, but so, too, have oil importers such as Ethiopia and Rwanda. Not all states have benefited equally, of course; fragile states such as Burundi and the Central African Republic, which are still struggling to recover from violent conflicts, have experienced only modest growth.

Africa's rebound has had many causes, including an increase in external assistance (partly from debt relief), a buoyant global economy until 2008, and high commodity prices. But the most significant has been an improvement in macroeconomic policies across all of sub-Saharan Africa, which has inspired confidence in investors and consumers. According to the World Bank's most recent annual "Country Policy and Institutional Assessment," the region's overall macroeconomic performance is now on par with that of developing countries in other regions. With stronger macroeconomic policies, African countries have taken advantage of the commodities boom that peaked before the global economic crisis and avoided a collapse when commodity prices plummeted. For example, in early 2008, when the international price of oil rose above $100 a barrel, some oil exporters in the region, such as Angola, Gabon, and Nigeria, planned their budgets as if oil prices were only $65 a barrel. When the price ultimately did fall to that level, in the fall of 2008, those countries were not caught off-guard and had a cushion to fall back on.

During the crisis, most countries continued with prudent economic policies; some even accelerated their reforms. Partly as a result of such efforts, African economies kept expanding throughout the global recession, and sub-Saharan Africa has maintained an average annual growth rate of nearly five percent since then, despite continued volatility in the global economy.

The Politics of Growth

In large part, the vast improvement in macroeconomic policy that began in the late 1990s can be traced to two factors. First, with the end of the Cold War, politics in Africa became freer, more vibrant, and more open to previously marginalized groups. As support from the United States or the Soviet Union diminished, autocratic regimes began to lose their monopolistic grips on power. Calls for multiparty democracy spread, and countries throughout the region held competitive elections. Such openings were limited, to be sure, but they provided a voice to many segments of African societies that had previously been marginalized, such as poor farmers in rural areas. Since the mid-1990s, those groups have benefited as politics has become more competitive, media have become freer, and communications technology has rapidly spread, especially since 2000. In several countries, including Ghana, Nigeria, Tanzania, and Uganda, these political changes brought to power more competent leaders, willing to place technocrats trained in modern economics in senior positions in the government, replacing the politically connected but less well-trained bureaucrats who often held similar posts in previous regimes.

Political liberalization also had a less direct but still profound effect on macroeconomic policy. In the past, many authoritarian African regimes kept their exchange rates artificially high, benefiting the small groups of urban elites on whom the regimes relied by making it easier for them to buy food and imported luxury goods. This policy amounted to a transfer of wealth from the rural poor to the urban rich, since the high exchange rates made it harder for farmers to export their crops. With the introduction of competitive elections, governments realized that they needed the support of the rural poor, who constitute a majority in most African countries, and so they allowed their countries' exchange rates to become more competitive. As a result, agricultural productivity and output rose as farmers received higher prices for their produce.

The second important factor that contributed to the improvement of African macroeconomic policy in the 1990s also involved the democratization of policymaking—spurred, in this case, by external intervention. When African countries were desperate for international aid in the 1980s, donors made their financial support contingent on the adoption of reform programs that African governments designed with input from the World Bank and the International Monetary Fund. But beginning in 1999, potential donors began to require African governments seeking debt relief to also consult with their own citizens—civil-society groups, businesses, community organizations—as they crafted policies to help the poor. This new process increased the chances that local citizens would buy into the policies. In the early 1990s, when international donors proposed changes to Zambia's system for pricing maize, the agriculture ministry rejected the changes, and they were never put in place, leading to periodic food shortages. A decade later, the government proposed similar reforms, but only after conducting consultations with a wide variety of Zambians whom the changes would affect. As a result, the public generally accepted the ideas; the reforms were implemented, and shortages were minimized.

Economic reforms, however, are not the only cause of Africa's growth surge. Three other factors have started to play a major role: demographic changes, urbanization, and technological advances. Since 1960, the dawn of the postcolonial era, the population of sub-Saharan Africa has grown rapidly, from fewer than 250 million people to around 900 million today. But around 2000, fertility rates began to decline, and so did child mortality rates. Consequently, working-age adults have come to constitute the fastest-growing segment of the region's societies. This shift has created a potential demographic dividend, since economies improve when there is a healthy ratio of working-age adults to dependents.

No country or region, meanwhile, has ever reached what the World Bank considers high-income status with low levels of urbanization. African populations have traditionally been mostly rural, but the cities of sub-Saharan Africa are growing at astonishing rates. The trend is such that by 2033, most of the region's inhabitants will live in cities—as most of the world's population already does. Firms have exploited this increased urban consumer base to enjoy economies of scale, benefiting themselves and consumers, who now have access to low-cost goods.

Perhaps the most visible sign of Africa's economic reemergence is the so-called mobile revolution. Cell phones have become ubiquitous, even in the poorest places. The change can be traced back to the reforms of the late 1990s, when several countries began opening up their telecommunications sectors. At the same time, technological breakthroughs have made low-cost cell phones affordable to a large number of Africans. In many African countries, the calling rates are among the lowest in the world. The explosion in mobile technology has spurred innovations such as M-Pesa, the mobile-money system widely embraced in Kenya and Tanzania, which allows users to make purchases and send cash transfers using their cell phones. In many countries, the spread of mobile devices has also allowed the information and communications sectors to become important parts of the economy; in Kenya, these industries are growing at an average of 20 percent each year, and in 2010, they accounted for five percent of the country's gdp.

Optimists have seized on all these trends to make the case that this African economic boom will prove sustainable. Much of the progress has resulted from political changes. But the remaining obstacles to a more lasting transformation of African economies will also depend on politics. And those problems might prove far more difficult to overcome.

More Money, More Problems

Africa faces a number of deep development challenges—in economic growth, poverty reduction, human development, and governance—that at the very least call into question the durability of the gains made during the last 15 years, and could even undermine them. Despite Africa's recent growth, there are few signs of what economists refer to as structural transformation: the shift from low-productivity agriculture to higher-productivity manufacturing and services. Sub-Saharan Africa's manufacturing sector remains dormant, and some countries, such as South Africa, have even experienced deindustrialization. And while there has been an increase in trade among the region's countries, their connections to the world economy remain weak and concentrated in just a few sectors, especially commodities and natural resources. These development challenges are the result of government failures, which helps explain their persistence amid rapid growth—but also points to possible solutions.

Perhaps none of these problems is more troubling than the seeming inability of African countries, including the fastest-growing economies, to convert growth into progress in fighting poverty. Despite years of significant oil revenues, the governments of Angola, Gabon, and Nigeria have not used their newfound wealth to significantly improve the welfare of their poor citizens. More troubling is the fact that during the past five years, some non-oil-producing countries, such as Burkina Faso, Mozambique, and Tanzania, have managed to reduce their poverty rates by only three or four percentage points, despite enjoying annual economic growth rates of around seven percent. That growth was very clearly driven by economic reforms, not the commodities boom. The persistence of poverty in those three countries is now providing rhetorical ammunition to the political elites who benefited from the misguided policies of the past, resisted reforms, and now want to reverse the changes. It also confirms the worst suspicions of critics of economic liberalization, who can point to these poverty numbers to argue that pro-trade reforms have simply made the rich richer and the poor poorer.

A more careful look at these countries, however, shows that the problem is not too much reform but too little. Specifically, the reforms have generated growth in only some sectors, especially services, with industries such as retail and wholesale trade, telecommunications, and public administration benefiting the most. But those industries provide relatively few jobs for low-skilled workers, and the reforms did not address the sectors in which the poor actually work. For example, in Mozambique, growth has come from large investment projects in mining that were made possible by changes in the country's foreign investment regulations. Such projects have increased aluminum exports and boosted GDP but created only 2,000 direct jobs. Most of Mozambique's labor force, meanwhile, is employed by small farms or household enterprises—parts of the economy in which productivity is growing very slowly.

In cases where there have been reforms in industries that employ the poor, corruption has sometimes prevented the benefits from accruing to the intended recipients. Tanzania, for example, has spent heavily to support its agriculture industry, especially on fertilizer subsidies. In 2009, to better target and streamline the subsidies, the government introduced a market-like system of vouchers: farmers could use government-issued vouchers to purchase fertilizers, and sellers would be reimbursed by the government. Unfortunately, local elected officials ended up gaining control of about 60 percent of the vouchers, making it difficult for poor farmers to access the government support.

If you Build it, Will They Come?

Even in countries that have achieved both rapid growth and poverty reduction, such as Ethiopia, Ghana, and Rwanda, there has been remarkably little structural transformation. The share of GDP represented by manufacturing, for example, is scarcely higher than it was before these countries started enjoying serious growth. There are many reasons why competitive manufacturing has not taken off in Africa, but most of them revolve around the high costs of production. Even though per capita incomes in Africa are among the lowest in the world, wages are relatively high and unit labor costs are even higher.

A major explanation for these high costs is the poor state of infrastructure. All across sub-Saharan Africa, anyone trying to do business is constantly stymied by power cuts, impassable roads, and leaky water pipes. Behind each of these infrastructure problems is a government failure that, although harmful to the economy, reflects a political equilibrium that will be difficult to undo simply by building new infrastructure.

Road transportation offers a good illustration of this problem. Exporters in the region face some of the highest transport prices in the world, especially when trying to ship goods from landlocked countries to a port. But a 2009 study published by the World Bank showed that vehicle operating costs along the four main transport corridors in sub-Saharan Africa are no higher than those in France. The difference between prices and vehicle operating costs is explained by the massive profit margins enjoyed by trucking companies in sub-Saharan Africa, some of which are close to 100 percent. The companies are able to charge a hefty premium thanks to regulations in most African countries that prohibit would-be competitors from entering the trucking industry. These regulations were introduced 40 years ago, when African governments, reflecting economic thinking at the time, viewed trucking as a natural monopoly because a single company could more easily ensure that trucks rode at full capacity. Not surprisingly, the outdated rules are now difficult to revoke because decades of high profits have provided the trucking industry with plenty of funds to pay for lobbying

to maintain the status quo. This problem is especially acute in places where the trucking business is controlled by politically connected families.

The region's water and electricity deficits also stem from political problems. Governments typically set prices for water and electricity that are below cost, with the intention of protecting the poor. As a result, the water and electrical utilities require government subsidies to operate. This relationship allows politicians to find ways to influence how the utilities are run and who receives their services. Officials often give priority treatment to neighborhoods they favor, which are not necessarily where the poor live. Furthermore, the subsidies rarely cover costs, so the utilities neglect maintenance, leading to leaky pipes and power outages. The rich opt out of the shoddy system altogether and use their own water tanks and electricity generators. The poor in underserved areas must rely on candles for lighting and buy water from private vendors, which costs multiple times the metered rates. One result of this political distortion is that since 2000, the percentage of households with access to water has declined in almost every urban area of Africa.

In addition to these deficiencies in infrastructure, a host of other factors serve to drive up the cost of doing business in the region, including the fact that African countries have some of the most complex and least transparent business regulations in the world. Like the distortions that shape transportation and infrastructure, these regulations did not come about by accident, nor is their persistence due to a lack of government capacity: they exist in order to serve specific political interests. If these interests are sufficiently powerful, they can block attempts at reform.

But simply improving the business climate will not lead to structural transformation. The reason is that business regulations mainly affect those who work in the private wage-employment sector, a group that accounts for less than ten percent of the region's labor force. Most Africans work for small farms or household enterprises, in what is often called the informal sector. This is unlikely to change in the medium term: in Uganda, for instance, even under the most optimistic assumptions, over 70 percent of the labor force will still be in the informal sector by 2020.

For that reason, structural transformation will depend not only on creating more wage and salary jobs but also on increasing the productivity of the informal sector. Improving infrastructure and reforming regulations will help to some extent. But more important are measures that can improve the skills of workers in the informal sector, in which those with barely any education are disproportionately concentrated. By increasing the skills of such workers, African governments can increase the productivity of small farms and household enterprises—and the incomes of the people who work there.

Raising Human Capital

Without a doubt, it will prove difficult to improve the skills of Africa's labor force enough to propel structural transformation. The fact is that despite some catch-up over the last decade, the countries of sub-Saharan Africa still have the lowest levels of human capital in the world. In one sense, that is not surprising: after all, at the time they won independence, most of these countries had very few people with higher education. Africa also has been buffeted by an onslaught of public health crises, including the world's worst manifestation of the HIV/AIDS pandemic.

The region's lack of sufficiently educated, skilled, healthy workers is even more distressing because for decades, donors and African taxpayers alike have spent considerable resources on health and education; yet they have little to show for it. Even in places where governments and foreign donors have improved access to schools and health clinics, there has been limited improvement in quality. Postapartheid South Africa, for instance, has increased its public spending on schools to redress the inequitable allocations of the past. Enrollment rates have risen dramatically, but learning outcomes have hardly changed, and only two in five young adults complete secondary school.

At least three factors explain this phenomenon. First, resources allocated to addressing the problems of poor people do not always reach their intended recipients. A landmark 2001 World Bank study on public spending showed that in Uganda, only 13 percent of the nonwage resources allocated to public primary education actually found their way to schools. Similarly, a 2009 study on health spending in Chad showed that less than one percent of nonwage spending ever arrived at primary clinics. Second, even when resources do reach schools or clinics, there are often no teachers or doctors there to use them. A recent report by the African Economic Research Consortium found that health workers in Senegal and Tanzania were absent 20 percent and 21 percent of the time, respectively. Finally, even when providers are present, the quality of their services is exceedingly poor. According to a 2009 World Bank review of public expenditures, teachers in Uganda spend less than 20 percent of class time teaching. Teachers in Tanzania spend slightly more time on instruction, but only 11 percent of them have what education experts consider to be the minimum level of language skills required for the job. The situation in the health sector is worse: in Tanzania, the average total amount of time doctors spend seeing patients is only 29 minutes per day.

These failures to deliver services are not simply the result of unprofessional conduct; underlying them is the fact that basic public services have been stolen by or diverted to political elites. The leakage of public funds intended for education and health care is the most straightforward example. Since these are expenditures for things other than salaries, officials are easily able to alter the amount of funding that is actually distributed. As the economists Ritva Reinikka and Jakob Svensson showed in a 2004 study, the amount of funding an African school receives likely depends on the principal's ties to a government bureaucrat or a local politician. The poor performance of service providers is similarly bound up in this form of patronage. Many teachers, for example, also serve as political operatives: relatively well-educated people who run election campaigns for local politicians and are then rewarded with teaching jobs, positions for which they are not necessarily qualified and that they do not always take very seriously.

The way political forces can thwart the delivery of services was illustrated in a recent study published by the Center for Global Development. The study analyzed the results of an experiment in Kenya that aimed to reduce teacher absenteeism by replacing salaried teachers with contract workers. In some cases, the plan was administered by a nongovernmental organization; in others, the government handled the hiring. Student learning outcomes improved when the plan was implemented by nongovernmental organizations but did not in the government-run cases. The study's authors concluded that the difference stemmed from the ability of teachers' unions to lobby the government to weaken the plan in various ways: for example, by delegating oversight to district officials who were not ultimately accountable to the government. The nongovernmental organizations did not succumb to the same pressure. The larger lesson is that efforts to solve problems such as teacher absenteeism with technical solutions, such as introducing contract teachers or electronic monitoring, will not succeed if the political system is not aligned with the ultimate goal.

Reasons for Optimism

It can be hard to stay optimistic about Africa's future when one considers the political pathologies that stand in the way of improving its human capital. But it is crucial to recall that the recent growth in sub-Saharan African economies resulted from fixing distorted macroeconomic policies that seemed irredeemable only 15 years ago. Triggered by reactions to the debt crises of the 1980s, the collapse of the Soviet Union, and the political liberalization of the 1990s, a regional consensus formed in favor of prudent macroeconomic policies. Those policies delivered growth, which created political support for further reforms, even during the global economic crisis of recent years.

The region now finds itself at another inflection point. Luckily, today, the combination of democratization, demographic change, rapid urbanization, and increasing levels of education has substantially altered policymaking processes, mostly for the better. There is now more political space to voice alternative views and challenge government policies. Even those who are opposed to reforms are less likely to resist if they feel they have been consulted. Moreover, thanks to better economic policies, foreign donors are less compelled to impose reforms from the outside, which creates even more space for homegrown reform efforts.

The almost complete connectedness of the region through cell phones will also aid reforms and structural transformation. Cell phones, by helping spread information of all kinds more quickly, enable poor people to learn about such issues as the regressive nature of government subsidies and the anti-poor bias of infrastructure spending. They also allow people to find out what their peers are thinking, greatly lowering the costs of mobilizing collective action. The spread of communications technology has also made it easier for politicians to discover what citizens are thinking—whether they want to or not—meaning that the voices of people living in marginalized areas will be heard more clearly in national capitals.

Whether one sees Africa's glass as half-full or half-empty depends on one's belief in the possibility of political change. The obstacles to durable growth in the region are primarily

Africa's Economic Boom: Why the Pessimists and the Optimists Are Both Right by Devarajan Shantayanan and Wolfgang Fengler

159

political. That hardly means that they will be easy to solve, as even a cursory glance at the troubled record of governance in postindependence Africa makes clear. But it does mean that they are not intractable. Sub-Saharan Africa's recent history of political change and reform leading to growth justifies a positive outlook. Believing in a more prosperous African future requires a healthy dose of optimism, but not a leap of faith.

Critical Thinking

1. Is poliitcal instability still a problem in Africa? Why or why not?

2. Why is the optimistic view of African growth closer to the mark?

3. Why have African states not been able to convert economic growth into progress in fighting poverty?

Create Central

www.mhhe.com/createcentral

Internet References

African Development Bank
www.afdb.org
African Union
www.africa-union.org
UN Economic Commission for Africa
www.uneca.org

SHANTAYANAN DEVARAJAN is chief economist of the World Bank's Africa Region. WOLFGANG FENGLER is the World Bank's lead economist for Eritrea, Kenya, and Rwanda.

Prepared by: Robert Weiner, *University of Massachusetts/Boston*

Article

The Crisis of Europe: How the Union Came Together and Why It's Falling Apart

Timothy Garton Ash

Learning Outcomes

After reading this article, you will be able to:

- Explain why the euro crisis endangers the entire European project.
- Discuss the central role of Germany in the European union.

May 10, 1943: German forces are destroying the Warsaw ghetto. Facing armed resistance from Polish Jewish fighters, they set fire to it house by house, burning some inhabitants alive and driving others out from the cellars. "Today, in sum 1,183 Jews were apprehended alive," notes the official report by the SS commander Jurgen Stroop. "187 Jews and bandits were shot. An indeterminable number of Jews and bandits were destroyed in blown-up bunkers. The total number of Jews processed so far has risen to 52,683." An appendix to this document contains the now-famous photograph of a terrified small boy in an outsize cloth cap, his hands held high in surrender. Marek Edelman, one of very few leaders of the Warsaw ghetto uprising to survive, concluded a memoir published immediately after the war with these words: "Those who were killed in action had done their duty to the end, to the last drop of blood that soaked into the pavements. . . . We, who did not perish, leave it up to you to keep the memory of them alive—forever."

Fast-forward exactly 60 years, to May 10, 2003, a month before Poland holds a referendum on whether to join the European Union. At a "yes" campaign rally in Warsaw, a banner in Poland's national colors, red and white, proclaims, "We go to Europe under the Polish flag." Outside the rebuilt Royal Castle, a choir of young girls in yellow and blue T-shirts—echoing the European flag's yellow stars on a blue background—breaks into song. To the music of the EU'S official anthem, which is drawn from the final movement of Beethoven's Ninth Symphony, they sing, in Polish, the words of the German poet Friedrich

Schiller's "Ode to Joy." Soon these young Poles will be able to move at will across most of a continent almost whole and free, to study, work, settle down, marry, and enjoy all the benefits of a generous European welfare state, in Dublin, Madrid, London, or Rome. "Be embraced, ye millions! This kiss to the entire world! Brothers, a loving father must live above that canopy of stars!"

To understand how a predicted crisis of European monetary union became an existential crisis of the whole post-1945 project of European unification, you have to see Europe's unique trajectory from one May 10 to the other. Both the memories of World War II and the exigencies of the Cold War drove three generations of Europeans to heights of peaceful unification that were unprecedented in European history and unmatched on any other continent. Yet that project began to go wrong soon after the fall of the Berlin Wall, as western European leaders hastily set course for a structurally flawed monetary union.

While many governments, companies, and households piled up unsustainable levels of debt, young Europeans from Portugal to Estonia and from Finland to Greece came to take peace, freedom, prosperity, and social security for granted. When the bubble burst, it left many feeling bitterly disappointed and led to excruciating divergences between the experiences of different nations. Now, with the current crisis still unresolved, Europe lacks most of the motivating forces that once propelled it toward unity. Even if a shared fear of the consequences of the eurozone's collapse saves it from the worst, Europe needs something more than fear to make it again the magnetic project it was for a half century. But what can that something be?

War on the Mind

Historians have identified many factors that contributed to the process of European integration, including the vital economic interests of European nations. Yet the single most important driving force across the continent was the memory of war. Among those parading down the streets of Warsaw in May

2003 was the bearded professor Bronislaw Geremek, who, as a ten-year-old Polish Jewish boy, had seen the Warsaw ghetto burning before his eyes. It was no accident that he became one of Poland's most ardent advocates of European integration, as a leader of the Solidarity movement, the Polish foreign minister, and then a member of the European Parliament.

To be sure, the Warsaw ghetto survivor, the Nazi soldier, the British officer, the French collaborator, the Swedish business-man, and the Slovak farmer had very different wars. Yet from all their throats rose the same passionate cry: "Never again!" For all the differences in national and subnational experiences across a hugely diverse continent, the historian Tony Judt could still title a history of Europe that covers the 60 years up to 2005 with a single word: Postwar. In this respect, if in no other, the European Union's favorite catch phrase, "Unity in diversity," was strictly accurate.

Those memories played an important role for those British Conservatives, most of them World War II veterans, who took the United Kingdom into the European Economic Community, the precursor to the European Union, in 1973. But above all, personal experience motivated those continental Europeans, up to and including French President Francois Mitterrand and German Chancellor Helmut Kohl, who created the EU of today. In a conversation I had with him after German reunification, Kohl delivered a line I will never forget. "Do you realize," he asked, "that you are sitting opposite the direct successor to Adolf Hitler?" As the first chancellor of a united Germany since Hitler, he explained, he was profoundly conscious of his historical duty to do things differently.

European integration has rightly been described as a proj-ect of the elites, but Europe's peoples shared these memories. When the project faltered, as it did many times, the elites' reac-tion was to seek some way forward, however complicated. Until the 1990s, when the custom of holding national referendums on European treaties began to spread, Europeans were seldom asked directly if they agreed with the solutions found, although they could periodically vote in or out of office the politicians responsible for finding them. Nonetheless, it is fair to say that for about 40 years, the project of European unification could rely on at least a passive consensus among most of Europe's national publics.

These 40 years were those of the Cold War, the other conflict that shaped the EU. From the 1940s through the 1970s, a cen-tral argument for Western European integration was to counter the Soviet threat, visible for all to see in the presence of the Red Army in East Germany and divided Berlin. Beside the memo-ries of Europe's own self-inflicted barbarism, there were, so to speak, the barbarians at the gate. Soviet leaders from Joseph Stalin to Leonid Brezhnev should be awarded posthumous medals for their service to European integration.

Cold War competition also goes a long way to explaining why the United States lent such strong support to European unification, from the Marshall Plan of the 1940s to the diplo-macy surrounding the reunification of Germany and the disso-lution of the Soviet Union in 1989–91.

For the half of Europe stuck behind the Iron Curtain—what the Czech writer Milan Kundera called "the kidnapped West"—the will to "return to Europe" went hand in hand with the struggle for national and individual freedom. The growing prosperity of Western Europe had a magnetic effect on those who saw it, whether at first hand or on Western television.

It is the most elementary historical fallacy to suggest that an event was caused by one that occurred after it, yet something that was only to happen in 1992 was a contributing cause of the velvet revolutions of 1989. The target year 1992, the widely trumpeted deadline that the European Economic Community had given itself for completing its single market, conveyed an urgent sense of being left ever-further behind, not just to the peoples of Eastern Europe but also to reform-minded Soviet-bloc leaders, including Mikhail Gorbachev.

This brings us to the last great motor of European integra-tion until the 1990s: West Germany. The West Germans, both the elites and a large part of the populace, demonstrated an exceptional commitment to European integration. They did this for two very good reasons: because they wanted to, and because they had to. They wanted to show that Germany had learned from its terrible pre-1945 history and wished to rehabil-itate itself fully in a European community of values, even to the point of surrendering much of its own sovereignty and national identity. Having been the worst Europeans, the Germans would now be the best. (As a joke at the time went, if someone intro-duced himself just as "a European," you knew immediately that he was German.) But they also had a hard national inter-est in demonstrating that European commitment, for only by regaining the trust of their neighbors and international part-ners (including the United States and the Soviet Union) could they achieve their long-term goal of German reunification. As Hans-Dietrich Genscher, the former West German foreign min-ister, once observed, "The more European our foreign policy is, the more national it is." West German Europeanism was not simply instrumental—it reflected a real moral and emotional engagement—but nor was it purely idealistic.

After the two German states were reunited in 1990, many observers wondered whether what was essentially an expanded West Germany would continue this extraordinary commitment to European integration. Well before the crisis of the eurozone broke, the answer was already apparent. Reunited Germany had become what some participants in the post-Wall debate called a "normal" nation-state—a "second France," in the commenta-tor Dominique Moisi's striking phrase. Like France, the new Germany would pursue its national interests through Europe whenever possible, but on its own when it deemed it neces-sary—as it did, for example, when securing its energy needs bilaterally with Russia, notably in the Nord Stream gas pipeline deal of 2005. Its leaders, in Berlin now, not Bonn, would still try to be good Europeans, but they would no longer open the checkbook so readily if Europe called.

The Birth of a Malformed Union

The immediate origins of the malformed currency union that is at the epicenter of today's European crisis also lie in the tem-pestuous moment of German reunification and its aftermath. Following the fall of the Berlin Wall on November 9, 1989,

Mitterrand, alarmed by the prospect of German reunification, pushed hard to pin Kohl down to a timetable for what was then called economic and monetary union. That proposal had already been elaborated to help the European Economic Community complete its single market and address the difficulty of managing exchange rates within it. Mitterrand's general purpose was to bind a united Germany, if united those two Germanies really must be, into a more united Europe; his specific purpose was to enable France to regain more control over its own currency, and even win some leverage over Germany's.

In a remarkable conversation with Genscher, the West German foreign minister, on November 30, 1989, Mitterrand went so far as to say that if Germany did not commit itself to the European monetary union, "We will return to the world of 1913." Meanwhile, Mitterrand was stirring up British Prime Minister Margaret Thatcher to sound the alarm as if it were 1938. According to a British record of their private meeting at the crucial Strasbourg summit of European leaders in December 1989, Mitterrand said that "he was fearful that he and the Prime Minister would find themselves in the situation of their predecessors in the 1930s who had failed to react in the face of constant pressing forward by the Germans."

David Marsh, the best chronicler of the euro's history, concludes that the "essential deal" to proceed with monetary union was done at Strasbourg. Tough negotiations followed, and exactly two years later a treaty was agreed on in the small Dutch city of Maastricht, setting the basic terms of what would become today's eurozone. It is too simplistic to characterize this as a straight tradeoff: "the whole of Deutschland for Kohl, half the deutsche mark for Mitterrand," as one wit quipped at the time. But Germany's need for its closest European allies—above all, France—to support its national reunification had a decisive influence on both the timetable and the design of Europe's monetary union.

To be sure, Kohl was a deeply committed European. He never tired of repeating that German and European unification were "two sides of the same coin." So now, he told U.S. Secretary of State James Baker three days after the Strasbourg summit, he had even agreed to a European monetary union. What stronger proof could he offer of Germany's European credentials? Kohl "took this decision against German interests," the German minutes of that meeting record him telling Baker. "For example, the president of the Bundesbank was against the present development. But the step was politically important, since Germany needed friends." As one does, when one is trying to unite Germany without blood and iron.

The design of the resulting monetary union can also be understood, like so much else in the history of European integration, as a Franco-German compromise. At the insistence of Germany, and especially of the Bundesbank, the European Central Bank would be a Bundesbank writ large, fiercely independent of governments (unlike in the French tradition) and devoted with Protestant fervor to the one true god of price stability (lest the Weimar nightmare of hyperinflation return). To his credit, Kohl wanted the monetary union to be complemented by a fiscal and political union, so there could be control of public spending and coordination of economic policy among the states, and more direct political legitimation of the whole enterprise. "Political union is the essential counterpart to economic and monetary union," he told the Bundestag in November 1991. "Recent history, not only in Germany, teaches us that it is absurd to expect in the long run that you can maintain economic and monetary union without political union."

But France was having none of that. The point was for it to gain some control over Germany's currency, not for Germany to gain control over France's budget. So the discussion of a fiscal union withered away into a set of "convergence criteria," which required would-be members of the monetary union to keep public debt under 60 percent of GDP and deficits under three percent.

Thus, in the Sturm und Drang of the largest geopolitical change in Europe since 1945, a sickly child was conceived. Most Germans opposed giving up their treasured deutsche mark. But they would not be asked; the West German constitution did not envisage referendums. Kohl had no intention of changing that. Alexandre Lamfalussy, the head of the European Monetary Institute, the precursor to the European Central Bank, later recalled telling him, "I don't know how you will get the German people to give up the D-Mark." Kohl's reply: "It will happen. The Germans accept strong leadership."

In France, meanwhile, the Maastricht Treaty scraped through in a September 1992 referendum with a yes vote of just over 50 percent. The passive consensus for further steps of European integration, advancing ever closer to the heart of national sovereignty, was beginning to break down even in heartlands of the postwar project.

A Crisis Foretold

With a hat tip to Gabriel Garcia Marquez, a history of Europe's monetary union could be called Chronicle of a Crisis Foretold. By the time the eurozone's 11 founding member states were preparing to introduce a common currency on January 1, 1999, most of the problems that would beset the euro a decade later had been predicted.

Critics at the time questioned how a common currency could work without a common treasury, how a one-size-fits-all interest rate could be right for such a diverse group of economies, and how the eurozone could cope with economic shocks that varied from region to region—what economists call "asymmetric shocks." For Europe had neither the labor mobility nor the level of fiscal transfers between states that characterized the United States.

"Since 1989, we have seen how reluctant West German taxpayers have been to pay even for their own compatriots in the east," noted one article in these pages in 1998. "Do we really expect that they would be willing to pay for the French unemployed as well?" Reporting a widespread view that the monetary union would face a crisis sooner rather than later, and that this would catalyze the necessary political unification, the author cautioned, "It is a truly dialectical leap of faith to suggest that a crisis that exacerbates differences between European countries is the best way to unite them."

Since I was that author, I should add that I did not anticipate three important things. First, I did not expect that the monetary union would flourish for so long. For nearly a decade, the euro appeared to be strong, edging up toward the dollar as a global trading and reserve currency. For businesses, it removed the risk of exchange-rate fluctuations inside the eurozone. For the rest of us, it was a delight to be able to travel from one end of the continent to the other without having to change currencies. To visit Dublin, Madrid, or Athens was to see cities booming as never before. Small wonder that in 2003 those young Poles sang Schiller's "Ode to Joy" at the prospect of joining the happy Irish, Spaniards, and Greeks. And I, like others sympathetic to the project, was lulled into a false sense of security.

Because the crash came later than originally expected, it was worse when it came. Over time, enormous imbalances had built up between the core, mainly northern European countries (above all, Germany), and the peripheral, mainly southern European countries (especially Portugal, Ireland, Italy, Greece, and Spain, which have sometimes been unkindly labeled "the PIGS").

To be sure, the initial shocks that started the earthquake came from outside Europe, in the U.S. subprime mortgage market. In this sense, the travails of the eurozone are part of a broader crisis of Western financial capitalism.

Yet the second thing we did not fully anticipate in the 1990s was the extent to which the eurozone would generate its own asymmetric shocks. Whereas Germany, still staggering under the financial burden of German reunification, impressively massaged down its labor costs, trimmed its welfare spending, and became competitive again, many of the peripheral countries allowed their unit labor costs to soar.

While Germany and some other northern European countries maintained fiscal discipline and moderate levels of debt, many of the peripheral countries went on the mother of all binges. In some places, such as Greece, it was public spending that skyrocketed; in others, such as Ireland and Spain, it was private spending. The open sesame to both kinds of excess was the same: governments, companies, and individuals could borrow at unprecedentedly low interest rates thanks to the credibility that eurozone membership lent their countries. In effect, Greece, which had snuck into the eurozone in 2001 with the aid of falsified statistics, could borrow almost as if it were Germany.

When, therefore, Germany was asked to help bail out those countries, German voters were understandably indignant. Why should we work even harder and retire even later, they asked, so these feckless Greeks, Portuguese, and Italians can retire earlier than we do and go sun themselves on the beach? "Sell your islands, you bankrupt Greeks," snorted Bild, Germany's largest tabloid, in October 2010.

The Germans had a good point: they had demonstrated remarkable prudence; the peripheral countries had not. But

there was another side to the story. The moment the Stability and Growth Pact (the formalized successor to the convergence criteria) was revealed to be toothless was when Germany itself, along with France, violated the deficit limit of three percent of GDP in 2003–4. The penalties envisaged in the pact were not even enforced.

Moreover, Germany had fared so well partly because the peripheral countries had fared so badly. The peripheral eurozone countries could no longer compete with Germany on price by devaluing their own national currencies, and part of their binge spending went to buying more BMWS and Bosch washing machines. The euro also enabled German exporters to price their goods more competitively in markets such as China. (One study, by Nathan Sheets and Robert Sockin of Citigroup, estimated that Germany's lower real exchange rate, courtesy of the euro, has lifted its real trade surplus by about three percent of GDP annually.) As the economist Martin Feldstein noted in these pages, in 2011 Germany's $200 billion trade surplus roughly equaled the rest of the eurozone's combined trade deficit. Germany was to Europe what China is to the world: the exporter that requires others to consume.

In addition, Germany and other northern European countries with current account surpluses recycled those surpluses partly by lending to Greeks, Irish, Portuguese, and Spaniards. So when Germany bailed out the peripheral eurozone countries, it was also bailing out its own banks.

The third element few foresaw in the 1990s was the spiraling scale, speed, and folly of global financial markets. Most egregious, bond markets contributed to the burgeoning imbalances by mispricing sovereign risk in general and the differential risk between various eurozone government bonds in particular. Despite the presence of a "no bailout" clause in the Maastricht Treaty, bond traders acted as if the risk associated with lending to the Greek or Portuguese governments was only fractionally higher than that of lending to Germany or the Netherlands.

When belief in the solidity of the eurozone began to collapse, soon after its tenth birthday, the markets plunged to the other extreme. Again and again, they punished eurozone leaders' belated half measures with soaring bond yield spreads, so that country after country found its borrowing costs whizzing upward. At interest rates of five to eight percent, it becomes very difficult for a government to sustain its debt burden, even with the most exemplary German-style fiscal discipline and structural reform. There was only so much that even the wisest and most economically responsible leaders, such as Italian Prime Minister Mario Monti, could ask of their own people.

Europe's Dysfunctional Triangle

Structurally, Europe now finds itself caught in a dysfunctional triangle, between national politics, European policies, and global markets. Ever since the European Coal and Steel Community was founded, in 1951, integration has proceeded through the development of common European policies: from those on agriculture, fisheries, and trade, all the way to monetary policy. The democratic politics of the EU have, however, remained stubbornly national.

While the volcanic magma was heating up under the outwardly calm crust of the eurozone, European leaders spent much of this century's first decade engaged in an ambitious attempt to write what some called a constitution for Europe. To cope with both the deepening of the EU, through monetary union, and its widening, through the historic enlargement to eastern Europe, they proposed a new set of institutional arrangements for the EU's 27 states (since 2007) and 500 million people. But in referendums, voters in France and the Netherlands rejected even a watered-down version of these lofty plans. "The nations don't want it," commented Geremek, that passionate but also realistic European, shortly before he died in 2008.

So the mountain labored again, and brought forth a mouse. The Treaty of Lisbon, which came into force in 2009, did give more powers to the directly elected European Parliament. But decision-making in today's EU still consists mainly of national politicians cutting deals behind closed doors in Brussels. And the politics and media they worry about are national, not European. There are Europe-wide political groupings, based on those in the European Parliament, but there are no truly European politics. The average turnout for elections to the European Parliament has declined with every vote since direct elections began in 1979. Although there are some Europe-wide media outlets, watched and read by a happy few, there is no broader European public sphere.

The French historian Ernest Renan said that a nation is "an everyday plebiscite." Well, today's EU has an election almost every day, but these are national elections, conducted in different languages and in national media. Increasingly, the election campaigns feature parties that blame the country's current travails on other European nations, or on the EU itself, or on both. Visiting Maastricht earlier this year—a city now a little worried about its place in the history books—I was told how the anti-immigrant and anti-Islamic Dutch populist Geert Wilders has redirected his political fire against "Europe." That's where he thinks the votes are now.

At the same time, panicky global markets instantly impinge on both European policies and national politics. As country after country finds its credit rating cut and its borrowing costs going through the roof, governments tremble and call yet another emergency summit in Brussels. As the clock ticks into the early hours, exhausted national leaders are torn between their terror of what the markets will do to them when trading opens the next morning and their terror of what their national media, coalition partners, parliaments, and voters will do to them when they get back home.

As soon as the meeting ends, each leader will dash out from the conference room to brief his or her own national media, so that every time, there is not just one version of a European summit but 27 different ones—plus a 28th, the implausibly irenic conclave described by the EU's own clutch of institutional heads. This is Europe's political Rashomon, with 28 conflicting versions of the same event delivered in 23 languages. It is an odd way to run a continent.

The Missing Ingredients

Europe's monetary union was a bridge too far—meaning not a bridge that should never have been crossed but a bridge that was crossed too soon, before Europe was strategically prepared to defend it. To be sure, carrying on for another decade or two with a system of fixing the margins within which exchange rates could fluctuate—the so-called Exchange Rate Mechanism—would have been demanding. But it is hard to disagree with this retrospective judgment by the economic commentator Martin Wolf: "Consider how much better off Europe would have been if the exchange rate mechanism had continued, instead, with wide bands."

We also have to consider other roads not taken. What if, instead of introducing the euro, Europe had deepened its still-far-from-complete single market? What if the whole EU had concentrated on improving its competitiveness, as Germany did so impressively, and not merely paid lip service to that goal in a catalog of good intentions called "the Lisbon agenda"? What if it had used this time to develop a more effective foreign policy? But regret is futile. An old and now politically incorrect English joke has an American couple arriving at a crossroads, deep in the Irish countryside, and asking a tweed-clad farmer the way to Tipperary. "If I were you," says the Irishman, "I wouldn't start from here." Yet here is where we are.

At the end of June this year, the EU held yet another "save the euro" summit—by a rough count, the 19th of the crisis. Germany said it would allow special European funds to be used to help imperiled Spanish banks, and the eurozone states resolved to create a single banking supervisory structure run by the European Central Bank. Although nobody noticed, the summit communique was a reminder of useful things the EU continues to do. For example, European leaders reached agreement on a unitary European patent system, which is expected to lower patenting costs for European companies by as much as 80 percent. They also decided to open accession negotiations with Montenegro, a newly independent state that just 13 years ago was still embroiled in the wars of former Yugoslavia.

As of this writing, no one knows how the euro saga will end. The possibilities include a total, disorderly collapse of the eurozone, a continued muddling through, and, most optimistically, systemic consolidation into a genuine fiscal and political union. Yet even if the eurozone crab-marches toward a political union, it will still have to generate the solidarity among its citizens necessary to underpin it, a degree of European compatriotism that does not yet exist. Another open question is how a more united eurozone core, which would itself contain creditor and debtor nations with very different perspectives, would relate institutionally and politically to EU member states not in the zone, such as the United Kingdom, Sweden, and Poland.

According to one projection by analysts at ING, a total collapse of the eurozone could cause GDP to fall by more than ten percent over two years in all the leading European economies, including Germany. Coming on top of the hardships already endured, that could lead to dangerous political radicalization.

(Unlike in the 1930s, such radicalization, to the far right and the far left, has been remarkably limited so far, even in Greece—a tribute to the resilience of contemporary European democracies.) But even if the eurozone falls apart, there will still be a place called Europe and probably a set of institutions called the European Union. And there will be a new yet also familiar historic challenge for Europeans: to pick themselves up from the ruins and rebuild.

Today's crisis is the greatest test yet of what has been called "the Monnet method" of unification, after Jean Monnet, a founding father of European integration. Monnet proposed moving forward, step by step, with technocratic measures of economic integration, hoping that these would catalyze political unification—not least through moments of crisis. "Crises are the great unifier!" he once explained. Yet even in the first 40 years of European integration, crises sometimes pulled Europe together and sometimes did not. If they tended more often to promote unity than division, that was in large part thanks to wartime memories and Cold War imperatives. So where are the drivers of integration now? Go back down the list.

A single market of 500 million consumers remains a powerful economic attraction for most European countries. However, it no longer seems as evident as it once did that Europe brings steadily growing prosperity and welfare to all its citizens. Exporting nations, especially Germany, and global service providers, such as the United Kingdom, are increasingly looking to emerging markets, where the growth is.

Unlike during the Cold War, there is no obvious external threat in Europe's front yard. Try as he might, Vladimir Putin just does not match up to Stalin, or even Brezhnev. Could China step into that role? Without stigmatizing China as an enemy, the most compelling new rationale for European unification is indeed the rise of non-Western great powers: China, mainly, but also India, Brazil, and South Africa.

One cannot simply extrapolate from current economic and demographic trends, but in any likely world of 2030, even Germany will be a small to medium-sized power. Then, the only effective way to defend the freedoms and advance the shared interests of all Europeans will be to act together and speak with one voice. Intellectually, this argument is persuasive. But emotionally, to sway a wider public, it does not compare with the visible presence of the Red Army at the heart of Europe.

If Russia no longer fits the bill for an external threat, the United States no longer plays the part of active external supporter. Already in 2001, President George W. Bush could ask, in a private meeting, "Do we want the European Union to succeed?" Part of his administration, at least in his first term, was inclined to answer no. President Barack Obama would definitely answer yes, but until the eurozone crisis threatened the U.S. economy, and hence his reelection prospects, it was hardly a priority. His administration has taken Europe as it has found it and dealt pragmatically with Brussels or with individual countries—whatever worked. Its geopolitical focus has been on China and Asia more generally, not Russia and Europe.

Conceivably, the United States' attitude could change if China really came to be seen as the new Soviet Union, a global geopolitical threat to the West. Then one option would be for Washington to seek a closer strategic partnership with a more united Europe, including, for example, a transatlantic free-trade area. Old Europe and its cousins across the water would work toward what Edouard Balladur, the former French prime minister, has imagined as a "Western Union." But there is scant evidence of such thinking at the moment. Rather, both the United States and Europe are making their own tense accommodations with China.

Another past driver of integration, eastern European yearnings, still has some traction today. Eastern Europeans have more recent memories than other Europeans do of dictatorship, hardship, and war. Many appreciate the new freedoms they enjoy in the EU; for some, belonging to the same club as western Europeans is the realization of a centuries-old dream. One Polish economist explains why Poland still aspires to join the eurozone thus: "We want to be on board the ship, even if it is sinking!" Of course, they would rather the ship stays afloat. Last fall, in a speech in Berlin, Radoslaw Sikorski, the Polish foreign minister, memorably observed, "I will probably be the first Polish foreign minister in history to say so, but here it is: I fear German power less than I am beginning to fear German inactivity."

European Germany, German Europe

Germany is the key to Europe's future, as it has been, one way or another, for at least a century. The irony of unintended consequences is especially acute here. If Kohl was the first chancellor of a united Germany since Hitler, Francois Hollande is the first Socialist president of France since Mitterrand, and it is Mitterrand's legacy he has to wrestle with. Monetary union, the method through which Mitterrand intended to keep united Germany in its proper place—co-driver with France, but still deferential to it—has ended up putting Germany at the wheel, with France as an irate husband flapping around in the passenger seat ("Turn left, Angela, turn left!").

At the time of German reunification, German politicians never tired of characterizing their goal in the finely turned words of the writer Thomas Mann: "Not a German Europe but a European Germany." What we see today, however, is a European Germany in a German Europe. This Germany is an exemplary European country: civilized, democratic, humane, law-abiding, and (although Mann might not have rated this one) very good at soccer. But the "Berlin Republic" is also at the center of a German Europe. At least when it comes to political economy, Germany calls the shots. (The same is not true in foreign and defense policy, where France and the United Kingdom are more important.) This is not a role Germany sought; leadership has been thrust upon it.

Moreover, if the need to win support for German reunification drove Kohl to accept European monetary union on a tight timetable, and without the political union he thought essential to sustain it, German reunification has changed the German attitude to the European project. The very same set of closely linked historical developments that has now produced, 20 years on, the need for a special German contribution to Europe has in the meantime reduced both the country's idealistic desire and its instrumental need to offer that contribution.

Were he still chancellor, Kohl would surely insist that the euro must be saved by moving decisively toward a political union. Merkel and her compatriots have reacted very differently, reluctantly doing the minimum needed to prevent collapse. The modest and plain-speaking Merkel is in many ways the personification of the civic, modern European virtues of this new Germany. She is also a brilliant and ruthless domestic political tactician. Whatever her personal convictions, she knows she faces what may be called the four Bs: the Bundestag (the lower house of the national parliament, from which Germany's most pro-European politicians have largely migrated to the European Parliament, another unintended consequence of that well-intended institution), the Bundesverfassungsgericht (the country's constitutional court, deliberately established after 1945 to be a U.S.-style check on a leader's power), the Bundesbank (still very influential in the German debate), and, last but by no means least, the populist tabloid Bild.

Many Germans resent the idea of bailing out Greeks and Spaniards and recall that they were given no say on Kohl's decision to give up the deutsche mark. In a German opinion poll conducted in May 2012, no less than 49 percent of respondents said it had been a mistake to introduce the euro. So far, the benefits they have derived from the euro have not been adequately explained. Yet this European Germany is a free country, open to argument, and some are now making the attempt.

Memory, Fear, and Hope

The greatest single driving force of the European project since 1945, personal memories of war, has disappeared. Where individual memory fades, collective memory should step in. Remember Edelman's appeal: "We, who did not perish, leave it up to you to keep the memory of them alive—forever." Yet most young Europeans' consciousness of their continent's tortured history is shallow. Their formative experiences have been in a Europe of peace, freedom, and prosperity. Even younger eastern Europeans from states such as Estonia, which did not exist on most maps just 22 years ago, have come to take these hard-won achievements for granted. In this sense, the deepest problem of the European project is the problem of success.

Over the last decade, European peoples with historical complexes about being consigned to the periphery of Europe felt themselves to be at last entering the core. Eastern Europeans joined the EU. Southern Europeans thought they were flourishing in the eurozone. In Athens, Lisbon, and Madrid, there was a sense of a leveling up of European societies, of a new, not merely formal equality among nations.

Now that illusion has been shattered. In Greece, the homeless line up at soup kitchens, pensioners commit suicide, the sick cannot get prescription medicines, shops are shuttered, and scavengers pick through dustbins—conditions almost reminiscent of the 1940s. In Spain, every second person under the age of 25 is unemployed; across the eurozone, the average is nearly one in four. But the pain is unevenly spread. In Germany, youth unemployment is comfortably under ten percent. There is a new dividing line across Europe, not between east and west but between north and south. Now, and probably for years to come, it will be a very different experience to be a young German or a young Spaniard, a young Pole or a young Greek.

Think back to those two May 10 moments in Warsaw. Someone whose formative teenage experience was of the terrors of 1943 would find today's crisis shocking, but still not half as bad as what he remembered—and he would insist that Europe must never fall back to that. The teenager of 2003 has a different mental lens: this is terrible, she thinks, and not what she was led to expect.

Europeans such as Geremek and Kohl witnessed Europe tear itself apart, and then dedicated themselves to building a better one. The generation of Spain's indignados, young protesters who have rallied across the country since May 2011, grew up in that better Europe, and have now been thrown backward. The trajectory of those who were, say, 15 years old in 1945 went from war to peace, poverty to prosperity, fear to hope. The trajectory of those who were 15 in 2003, especially in the parts of the continent now suffering the most, has arched in the opposite direction: from prosperity to unemployment, convergence of national experiences to divergence, hope to fear.

Could this very discontent provide the psychological basis for a popular campaign to save Europe? The signs are not promising. Popular movements have arisen during the crisis, but they have pointed in other directions. One of the largest was against the Anti-Counterfeiting Trade Agreement, which many young Europeans saw as a threat to their online freedom. The indignados of all countries, Europe's counterparts to the Occupy Wall Street movement, rail against bankers, politicians, and baby boomers, whom they see as having stolen their future. An interview-based survey of activists in these diverse campaigns, coordinated by Mary Kaldor and Sabine Selchow of the London School of Economics, found that the EU is either invisible among them or viewed somewhat negatively.

Fear should not be underestimated as a motivating force in politics. When, in a repeat election this June, the Greeks narrowly voted for parties that were serious about keeping the country in the eurozone, the Swiss cartoonist Patrick Chappatte drew a weary-looking man standing next to a ballot box in the shadow of the Acropolis and exclaiming, "Good news! Fear triumphed over despair." Adapting a famous phrase of U.S. President Franklin Roosevelt, one might almost say that today Europe has nothing to put its hope in but fear itself. The fear of collapse, the Monnet-like logic of necessity, the power of inertia: these may just keep the show on the road, but they will not create a dynamic, outward-looking European Union that enjoys the active support of its citizens. Without some new driving forces, without a positive mobilization among its elites and peoples, the EU, while probably surviving as an origami palace of treaties and institutions, will gradually decline in efficacy and real significance, like the Holy Roman Empire of yore. Future historians may then identify some time around 2005 as the apogee of the most far-reaching, constructive, and peaceful attempt to unite the continent that history has ever seen.

Critical Thinking

1. Why does the author consider the European union to be dysfunctional?

2. Why is Germany the key to Europe's future?

3. Should the United Kingdom withdraw from the European Union? Why or why not?

Create Central

www.mhhe.com/createcentral

Internet References

Europa
   europa.eu/int
Deutsche Bundesbank
   www.bundesbank.de/Navigation/DE/Home/home_node.html
Maastricht Treaty
   http://europa.eu/legislation_summaries/institutional_affairs/treaties/treaties_maastricht_en.

TIMOTHY GARTON ASH is Professor of European Studies at Oxford University and a Senior Fellow at the Hoover Institution at Stanford University.

Article

Prepared by: Robert Weiner, *University of Massachusetts/Boston*

Mutual Assured Production: Why Trade Will Limit Conflict Between China and Japan

RICHARD KATZ

Learning Outcomes

After reading this article, you will be able to:

- Explain why economic interdependence will reduce tensions between Japan and China.

- Explain why Pax Economica offers an alternative to the realist model of international relations.

Today, even though tensions between China and Japan are rising, an economic version of mutual deterrence is preserving the uneasy status quo between the two sides. Last fall, as the countries escalated their quarrel over an island chain that Japan has controlled for more than a century, many Chinese citizens boycotted Japanese products and took to the streets in anti-Japanese riots. With no resolution in sight, those who fear an escalation can nonetheless take solace in the fact that China and Japan stand to gain far more from trading than from fighting. For now, China is sending mixed signals as to whether it will escalate or calm these tensions, something the Japanese interpret as a lack of a policy consensus. Even though some officials in both capitals are looking for ways to save face and step back from the precipice, an unintentional conflict is still possible.

During the Cold War, the United States and the Soviet Union carefully avoided triggering a nuclear war because of the assumption of "mutual assured destruction": each knew that any such conflict would mean the obliteration of both countries. Today, even though tensions between China and Japan are rising, an economic version of mutual deterrence is preserving the uneasy status quo between the two sides.

Last fall, as the countries escalated their quarrel over an island chain that Japan has controlled for more than a century, many Chinese citizens boycotted Japanese products and took to the streets in anti-Japanese riots. This commotion, at times encouraged by the Chinese government, led the Japanese

government to fear that Beijing might exploit Japan's reliance on China as an export market to squeeze Tokyo into making territorial concessions. Throughout the crisis, Japan has doubted that China would ever try to forcibly seize the islands-barren rocks known in Chinese as the Diaoyu Islands and in Japanese as the Senkaku Islands—if only because the United States has made it clear that it would come to Japan's defense. Japanese security experts, however, have suggested that China might try other methods of intimidation, including a prolonged economic boycott.

But these fears have not materialized, for one simple reason: China needs to buy Japanese products as much as Japan needs to sell them. Many of the high-tech products assembled in and exported from China, often on behalf of American and European firms, use advanced Japanese-made parts. China could not boycott Japan, let alone precipitate an actual conflict, without stymieing the export-fueled economic miracle that underpins Communist Party rule.

For the moment, the combination of economic interdependence and Washington's commitment to Japan's defense will likely keep the peace. Still, an accidental clash of armed ships around the islands could lead to an unintended conflict. That is why defense officials from both countries have met with an eye to reducing that particular risk. With no resolution in sight, those who fear an escalation can nonetheless take solace in the fact that China and Japan stand to gain far more from trading than from fighting.

The Ties that Bind

Although China first claimed the Diaoyu/ Senkaku Islands in 1971, it never did much to pursue its claim until recently. On the contrary, in the interests of improving economic and political ties, Chinese Premier Zhou Enlai and Japanese Prime Minister Kakuei Tanaka agreed in 1972 to shelve the issue indefinitely. Beijing even stopped Chinese nationalist activists from trying to land on the islands and prevented articles that asserted

China's claim to them from appearing in the Chinese press. In the last few years, however, China has reversed course and started to back up its claims with actions. In 2010, for example, a Chinese fishing boat rammed a Japanese coast guard ship in the waters around the islands. When the coast guard personnel arrested the fishing boat's captain, Beijing declared that Japan had no jurisdiction in "Chinese territory" and cut off supplies to Japan of vital rare-earth minerals until he was released.

It took until July 2012 for the issue to explode. That month, then Japanese Prime Minister Yoshihiko Noda announced that his government intended to buy some of the islets from their private Japanese owner. Noda's aim was to prevent them from being sold to the right-wing governor of Tokyo, who had revealed plans for the islands that would certainly have provoked China. But Beijing told Noda that it would see even the government's purchase as an unacceptable change in the status quo. The Noda administration ignored warnings from both Beijing and the U.S. State Department and deluded itself that China would acquiesce to the purchase.

Then came the riots and the boycotts. For several weeks in August and September, Chinese protesters caused a ruckus, damaging Japanese-made cars, vandalizing stores selling Japanese products, and setting a Panasonic factory on fire. The police vacillated between encouraging and suppressing the riots, and some Chinese state media outlets listed Japanese brands to boycott. By the time the dust settled, Japanese firms operating in China had suffered about $120 million in property damage, and for a few months thereafter, sales of Japanese cars fell by approximately 40–50 percent.

The riots have stopped, but the larger conflict shows no signs of fading. China regularly sends armed surveillance boats into the islands' territorial waters, and the Chinese Foreign Ministry now calls the islands a "core interest," a term limited to the most sensitive areas regarding China's sovereignty, such as Taiwan and Tibet. China's Commerce Ministry has hinted at the possibility of a prolonged boycott to get Tokyo to concede that China has legitimate claims to the islands. It warned last September that Noda's purchase of the islands would "inevitably affect and damage . . . Sino-Japanese economic and trade relations."

A boycott would indeed prove disastrous for Japan's export-dependent economy. From 2002 to 2007, a third of Japan's gdp growth came from an increase in its trade surplus, and another third came from capital investment, much of which was tied to exports. And China stands at the center of this picture. From 1995 to 2011, increased shipments to China accounted for 45 percent of the overall growth in Japanese exports. Since the crisis erupted last July, however, Japan's price-adjusted exports to China have fallen by 20 percent, compared with an 11 percent drop in its global exports (as of March).

Japan's dependence on China helps explain why the new Japanese prime minister, Shinzo Abe, has not followed through on the hawkish positions he touted during last December's election campaign, such as his plan to place personnel and facilities on the Senkakus. Abe knows that his popularity hinges on Japan's economic recovery, and lest he forget it, Japanese businesses have been urging him to refrain from any provocations

while still seeking a resolution that maintains the country's sovereignty over the islands.

But it is not only Tokyo's behavior that has been tempered by economic interdependence. This year, Chinese censors have blocked the phrase "Boycott Japan" from Weibo, China's equivalent of Twitter. During February's New Year's celebrations, Beijing banned sales of the popular "Tokyo Big Bang" fireworks, which simulate the burning of the Japanese capital. In late March, China even joined Japan and South Korea in long-anticipated talks aimed at forming a trilateral free-trade agreement.

Meanwhile, Chinese provincial governments, hungry for jobs and tax revenue, keep imploring Japanese companies to expand their operations in China. In February, the city of Chongqing hired the Mitsui Group, a Japanese conglomerate, to develop an industrial park aimed at attracting foreign investment. At a March conference of the Japan-China Economic Association in Beijing, China's new vice president, Li Yuanchao, may have insisted that the media not photograph him shaking hands with Japan's top business leaders, but he nevertheless asked those leaders to step up their investments. Even on the national level, China is far more pluralistic than it used to be. The Communist Party- owned Global Times published both pro- and anti-boycott op-eds last fall.

Nor do most Chinese consumers seem interested in a boycott. The Japanese products that have lost the most sales during these latest tensions have been the highly visible ones, which are vulnerable to social pressure. Last fall and winter, sales of the popular cosmetics and skin-care products made by the Japanese company Shiseido tumbled, partly because many customers refrained from sending them as holiday gifts. Some stores temporarily displayed Shiseido products less prominently, but very few stopped carrying them altogether.

The worst-hit Japanese products have been cars, since many were vandalized by hooligans during the riots. But sales are recovering. Last fall, Nissan, the Japanese automaker, offered its Chinese customers its new "Promise for Car Security" program, a guarantee of free repairs for vehicles damaged in anti-Japanese riots. That's one reason why in March, sales at Nissan dealerships finally rose above the previous year's levels. And as China's infamous air pollution has worsened, February sales of air purifiers by Panasonic doubled from the year before, and sales of those made by Daikin quadrupled. Both are Japanese companies.

Made in China, with Help from Japan

Why does China, the world's largest exporter, so badly need what Japan is selling? Put simply, China's export driven economic miracle depends on imports. Around 60–70 percent of the goods China imports from Japan are the machinery and parts needed to make China's own exports. China cannot cut off this flow, or risk disrupting it through conflict, without crippling its economy. That is why, during the height of the island crisis last fall, the same Chinese customs officials who sometimes delayed shipments of Japanese consumer goods let industrial parts pass through.

For years, Japan has been China's single largest source of imports. A 2012 International Monetary Fund report calculated that for every one percent of growth in China's global exports, its imports from Japan rise by 1.2 percent. Take away those imports, and China's exports collapse.

Consider the iPhone and the iPad. Although Apple hires Taiwan's Foxconn to assemble these products in China, they contain Japanese parts, including Toshiba flash memory drives and Sharp lcd screens. The case of Apple is instructive for two reasons. First, as China has increasingly begun to export high-tech products, it has needed to rely more and more on imported parts. During the middle of the last decade, imported parts accounted for 22 percent of the value of low-technology goods exported from China. For information and communications equipment, however, that number was almost 50 percent. And it is precisely these import-intensive machinery and electronics products that are becoming more important to China's economy, rising from just 22 percent of all Chinese exports in 1992 to 63 percent in 2006 and presumably even more today.

Second, China's modernization depends on a host of multinational corporations using China as their workshop. In 2010, foreign companies and joint foreign-Chinese ventures accounted for more than 25 percent of China's entire industrial output, 39 percent of its apparel exports, and 99 percent of its computer exports. And these companies rely on imports from Japan. China cannot single out Japanese products without damaging and alienating the network of multinational companies that are fueling China's march up the value chain and toward higher living standards.

Today, China's cheap but high-quality labor and its fantastic infrastructure make the country an attractive production base. But if nasty international politics get in the way of those advantages, multinational corporations have other places to go. Only $4 of the value of the 2005 vintage iPod, priced at $299, came from China, and most of that was assembly work that could have been done elsewhere. Rising wages and worsening pollution have already led some electronics firms to shift their assembly operations to Southeast Asia. China cannot take foreign investment for granted.

Japanese firms know this, which explains why they are not fleeing despite the recent tensions. Japan remains the largest source of foreign investment in China today. According to Masaki Yamazaki of the Japan Center for Economic Research, some Japanese firms are considering a "China plus one" strategy, a way of diversifying their risks and finding another large market to invest in and export from. But, he added, "it's unrealistic to think that all the Japanese companies will rush away from China." As long as multinationals want to assemble products in China, Japanese suppliers need to be there as much as China needs to host them. What is more, China's bulging middle-class market is too big to be ignored by Japanese companies that produce consumer products and are plagued by low growth at home.

Indeed, in a survey conducted by the Japan External Trade Organization last November and December, just after the spate of violence, only 6 percent of Japanese companies in China said they were going to leave or downsize. Conversely, 52 percent planned on expanding, and 42 percent indicated that they would keep their operations at the same level while monitoring the situation. In 2012, a year in which global foreign direct investment in China fell by 3.7 percent, Japanese investment rose by 6 percent. As one Japanese business consultant told me, "Japanese firms see no early end to the territorial tensions. Some feel that they may just have to accept that, every few years, there will be an outburst of boycotts and even some violence."

Pax Economica

As World War I cruelly demonstrated, economic self-interest does not always override nationalist emotions. But it does raise the costs of letting passions dominate foreign policy.

For most of the past three decades, in recognition of those costs, China has sought what its leaders term a "peaceful rise." In the past few years, however, Beijing has shifted to a far more abrasive posture toward several countries in Asia. Some observers speculate that China's newfound assertiveness is a response to political dysfunction and the financial crisis in the West, which have led Beijing to doubt the United States' staying power and overestimate its own strength. Whatever the reasons, the new approach is reportedly disdained as self-defeating by many of China's business leaders and, according to Kiyoyuki Seguchi, research director at the Canon Institute for Global Studies, even by certain elements of its military.

Mao once observed that "political power grows out of the barrel of a gun." But in today's China, it is trade and globalization that pay for that gun. Despite Beijing's increasingly assertive stance, many Chinese officials recognize the costs of threatening the country's economic ties. As an op-ed in China Daily, an official Communist Party paper, put it last August, "Blindly boycotting Japanese goods by giving way to sentiments could harm our own industries and exports, and reduce employment."

Unfortunately, Beijing is not the only capital where politicians are ginning up nationalist fervor. In April, Abe began to show his more hawkish side. That month, some of his cabinet ministers and a record 168 Diet members visited the Yasukuni Shrine, which pays tribute to Japan's war dead and is controversial because it includes World War II war criminals—such a trip understandably outrages China, South Korea, and others. Then, Abe publicly questioned whether Japan's actions in China and Korea during World War II and earlier really constituted "aggression" and indicated that he might partially retract Tokyo's 1995 apology for Japan's behavior. Some fear that if Abe's party wins the July elections for the upper house of the Diet, Abe will further unleash his nationalist instincts.

For now, China is sending mixed signals as to whether it will escalate or calm these tensions, something the Japanese interpret as a lack of a policy consensus. Even though some officials in both capitals are looking for ways to save face and step back from the precipice, an unintentional conflict is still possible.

But there is at least one reason for hope: although money is said to be the root of all evil, it may also be what ultimately tips the balance of forces within each country back toward those who can prevent war and eventually put the conflicting territorial claims back on the shelf.

Critical Thinking

1. Why should the U.S. defend Japan in its dispute with China over the sovereignty of some islands in the South China sea?

2. What does China mean by core interest?

3. What is the nature of Chinese economic dependance on Japan?

Create Central

www.mhhe.com/createcentral

Internet References

The Foreign Ministry of China
www.mfa.gov.cn/eng

Japan's Ministry of Foreign Policy
www.mofa.go.jp

RICHARD KATZ is editor of the semiweekly The Oriental Economist Alert and the monthly The Oriental Economist Report, both reports on Japan.

Unit 7

UNIT

Prepared by: Robert Weiner, *University of Massachusetts/Boston*

Global Environmental Issues

Climate change is certainly the most important global environmental problem that the international community faces and one that the global environmental governance mechanisms have not yet been able to resolve. The scientific community has developed different models that attempt to predict the effects of the warming climate, which is caused by the release of greenhouse gases into the atmosphere. The emission of greenhouse gases into the atmosphere is a major factor that contributes to the warming of the climate. The deforestation of the Amazon Basin in Brazil has contributed to this phenomenon, as the tropical rainforest in Brazil had served as a carbon sink, soaking up carbon dioxide. Climate warming has also had a generally negative effect on the production of food, especially global food security. Food production continues to be threatened by the effects of climate change, with more and more extreme weather events in the offing.

The warming of the climate has had the effect of melting glaciers and ice sheets in Greenland and Antarctica. The melting of the Arctic Ocean has geopolitical implications and will have effects on the exploitation of oil in the Arctic. Arctic states such as Russia are staking out claims to adjacent continental shelves and seabeds, such as the Lomonosov ridge, in order to exploit the large amounts of energy that are available there. Besides Russia, some of the other important Arctic states are the United States, Canada, Denmark, and Norway. It is estimated that the energy reserves in the Arctic may amount to about 25% of the world's total reserves. A vast amount of scientific data, such as oceanographic data and satellite photos, show the melting of the Arctic sea ice.

The melting of the ice in Canada's fabled Northwest Passage will allow goods to be shipped from Europe to Asia and Asia to Europe, subtracting thousands of miles from the traditional sea routes. States and multinationals have expressed an interest in using the shorter routes, when they open up, perhaps by 2050 or earlier, to ship goods such as oil between East and West, as there would obviously be significant savings of shipping costs. For example, supertankers that cannot use the Panama Canal because they are too big would be interested in using the Northwest Passage, if feasible. The melting of the ice in the Northwest Passage has also opened up security issues, because it means that vessels of states other than Canada that do not accept Canadian claims to sovereignty over the waterway might be tempted to test Canada's ability to assert its sovereignty over the area. For example, the main difference between Canada and the United States over the Northwest Passage is that Canada views the Northwest Passage as lying within its internal waters, whereas the United States views the Northwest Passage as an international strait. The Canadian position on the Northwest Passage means that those states

wishing to use it must seek the permission of the Canadian government. According to this line of reasoning, Canada, then, should have the capacity to control the Northwest Passage if it wants its claim to sovereignty upheld. On the other hand, the U.S. position is that the Northwest Passage is an international waterway, and therefore can be used for the purposes of innocent passage.

In the final analysis, however, the potential of the vast energy resources opening up in the Arctic because of the melting of the Arctic sea ice is also resulting in the increasing militarization of the Arctic Ocean states, especially Canada.

The economic development of emerging markets such as China and India has resulted in an intensified competition with the United States around the globe for energy and mineral resources, especially as China scours the globe for oil. Moreover, the catastrophic effects of a lack of an adequate energy infrastructure were seen in India in the summer of 2012, when more than 600 million people were hit by a massive blackout.

Furthermore, energy security is a critical issue for the United States, as the rise in the price of oil due to the turmoil in the Middle East in 2013 could upset the efforts on the part of a country like the United States to recover from the recession of 2009. Most of the world's oil is located in the Middle East and is controlled by a cartel known as the Organization of Petroleum Exporting Countries (OPEC), which was formed in the 1960s. The United States, in an effort to reduce its dependency on foreign oil from regions such as the Middle East, has diversified the sources from which it imports this commodity. The United States is also taking advantage of new developments in energy technology, for example, to extract gas from shale as well as to exploit oil from wells that have been thought to have been depleted as the United States focuses more on the development of domestic sources of energy.

The growth of the world's population is also placing an enormous strain on Earth's resources and the carrying capacity of the planet to sustain the increased numbers, raising the spectra of a Malthusian nightmare. In 2012, the world's population reached 7 billion, and a population of 9 billion is projected by some experts by 2050. The greatest increase in global population is taking place in the developing or southern part of the world, therefore increasing the economic gaps and disparities that exist between the rich, industrialized, northern states located in regions such as North America, Western and Eastern Europe, and Russia and the states located in the southern part of the globe. This means that in the age of globalization, a significant number of the world's population will continue to exist on the periphery of the industrialized core of world society, eking out an existence of abject poverty.

Article

Prepared by: Robert Weiner, *University of Massachusetts/Boston*

Too Much to Fight Over

Arctic countries have decided to join hands and gorge on Arctic resources.

JAMES ASTILL

Learning Outcomes

After reading this article, you will be able to:

- Explain why the prediction of conflict over the resources of the Arctic Sea is exaggerated.
- Discuss the important role that the Arctic Council plays in promoting the cooperation of the Arctic Sea states.

The Geopolitics of the new Arctic entered the mainstream on August 2nd 2007. Descending by *Mir* submersible to a depth of over 4km, a Russian-led expedition planted a titanium Russian flag beneath the North Pole. The news shocked the world.

The Lomonosov ridge under the pole, which is probably rich in minerals, is claimed by Russia, Canada and Denmark. The Russians, it was assumed, were asserting their claim, perhaps even launching a scramble for Arctic resources. One of their leaders, Artur Chilingarov, Russia's leading polar explorer and a Putin loyalist, fanned the flames. "The Arctic has always been Russian," he declared. Yet the expedition turned out to have been somewhat international, initiated by an Australian entrepreneur and a retired American submarine captain, and paid for by a Swedish pharmaceuticals tycoon.

Even so, fears of Arctic conflict have not gone away. In 2010 NATO's top officer in Europe, James Stavridis, an American admiral, gave warning that "for now, the disputes in the north have been dealt with peacefully, but climate change could alter the equilibrium". Russia's ambassador to NATO, Dmitry Rogozin, has hinted at similar concerns. "NATO", he said, "has sensed where the wind comes from. It comes from the north." The development of the Arctic will involve a rebalancing of large interests. The Lomonosov ridge could contain several billion barrels of oil equivalent, a substantial prize. For Greenland, currently semi-autonomous from Denmark, Arctic development contains an even richer promise: full independence. That would have strategic implications not only for Denmark but also for the United States, which has an airbase in northern Greenland.

There are also a few Arctic quirks that turn the mind to confrontation. Most countries in the region (the United States being the main exception) have powerful frontier myths around their northern parts. This is truest of the biggest: Russia, for which the Arctic has been a source of minerals and pride in the feats of Russian explorers, scientists and engineers since the late 19th century; and Canada, which often harps on Arctic security, perhaps as a means of differentiating itself from the United States.

During the cold war the Arctic bristled with Soviet submarines and American bombers operating from airbases in Iceland and Greenland. The talk of Arctic security risks sometimes betrays a certain nostalgia for that period. Some people also worry about Arctic countries militarising the north. Canada conducted its biggest-ever military exercise in the north, involving 1,200 troops, in the Arctic last year.

The risks of Arctic conflict have been exaggerated. Far from violent, the development of the Arctic is likely to be uncommonly harmonious.

Yet the risks of Arctic conflict have been exaggerated. Most of the Arctic is clearly assigned to individual countries. According to a Danish estimate, 95 percent of Arctic mineral resources are within agreed national boundaries. The biggest of the half-dozen remaining territorial disputes is between the United States and Canada, over whether the north-west passage is in international or Canadian waters, hardly a *casus belli*.

Far from violent, the development of the Arctic is likely to be uncommonly harmonious, for three related reasons. One is the profit motive. The five Arctic littoral countries, Russia, the United States, Canada, Denmark and Norway, would sooner develop the resources they have than argue over those they do not have. A sign of this was an agreement between Russia and Norway last year to fix their maritime border in the Barents Sea, ending a decades-long dispute. The border area is probably rich in oil; both countries are now racing to get exploration started.

Another spur to Arctic co-operation is the high cost of operating in the region. This is behind the Arctic Council's first binding agreement, signed last year, to co-ordinate search-and-rescue efforts. Rival oil companies are also working together, on scientific research and mapping as well as on formal joint ventures.

The third reason for peace is equally important: a strong reluctance among Arctic countries to give outsiders any excuse to intervene in the region's affairs. An illustration is the stated willingness of all concerned to settle their biggest potential dispute, over their maritime frontiers, according to the international Law of the Sea (LOS). Even the United States accepts this, despite its dislike for treaties—though it has still not ratified the United Nations Convention on the Law of the Sea, an anomaly many of its leaders are keen to end.

The LOS entitles countries to an area of seabed beyond the usual 200 nautical miles, with certain provisos, if it can be shown to be an extension of their continental shelf. Whichever of Russia, Canada and Denmark can prove that the Lomonosov ridge is an extension of its continental shelf will therefore have it. It will be up to the countries themselves to decide this: the UN does not rule on disputed territories. The losers will not do too badly, though: given the Arctic's wide continental shelves, the LOS guarantees each a vast amount of resource-rich seabed.

The 2007 furore over the Russian flag led to an important statement of Arctic solidarity, the Ilulissat Declaration, issued by the foreign ministers of the five countries adjoining the Arctic Ocean (to the chagrin of the Arctic Council's other members, Sweden, Iceland and Finland). This expressed their commitment to developing the Arctic peacefully and without outside interference. Possible defence co-operation between Arctic countries points in the same direction. Their defence chiefs met for the first time in Canada in April in what is to become an annual event.

A Warm Atmosphere

The Arctic Council, founded in 1996, was not designed as a regional decision-making forum, though outsiders often see it that way. Its mission was to promote conservation, research and sustainable development in the Arctic. The fact that six NGOS representing indigenous peoples were admitted to the club as non-voting members was evidence of both this ambition and the countries' rather flaky commitment to it. But since 2007, under Danish, Norwegian and now Swedish chairmanship, the council has become more ambitious. Next year it will open a permanent secretariat, paid for by Norway, in the Norwegian city of Tromso. A second binding pact, on responding to Arctic oil spills, is being negotiated; others have been mooted.

Russia, which has at least half of the Arctic in terms of area, coastline, population and probably mineral wealth, is in the thick of the new chumminess. It has a reputation for thinking more deeply about Arctic strategy—in which Mr Putin and his prime minister, Dmitry Medvedev, are both considered well-versed—than any other power, and appears to have concluded that it will benefit more from collaboration than from discord. Indeed its plans for the Northern Sea Route may depend upon international co-operation: Norway and Iceland both have ambitions to provide shipping services in the region.

Russia's ambassador for Arctic affairs, Anton Vasiliev, is one of the council's most fluent proponents of such collaborations. At a recent conference in Singapore, convened by *The Economist,* he surprised many by declaring Russia eager to standardise safety procedures for Arctic oil and gas production. "The Arctic is a bit special for civility," he says, "You cannot survive alone in the Arctic: this is perhaps true for countries as well as individuals."

The United States is less prominent in Arctic affairs, reflecting its lesser interest in the region and lukewarm enthusiasm for international decision-making. Although its scientists lead many of the council's working groups on subjects such as atmospheric pollution and biodiversity, it only hesitantly supports the council's burgeoning remit.

Frustrated advocates of a more forthright American policy for the Arctic, mostly from Alaska, lament that the United States hardly sees itself as an Arctic country, a status it owes to its cut-price $7.2m purchase of Alaska (Russian America as was) in 1867. A common complaint is the United States' meagre ice-breaking capability, highlighted last winter when an ice-capable Russian tanker had to be brought in to deliver fuel to the icebound Alaskan town of Nome.

The African Arctic

As governments wake up to the changing Arctic, global interest in the region is booming. A veteran Scandinavian diplomat recalls holding a high-level European meeting on the Arctic in the early 1990s to which only her own minister turned up. "Now we're beating countries away," she says. "I've had a couple of African countries tell me they're Arctic players."

Asia's big trading countries, including strong exporters like China and Japan, shipbuilders like South Korea and those with shipping hubs, like Singapore, make a more convincing case for themselves. All have applied to join the council as observers, as have Italy and the EU. Half a dozen European countries with traditions of Arctic exploration, including Britain and Poland, are observers already.

Some council members are reluctant to expand their club. Canada is especially wary of admitting the EU because the Europeans make a fuss about slaughtering seals; Russia has a neurotic fear of China. Even the relaxed Scandinavians are in no hurry to expand the council. Yet the disagreement has been overblown. If the EU, China and others were to be denied entry to the council, they would no doubt try to raise Arctic issues elsewhere, probably at the UN, which is a far more dreadful prospect for Arctic countries. So by the end of Sweden's chairmanship, in May 2013, these national applicants are likely to be admitted.

But Greenpeace, which also wants to be an observer, may not be, even though another green NGO, the World Wildlife Fund (WWF), already is. Several Arctic governments have been put off by Greenpeace's aggressive methods. Greens against governments, not country against country, looks likely to be the most serious sort of Arctic conflict. That is progress of a sort.

Critical Thinking

1. What is the relationship between the Law of the Sea and rival claims by Arctic Sea states to the energy resources of the Arctic Ocean?

2. Can Canada successfully uphold its claim to sovereignty over the Northwest passage? How?

3. How can the Arctic Council promote cooperation between the Arctic Sea states?

Create Central

www.mhhe.com/createcentral

Internet References

The Arctic Council
www.arctic-council/org

Arctic Map
http://geology.com/world/arctic-ocean-map-shtml

Article

Prepared by: Robert Weiner, *University of Massachusetts/Boston*

A Light in the Forest: Brazil's Fight to Save the Amazon and Climate-Change Diplomacy

JEFF TOLLEFSON

Learning Outcomes

After reading this article, you will be able to:

- Explain how Amazonian deforestation affects the global climate.

- Discuss what can be done to stop the deforestation of the Amazon.

Across the world, complex social and market forces are driving the conversion of vast swaths of rain forests into pastureland, plantations, and cropland. Rain forests are disappearing in Indonesia and Madagascar and are increasingly threatened in Africa's Congo basin. But the most extreme deforestation has taken place in Brazil. With the resulting increase in arable land, Brazil has helped feed the growing global demand for commodities, such as soybeans and beef. Brazil has dramatically slowed the destruction of its rain forests, reducing the rate of deforestation by 83% since 2004, primarily by enforcing land-use regulations, creating new protected areas, and working to maintain the rule of law in the Amazon. At the same time, Brazil has become a test case for a controversial international climate-change prevention strategy known as REDD, short for "reducing emissions from deforestation and forest degradation," which places a monetary value on the carbon stored in forests.

Across the world, complex social and market forces are driving the conversion of vast swaths of rain forests into pastureland, plantations, and cropland. Rain forests are disappearing in Indonesia and Madagascar and are increasingly threatened in Africa's Congo basin. But the most extreme deforestation has taken place in Brazil. Since 1988, Brazilians have cleared more than 153,000 square miles of Amazonian rain forest, an area larger than Germany. With the resulting increase in arable land, Brazil has helped feed the growing global demand for commodities, such as soybeans and beef.

But the environmental price has been steep. In addition to providing habitats for untold numbers of plant and animal species and discharging around 20 percent of the world's fresh water, the Amazon basin plays a crucial role in regulating the earth's climate, storing huge quantities of carbon dioxide that would otherwise contribute to global warming. Slashing and burning the Amazon rain forest releases the carbon locked up in plants and soils; from a climate perspective, clearing the rain forest is no different from burning fossil fuels, such as oil and gas. Recent estimates suggest that deforestation and associated activities account for 10–15 percent of global carbon dioxide emissions.

But in recent years, good news has emerged from the Amazon. Brazil has dramatically slowed the destruction of its rain forests, reducing the rate of deforestation by 83 percent since 2004, primarily by enforcing land-use regulations, creating new protected areas, and working to maintain the rule of law in the Amazon. At the same time, Brazil has become a test case for a controversial international climate-change prevention strategy known as REDD, short for "reducing emissions from deforestation and forest degradation," which places a monetary value on the carbon stored in forests. Under such a system, developed countries can pay developing countries to protect their own forests, thereby offsetting the developed countries' emissions at home. Brazil's preliminary experience with REDD suggests that, in addition to offering multiple benefits to forest dwellers (human and otherwise), the model can be cheap and fast: Brazil has done more to reduce emissions than any other country in the world in recent years, without breaking the bank.

The REDD model remains a work in progress. In Brazil and other places where elements of REDD have been applied, the funding has yet to reach many of its intended beneficiaries, and institutional reforms have been slow to develop. This has contributed to a rural backlash against the new enforcement measures in the Brazilian Amazon—a backlash that the government is still struggling to contain. But if Brazil can consolidate its early gains, build consensus around a broader vision for development, and follow through with a program to overhaul the economies of its rainforest regions, it could pave the way for a new era of environmental governance across the tropics. For the first time, perhaps, it is possible to contemplate an end to the era of large-scale human deforestation.

Lula Gets Tough

The deforestation crisis in Brazil ramped up in the 1960s, when the country's military rulers, seeking to address the country's poverty crisis, encouraged poor Brazilians to move into the Amazon basin with promises of free land and generous government subsidies. In response, tens of thousands of Brazilians left dry scrublands in the northeast and other poor areas for the lush Amazon basin—a mass internal migration that only increased in size throughout the 1970s and beyond.

But the government did not properly plan for the effect of a population explosion in the Amazon basin. The result was a land rush, during which short-term profiteering from slash-and-burn agriculture prevented anything resembling sustainable development. Environmental and social movements arose in response to the chaotic development, but it was not until the 1980s, when scientists began systematically tracking Amazonian deforestation using satellite imagery, that the true scale of the environmental destruction under way in the Amazon became apparent. The end of military rule in 1985 and Brazil's transition to democracy did nothing to slow the devastation; the ecological damage only worsened as road-building projects and government subsidies for agriculture fueled a real estate boom that wiped out forests and threatened traditional rubber tappers and native peoples. Meanwhile, the total population of the Amazon basin increased from around six million in 1960 to 25 million in 2010 (including some 20 million in Brazil), and agricultural production in the Amazon region ramped up as global commodity markets expanded.

Things began to change in 2003, when Luiz Inácio Lula da Silva, the newly elected Brazilian president, known as Lula, chose Marina Silva as his environment minister. A social and environmental activist turned politician, Silva hailed from the remote Amazonian state of Acre and had worked alongside Chico Mendes, a union leader and environmentalist whose murder in 1988 at the hands of a rancher drew global attention to the issue of the Amazon's preservation. With Lula's blessing, Silva immediately set about doing what no Brazilian government had previously attempted: enforcing Brazil's 1965 Forest Code, which had set forth strong protections for forests and established strict limits on how much land could be cleared. Doing so represented a major shift in domestic policy and was equally striking at the international level: Brazil chose to act at a time when most developing countries were resisting any significant steps to combat global warming absent the industrialized world's own more aggressive actions and provision of financial aid.

After peaking in 2004, when an area of rain forest roughly the size of Massachusetts was mowed down in a single year, Brazil's deforestation rate began to fall. Then, in late 2007, scientists at Brazil's National Institute for Space Research warned that the rate of deforestation had spiked once again. The increase coincided with a sudden rise in global food prices, which created an incentive for landowners in the Amazon to illegally clear more forest for pasture and crops. This suggested that the earlier decline in the rate of deforestation might have been driven by market forces as much as by government intervention, but Lula nevertheless doubled down on enforcement. The government deployed hundreds of Brazilian soldiers in early 2008 to crack down on illegal logging, issuing fines to those who broke the law and in some instances hauling lawbreakers to jail.

The following year, Brazil announced that its rate of deforestation had hit a historic low, and Lula pledged that by 2020 the country would reduce its deforestation to 20 percent of the country's longterm baseline, then defined as the average from 1996 to 2005. His plan to achieve that goal was based on one version of the REDD model, which had vaulted onto the international agenda several years earlier as scientists made advances in quantifying the impact of tropical deforestation on climate change.

Green-Lighting REDD

Politicians and commentators usually describe global warming as a long-term threat, but scientists also worry about transgressing invisible thresholds and thus provoking potentially rapid and irreversible near term changes in the way environmental and biological systems function. During the past decade, based in part on the results of intensive climate modeling, some scientists began to grow concerned that the Amazon could represent one of the clearest examples of such tipping points.

Think of the rain forest not as a collection of trees but as a hydrologic system, a massive machine for transporting and recycling water in which trees act as pumps, pulling water out of the ground and then injecting it, through transpiration, into the air. This process ramps up as the sun rises over the Amazon each day; as the forest heats up, evaporation increases, and trees transpire water to stay cool, simultaneously increasing the amount of water they take up through their roots. By constantly replenishing the atmosphere with water vapor, the Amazon helps create its own weather on a grand scale.

Humans interfere with this process whenever they chop down rain forests, and at some point, the system will begin to shut down. And this is not the only threat. Studies suggest that the Amazon could also be susceptible to rising temperatures and shifting rainfall patterns due to global warming. The nightmare scenario is known as "Amazon dieback," wherein the rains decrease and open savannas encroach on an ever-shrinking rain forest. The resulting loss of fresh water could be catastrophic for communities, agriculture, and hydropower systems in the Amazon, and dieback would have drastic effects on biodiversity and the global carbon dioxide cycle. The Amazon stores some 100 billion metric tons of carbon, equivalent to roughly a decade of global emissions. Converting carbon-rich rain forests into open savannas would pump massive quantities of carbon dioxide into the atmosphere, making it even harder for humans to prevent further warming.

Roughly 20 percent of the Amazon has been cleared to date, and there is already evidence that precipitation and river-discharge patterns are changing where the deforestation has been most intense, notably in the southwestern portion of the basin. And some scientists fear that the shifting climate may already be exerting an influence. In the past seven years, the Amazon has suffered two extremely severe droughts; normally, such droughts would be expected to occur perhaps once a century. One of the most comprehensive modeling studies to date, conducted in 2010 under the auspices of the World Bank, suggests that even current levels of deforestation, when combined with the impacts of increasing forest fires and global warming, are making the Amazon susceptible to dieback.

Such projections have heightened the sense of urgency in climate policy circles and helped focus attention on the REDD model. The concept has been around in some form for more than 15 years, but it was first placed on the international agenda in 2005 by the Coalition for Rainforest Nations, a group of 41 developing countries that cooperates with the UN and the World Bank on sustainability issues. At the core of the model is the belief that it is possible to calculate how much carbon is released into the atmosphere when a given chunk of forest is cut down. Fears that this would prove impossible helped keep deforestation off the agenda when climate diplomats signed the Kyoto Protocol in 1997. Scientists are steadily improving their methods for estimating how much carbon is stored in forests, however, and most experts agree that carbon dioxide can be tracked with enough accuracy to calculate baseline figures for every country.

Under various proposed versions of the REDD model, wealthy countries or businesses seeking to offset their own impact on the climate would pay tropical countries to reduce their emissions below their baseline levels. There is no consensus about the best way to design such a system of payments; since REDD was formally adopted as part of the agenda for climate negotiations at the UN Climate Change Conference in Bali, Indonesia, in 2007, dozens of countries and nongovernmental organizations have put forward a range of ideas. Most of these call for the creation of a global market that, like the European carbon-trading system, would allow industrial polluters to purchase carbon offsets generated by rain-forest preservation. Some environmentalists and social activists worry about the validity and longevity of such credits, as well as the prospect of banks and traders entering the conservation business. One fear is that "carbon cowboys," a new class of entrepreneurs specializing in the development of carbon-offset projects, would sweep through forests, trampling the rights of indigenous and poor people by taking control of their lands and walking away with the profits. This concern is valid, as there is always a danger of bad actors. But civil-society groups and governments, including Brazil's, are aware of the problem and are working on safeguards.

Brazilian officials have also expressed worries that the ability to simply purchase unlimited offsets would allow wealthy countries to delay the work that needs to be done to reduce their own emissions. An alternative backed by Brazil's climate negotiators and others would be a state-based funding system, in which money would flow from governments in the developed world to governments in the developing world, which would guarantee emissions reductions in return.

Norwegian Wood

In 2008, Lula, perhaps hoping to preempt an interminable debate over how best to design a global REDD system, announced the establishment of the Amazon Fund, calling on wealthy countries to contribute some $21 billion to directly fund rain-forest-preservation measures. The proposal went against the market-based approach being pushed by the Coalition for Rainforest Nations. Based on a more conventional system of government donations, the Amazon Fund would allow Brazil to control the money and manage its forests as it saw fit. To the fund's backers, the resulting reductions in emissions would represent offsets of a sort.

Only one country decided to take up Lula's challenge: Norway, which stepped forward with a commitment of up to $1 billion. Coming well in advance of any formal carbon market and the international treaty that many hoped would be signed at the UN climate summit in Copenhagen in 2009, Norway's pledge was largely an altruistic vote of confidence in Brazil's approach, with donations conditioned on measurable progress. Since 2010, when the funding began, the Brazilian Development Bank, which manages the fund, has undertaken 30 projects, costing nearly $152 million. These projects include direct payments to landowners in return for preserving forests and initiatives to sort out disputes over landownership, educate farmers and ranchers about sustainability, and combat forest fires.

Although environmentalists and scientists have criticized some delays in the program, Brazil's deforestation rate has continued to plunge. Each year from 2009 to 2012, the country registered a new record low for deforestation; in 2012, only 1,798 square miles of forest were cleared. That is 76 percent below the long-term baseline, leaving Brazil just four percent shy of its Copenhagen commitment with eight years to go. Recent calculations by Brazilian scientists suggest that the cumulative release of carbon dioxide expected as a result of deforestation in the Brazilian Amazon dropped from more than 1.1 billion metric tons in 2004 to 298 million in 2011—roughly equivalent to the effect of France and the United Kingdom eliminating their combined carbon dioxide emissions for 2011.

REDD remains a distant promise for most landowners and communities, and the precipitous drop in deforestation in Brazil is more a function of broader government policy than the result of any individual project. Still, the Amazon Fund is demonstrating the promise and practicality of the REDD model. Although the actual cost of preventing emissions remains unclear, Brazil is offering donors carbon offsets at a discounted price of $5 per metric ton of carbon dioxide, intentionally underestimating how much biomass its forests contain in order to avoid arguments over the price. Of course, implementing the REDD model could prove significantly more expensive elsewhere. But the price would nonetheless be significantly cheaper than for many other methods of cutting emissions, such as capturing carbon dioxide from a coal-fired power plant and pumping it underground, which could cost upward of $100 per metric ton in the initial stages.

Rousseff and the Ruralistas

Lula was succeeded by his protégé and former chief of staff, Dilma Rousseff, in 2011. Although environmentalists have been critical of her broader development agenda in the Amazon and beyond, Rousseff has upheld Lula's deforestation policies. And she has done so despite intense pressure from the so-called ruralista coalition of landowners and major agricultural interests, which currently exercises tremendous influence in Brasília.

In the spring of 2012, the Brazilian Congress passed a bill that would have eviscerated the country's vaunted Forest Code by scaling back basic protections for land alongside rivers and embankments and offering outright amnesty to companies and landowners who had broken the law. Rousseff fought back, and a prolonged tussle ensued. The final result was a law that is generally more favorable to agricultural interests but that nonetheless retains minimum requirements for forest protection and recovery on private land.

More troubling than the new law itself, perhaps, is the political polarization that accompanied its passage. Brasília now seems divided into rigid environmentalist and agricultural factions. Fierce opposition to Brazil's rain-forest-preservation efforts is sure to persist, and many observers fear that landowners, impatient with the slow pace of progress on REDD, will ultimately begin to test the limits of the newly revised Forest Code. As if on cue, last September, Brazilian scientists announced that deforestation was 220 percent higher in August than it had been in August of 2011. But it is too early to tell what this latest outbreak might mean. After all, prior spikes have incurred a government response, and each time the damage has been contained.

It is also worth noting that not only has Brazilian deforestation decreased overall, but the size of the average forest clearing has also decreased over time. The powerful landowners and corporate interests responsible for large-scale deforestation have apparently decided that they can no longer cut down rain forests with impunity. The upshot is that for the first time ever, in 2011, the amount of land cleared in the Brazilian Amazon dropped below the combined amount cleared in the surrounding Amazon countries, which make up 40 percent of the basin. In those countries, the trend is not so encouraging: deforestation in the non-Brazilian Amazon increased from an estimated annual average of 1,938 square miles in the 1990s to 2,782 square miles last year, according to an analysis published by the World Wildlife Fund.

Missing the Forest for the Trees?

There was very little progress on REDD at the most recent UN climate summit, in Doha, Qatar, last November. Negotiators left the door open to a full suite of REDD-style models, from government-to-government financial transfers to a privatized carbon market, but failed to agree on the details. Regardless of which particular models are codified in a hypothetical future treaty on climate change, countries need to focus on making

the money flow: some studies suggest that halving deforestation would cost $20–$25 billion annually by 2020. So far, governments have committed several billion dollars to forest protection through various bilateral and multilateral agreements. Through the UN, the industrialized countries have also made impressive commitments to combating climate change in the developing world, promising to contribute up to $100 billion annually by 2020, a portion of which could fund forest protection.

But it is not at all clear that this money will materialize, due in part to the current weakness of the global economy. And there is a limit to government largess. Advocates of rain-forest preservation are now trying to convince governments to commit money from revenue streams that do not depend on annual appropriations, which are more vulnerable to political and economic pressure. But that, too, is an uphill battle. Indeed, forest-preservation advocates cannot rely on governments alone; they will ultimately need to attract private-sector investment.

In the meantime, the fight against deforestation will rely on a patchwork of international partnerships and initiatives. Most significant, perhaps, Norway has transferred the model it developed with Brazil to Indonesia, which now ranks as the largest emitter of carbon dioxide from tropical deforestation. Just as in Brazil, the promise of REDD helped inspire some bold political commitments by Indonesian authorities, who have agreed to reduce their greenhouse gas emissions—most of which come from deforestation—by up to 41 percent by 2020 if international aid materializes. But Indonesia has neither the monitoring technology nor the institutional wherewithal of Brazil, so Norway's $1 billion commitment is aimed at helping the country build up its scientific and institutional capacity. Progress has been slow, but the advantage of a results-based approach, such as REDD, is that these initiatives cost money only if they yield positive results.

Brazil's experience offers some lessons for other tropical countries. The first is that science and technology must be the foundation of any solution. Brazil's progress has been made possible by major investments in scientific and institutional infrastructure to monitor the country's rain forests. Nations seeking to follow suit must invest in tools that will help them not only monitor their forests but also estimate just how much carbon those forests store. Working with scientists at the Carnegie Institution for Science, the governments of Colombia and Peru are deploying advanced systems for tracking deforestation from readily available satellite data. Combined with laser-based aerial technology that can map vast swaths of forest in three dimensions, these systems will be able to more accurately calculate and monitor stored carbon across an entire landscape—a feat that could allow these countries to leapfrog Brazil.

Brazil's Amazon Fund also shows that it is possible to move forward despite lingering scientific uncertainty about how to quantify the carbon stored in forests. Some critics of the REDD model have worried that it could draw attention away from the enforcement of existing forestry laws, ultimately increasing the cost of conservation and rewarding wealthy lawbreakers. But

Brazil's experience shows that the two approaches can go hand in hand. Indeed, most of Brazil's progress to date has come from simply enforcing existing rules. The government has also created formal land reserves, outlawing development on nearly half its territory, and environmental groups have played a role by rallying public opinion and partnering with industry groups to improve agricultural practices. Still, enforcement can go only so far with the smaller landholders and subsistence farmers who are responsible for an increasingly large share of the remaining deforestation. Brazil must focus the Amazon Fund and other government initiatives on projects that will create more sustainable forms of agriculture for these small-scale farmers and ranchers.

The government also needs to look ahead. Cities in the Amazon are booming, and larger populations will translate into additional demands for natural resources and food. The Brazilian government has sought to increase agricultural productivity across the basin, recognizing that there is more than enough land available to expand production without clearing more of it. But Brazil should also encourage more forest recovery, which would bolster the Amazon's ability to produce rain and absorb carbon dioxide from the atmosphere. Globally, forests currently absorb roughly a quarter of the world's carbon emissions, thanks to the regrowth of forests cut down long ago in places such as the United States, and they could provide an even larger buffer going forward. Roughly 20 percent of the areas once cleared in the Amazon are already regrowing as so-called secondary forest. Scientists have calculated that if the government can increase that figure to 40 percent, the Brazilian Amazon will transition from a net source of carbon dioxide emissions to a "carbon sink" by 2015, taking in more carbon dioxide than it emits.

Deforestation is just one of many challenges buffeting the Amazon region, and improvements on this front should not obscure the ongoing problems of poverty, violence, and corruption. But at a time when expectations for progress on climate change are falling, Brazil has given the world a glimmer of hope. In many ways, the hard work is just beginning, but the results so far more than justify continuing the experiment.

Critical Thinking

1. How has Brazil slowed the destruction of its rainforest in the Amazon?
2. What is the relationship between the international economy and the slowing deforestation in the Amazon rainforest?
3. How can Brazil serve as a model for other tropical countries?

Create Central

www.mhhe.com/createcentral

Internet References

Amazon Fund
www.amazonfund

Coalition for Rainforest Nations
http://rainforestcoaliton.org

Gateway to the United Nations' Systems Work on Climate Change
www.un.org/climatechange

UN Framework Convention on Climate Change Secretariat
www.unfcc.de

Real Climate
www.realclimate.org

JEFF TOLLEFSON is a U.S. correspondent for Nature, where he covers energy, climate, and environmental issues.

Article

Prepared by: Robert Weiner, *University of Massachusetts/Boston*

Climate Change and Food Security

BRUCE A. McCARL, MARIO A. FERNANDEZ, JASON P. H. JONES, AND MARTA WLODARZ

Learning Outcomes

After reading this article, you will be able to:

- Explain the relationship between climate change and food security.
- Explain how the increase of carbon dioxide in the atmosphere will affect food security.
- Explain policy strategy in dealing with the effects of climate change on food security.

Of the 10 warmest years in recorded history, 9 have appeared in the past 10 years, and all since 1998. Furthermore, 2012, the 9th-warmest year in history, was the 36th year in a row above the twentieth century average. Simultaneously, precipitation patterns are changing, with rainfall generally becoming more concentrated. Not surprisingly, the effects on agriculture from such climate change are proving significant and worldwide, including in the United States. The US National Oceanic and Atmospheric Administration estimated, for instance, that climate change made a 2011 drought in the American Southwest 20 times more likely to occur.

At the same time, the role of agriculture, a sector highly vulnerable to climate change, is changing globally. Not only does farming remain vital for food and fiber supplies; it is also growing in importance as a source of feedstock for energy production. It is frequently mentioned, too, as a possible source of offsets to the greenhouse gas emissions that contribute to global warming.

Climate trends, in short, raise critical questions for the future of agriculture. What influence is climate change having on agricultural yields? Does it imply that farming might be less able to supply future food needs, especially given the likely demands from a growing population and from populations with growing income? And what might nations do to lessen the disruptive influence of climate change on agriculture?

To help put these questions in perspective, it is worth mentioning a couple of climate change's fundamental characteristics. First, the preponderance of evidence indicates that it is likely to make conditions hotter and overall wetter, but with a more variable set of weather patterns. Second, climate change has not been observed to, nor is it projected to, have geographically uniform effects. In particular, while most every place is expected to be hotter with more variable conditions, some regions are likely to be drier while others will be wetter.

The Culprits

A changing climate certainly alters agricultural productivity. Ultimately, conditions involving extreme heat or extreme cold, as well as extreme wetness or extreme dryness, are unsuitable for raising crops. Crops fare best within narrow temperature and precipitation bands. Fortunately, temperature and precipitation conditions vary geographically. Conditions near the poles are generally too cold, while those near the equator can be too hot. Not all crops need the same ranges: Wheat, for example, fares best under comparatively colder conditions, and cotton or rice under hotter ones, while corn and soybeans need moderate conditions. This means a warmer climate will benefit certain crops and regions but harm others. It will also alter the geographic distribution of crop production, causing current crop ranges to move generally poleward.

Carbon dioxide is a related factor that will also affect agriculture. Considerable scientific evidence indicates that today's climate change is being driven in large part by increasing atmospheric greenhouse gas concentrations. Increases in carbon dioxide, the most abundant of greenhouse gases, stimulates the growth of certain classes of crops (so-called C3 crops such as rice, wheat, barley, oats, soybeans, potatoes, and most fruits), while the growth of others (so-called C4 crops like corn, sugarcane, sorghum, millet, and some grasses) is not greatly stimulated, but does better under drought conditions. Carbon dioxide effects on production are not strictly positive: Weed competition, for instance, also will be stimulated. However, carbon dioxide effects could partially offset yield losses that will occur solely based on temperature and precipitation changes.

And these are far from the only climate change factors with important effects on agriculture. Sea-level rise caused by ice melt and thermal expansion of the ocean could inundate substantial areas of agricultural land, particularly in low-lying producing countries such as Egypt, Bangladesh, India, and Vietnam. Pest populations are likely to be affected, and significant shifts have already been observed in pest extent and incidence. Observations show that weed and pest damages are greater in warmer areas, portending an expanding region of damage as the climate warms. Decreased frequency of extreme cold spells can also stimulate pest spread, as has been observed in North American forests with the wide spread of the destructive pine bark beetle.

Climatic extreme events—for example, droughts, floods, heat waves, and extreme cold—are projected to increase, and these can lead to lower, less stable agricultural yields, while also inducing greater incidence of famine and shifts in land use away from cropping. The Intergovernmental Panel on Climate Change (IPCC) recently published a report on extreme events suggesting that droughts may intensify in many parts of the world, including North and South America, Central Europe, and Africa. This in turn could reduce production and cause domestic and international food prices to increase, as was seen during 2012. Countries whose inhabitants already spend a large portion of their income on food will be most severely affected, resulting in increased malnourishment and poverty.

The IPCC report on weather extremes indicates the likelihood of more heat waves, which would stress water availability, crop production, and livestock production. They could as well decrease livestock disease resilience. The IPCC also provides evidence of an increase in the proportion of heavy rainfall events, relative to total rainfall. This would increase soil and fertilizer runoff, in turn causing water pollution and algae blooms.

Climate variation does not arise from a single source. Earth's climate has always exhibited strong natural variability on a seasonal, annual, and multiyear basis. Such variation originates from interactions within and among the atmosphere, ocean, land, sea ice, and glaciers, among other factors. One widely discussed cause of between-year climate variability is the El Niño Southern Oscillation (ENSO). Arising from interactions between the ocean and the atmosphere, ENSO causes shifts in the jet steam with effects on climate over large areas. For example, in Texas the occurrence of the La Niña phase of ENSO has been associated with the driest years in recorded history, including the record drought in 2011.

Many other major ocean-atmosphere interactions have been identified as contributing factors as well, including longer-term phenomena like the so-called North Atlantic Oscillation. Interestingly, some analysts have projected interaction between climatic change forces and the ocean phenomena with, for example, extreme ENSO events becoming more common and stronger. The jury is still out on whether this is likely to happen.

Crop and Livestock Yields

Certainly there is reason for concern given climate change effects and natural variability coupled with agriculture's enormous dependence on climate. And this concern is borne out in current agricultural production trends.

Recent years have witnessed substantial variability in agricultural yields. Consider data from the United States. During the 2011 drought in the Southwest, nearly 40 percent of the cotton crop was abandoned, with yields judged insufficient to merit harvest. Cattle were widely sold off. Irrigators in many areas found that they could not pump enough water to compensate

for the extremely dry conditions. The net loss was estimated at $7.4 billion. Then came a 2012 drought in America's Midwest, resulting in a corn crop estimated to be 25 percent smaller than expected, and a near-doubling of corn prices.

Increasing variability in yields is also evident in developing countries. In subsistence areas, dry conditions have led to widespread famine in some instances, while extremely wet and favorable conditions can cause an oversupply in markets not capable of moving the commodities, resulting in a collapse in prices.

Some of climate change's damaging effects on agricultural yields are offset by technological progress. Indeed, increases in yield stimulated by research investment and technology dissemination have been a key feature of agriculture for many years. In some areas of the world, food supply has grown faster than the population—leading to declining real prices and enhancing nations' ability to export more food. This also has allowed farmers to devote increasing amounts of land to bioenergy resources.

However, recent years have seen an overall decline in rates of yield growth. In the United States, corn yield growth until the 1970s exceeded 3 percent per year; now it is below 1.7 percent. Many complex factors have led to this result, including reductions in yield-enhancing investment levels. But certainly climate change has been a factor, and will be one in the future. This portends lower future growth in yields relative to demand growth, and perhaps may restrict agriculture's ability to meet the multiple demands now placed upon it. It also calls for larger levels of investment in productivity-increasing factors like research and technology dissemination.

The agricultural impacts of climate change and climate variation show considerable geographic differences, both within and across regions, due to differing soil characteristics, regional climates, and socioeconomic conditions. For example, according to projections reviewed in IPCC reports, rain-fed agricultural production in sub-Saharan Africa will decline by up to 50 percent by 2020. Maize production in Africa and Latin America is projected to fall by 10 percent to 20 percent by 2050. Yet the maize yield on China's Loess Plateau is projected to increase by around 60 percent during 2070–99. Wheat yields in southern Australia are projected to drop by 13.5 percent to 32 percent by 2050, yet over the same period winter wheat production in southern Sweden will increase by 10 percent to 20 percent.

In areas of Illinois and Indiana, due to an increase in daily maximum temperatures, some analysts project long-season maize yields will decline by 10 percent to 50 percent between 2030 and 2095. However, maize yields in the Great Plains area are projected to increase 25 percent by 2030 and 36 percent by 2095. A warming of 9 degrees to 11 degrees Fahrenheit by 2050 would cause a projected 10 percent decline in livestock yields, on average, in cow and calf and dairy operations in the Appalachian region, the Southeast (including the Mississippi Delta), and the southern plains.

Simultaneously, water is expected to become a growing issue. IPCC projections indicate that water availability within some dry regions at mid-latitudes and in the dry tropics will experience a reduction of 10 percent to 30 percent by 2050. The projections also show that, at higher latitudes and in some wet tropical areas, water supplies will increase by 10 percent to 40 percent over the same period. Also, the portion of river basins under severe water stress is expected to expand, with the ability to withdraw water either stabilizing or declining in 41 percent of global river basins. On the one hand, such impacts are expected to be more prevalent in developing countries than in industrialized ones. On the other hand, warming may well help in regions closer to the poles by limiting cold stress, even as it raises the heat stress in regions closer to the equator.

As already hot regions grow hotter, cows and pigs will not eat as much; the heat suppresses their appetites. This will negatively affect their growth performance. Additionally, evidence suggests that higher average temperatures cause lower birth rates and reduced milk and wool production. A study by the US Department of Agriculture (USDA) estimates that additional stress from heat will cause the beef industry to lose $370 million per year. This, coupled with altered feed availability, could cause large pole-ward shifts in regions of livestock production.

Forage properties are also at issue. Under hotter conditions in already hot areas, the quality of forage deteriorates and its protein content worsens. Also, grass and hay are projected to grow at a slower pace; thus livestock stocking rates per unit of land area will go down.

Livestock diseases and pests are projected to become more prevalent. For example, higher temperatures have been found to increase the probability of avian influenza outbreaks, raising threats to poultry as well as human populations. In Niger, an invasion of desert locusts in 2005–06 caused massive damage to pasture lands and was followed by an extreme food crisis, with around 4 million people facing chronic famine.

Collectively, the water and agricultural implications of climate change will add to the developmental challenges of ensuring food security and reducing poverty.

Adapt and Mitigate

The 2007 IPCC report identifies two basic forms of actions for addressing the impact of climate change on agriculture. First, society can alter agricultural production processes to accommodate the altered climate. Second, society can act to reduce greenhouse gas emissions in an effort to mitigate (or limit) the extent of future climate change, with farming playing a role in this effort. Climate change likely will affect agriculture negatively where societies do not find ways to adapt.

To prepare for changing climate conditions, policy makers require a clear picture of the risks that their country or region will face in the future. The extent of these risks is generally uncertain. Traditionally we have used historical climate behavior as a starting point for predictions. That is, we typically assumed that any climatic cycles or phenomena that occurred in the past will likely recur (for example, the 100-year flood). This was a reasonable approach in earlier times, but in a future with climate change the repeatability of the past is not likely to hold.

Climate change alters the variability of droughts, heat waves, and floods. Not only will it affect future average crop and livestock yields; it will also make more uncertain the year-to-year variations in production. Thus, it will not be appropriate to assume that, for example, an observed flood or drought of a particular severity that occurred once in the past hundred years will occur with such frequency in the future.

Agriculture can be adapted to climate change by altering the management and location of production. Indeed, adaptation is not a new concept in agriculture. Producers in any region are faced with local conditions in terms of climate, pests, water availability, demand, land suitability, environmental regulation, and market competition. In turn, they choose an appropriate mix of crops, livestock, and management techniques to accommodate those conditions. As we have noted, for instance, areas where rice and cotton are grown are generally hotter than areas where wheat grows. As climate heats up, relocation of negatively affected crops toward the poles is an effective adaptation.

At the same time, selection of animal, crop, and forage species or breeds that are more resistant to heat and drought might help, along with the provision of irrigation and shade for animals. These possibilities will aid agriculture's adaptation, but likely will not alleviate difficulties in particularly vulnerable regions. In these regions a lack of resources such as available capital, producer education and knowledge, and available information, together with the infeasibility of certain actions, might preclude full adaptation, leaving residual damages from climate change.

In general, adaptations can be private and autonomous or public in nature. Producers often undertake adaptations autonomously. For example, warmer conditions historically have caused crop shifts. In the United States, the geographic center of corn and soybean production in 1990 showed a northwestern shift of approximately 120 miles, in comparison with production locations in the early 1900s. More recent data show a further northwestern shift of more than 75 miles since 1990.

Policy Strategies

Public adaptations, on the other hand, encompass actions that are beyond the capabilities of individuals, or are far too costly for individuals to invest in, or once developed are not the kind of practices that an individual can patent and be paid for by other users. Public adaptations range from developing heat-resistant crop and livestock varieties, to disseminating climate-forecasting information to populations that need the knowledge in order to adapt.

For one example, the US National Aeronautics and Space Administration, the National Oceanic and Atmospheric Administration, and the US Geological Survey have created a famine early-warning system using satellite information on soil moisture levels and crop health. The system is designed to help farmers adapt to projected unfavorable climate change and to lower the cost of extreme events.

Publicly supported adaptations can also involve the development of institutions such as financial systems that reduce farmers' exposure to risk, or the implementation of a freer trade policy that more readily provides food to areas where climate change reduces production. However, in this regard there is a serious risk of public underinvestment. The World Bank estimates a current need for between $9 billion and $40 billion in annual climate change adaptation funding. The United Nations' Food and Agriculture Organization (FAO) indicates that in 2011, some $244 million was dispersed to all countries in total.

Agricultural damages from climate change impacts are expected to be greatest among countries with the least ability to adapt, primarily poor countries. When such nations face a prolonged drought or multiyear crop failures, their strained food supply could cause a collapse of rural production, large-scale out-migration, social unrest, and famine. The severity of impact is related to the limited human and physical resources available for investments in technological knowledge, human capital, water and food storage, processing, and distribution.

What's at Stake

There is increasing evidence that the welfare of current and future generations will depend heavily not only on atmospheric greenhouse gas concentration levels, but also on the actions taken to stop and reverse greenhouse gas accumulation. In 2012 carbon dioxide levels in the atmosphere were measured to be more than 40 percent higher than pre-industrial levels. Agriculture itself is the source of between 50 percent and 70 percent of methane and nitrous oxide emissions, and atmospheric concentrations of these greenhouse gases also have increased significantly.

Agriculture can play a role in reducing atmospheric greenhouse gases by increasing carbon storage (sequestration), increasing tree planting, easing tillage, converting croplands to grasslands, or otherwise managing to increase soil organic content. Agriculture can also help avoid emissions by reducing fossil fuel use, altering nitrogen fertilization practices, better managing ruminant livestock and manure, and reducing rice methane emissions, among other means. Finally, agriculture can provide substitute products that can be used in place of fossil-fuel-intensive products. For example, biomass-based feedstocks can be substituted for liquid energy or electricity production, and new building materials can replace steel and concrete.

In considering adaptation and mitigation, one must be cognizant of the fact that land use for some environmentally adaptive alternatives can come into competition with land use for the food supply. The recent corn ethanol boom in the United States is an important example: An expansion of ethanol consumption from roughly 6 percent to nearly 40 percent of the US corn crop between 2002 and 2012 has, coupled with other factors, led to increased land use, diverted production, higher food prices, and some degree of increased price instability.

Rising food prices are not the only problem caused by expanded mitigation activity. Increased biomass production and utilization (for example, removal of corn residues from fields) cause increases in pesticide use, ground water depletion, soil erosion, and biodiversity loss. Furthermore, the rise in commodity prices can induce expansions in domestic and international agricultural land use, possibly leading to greater rates of deforestation and losses in associated carbon sequestration.

FAO figures show that the world's agricultural production has more than doubled in the past 50 years, and in developing countries it has more than tripled. The amount of available food has grown steadily, allowing the fulfillment of basic nutritional requirements for an increasing share of a growing global population. In part, advances in farmers' management skills, fertilizers and pesticides, and irrigation supply have contributed to increasing crop productivity in formerly famine-prone areas, particularly in Africa.

Still, the USDA estimates that 850 million of the earth's inhabitants currently lack access to a secure food supply. Oxfam, an international organization for famine relief, recently projected a doubling of prices for the world's staple food products over the next 20 years, with half of the increase attributed to climate change. This would likely result in major food security issues, particularly in areas of Africa, India, and Southeast Asia.

Population growth also contributes to the problem. By 2050 the world is projected to have 3.3 billion more mouths to feed. The challenge is feeding them while also adapting to or mitigating climate change. The dual forces of population growth and climate change will exacerbate pressures on land use, water access, and food security.

It is likely that the impacts of climate change on agriculture will affect everyone. However, the degree of impact will vary depending on how or whether one's society chooses to adapt, and how or whether we act on a national and global basis to limit the extent of future impacts by mitigating atmospheric greenhouse gas concentrations. Both adaptation and mitigation require actions and investments that will compete with each other and with conventional production and consumption. Food security in some regions is certainly at stake.

Critical Thinking

1. What will the effect of climatic extreme events be on food security?
2. What are the negative and positive effects of a warmer climate on food security?
3. What can be done to reverse the concentration gases in the atmosphere?

Create Central

www.mhhe.com/createcentral

Internet References

Famine Early Warning System Network
www.fews.net/Pages/default.aspa
Food and Agricultural Organization
www.fao.org

BRUCE A. MCCARL, a professor of agricultural economics at Texas A&M University, has contributed to the work of the Intergovernmental Panel on Climate Change. MARIO A. FERNANDEZ, JASON P. H. JONES, and MARTA WLODARZ are graduate students at Texas A&M.

McCarl et. al., Bruce A. From *Current History*, January 2013, pp. 33–37. Copyright © 2013 by Current History, Inc. Reprinted by permission.